XVIIth Conference of PhD Students and Young Scientists 2017

E3S Web of Conferences Volume 29 (2018)

Szklarska Poreba, Poland
23 - 26 May 2017

Editors:

T.A. Przylibski
D. Kasza

ISBN: 978-1-5108-5722-3

Printed from e-media with permission by:

Curran Associates, Inc.
57 Morehouse Lane
Red Hook, NY 12571

Some format issues inherent in the e-media version may also appear in this print version.

Printed by Curran Associates, Inc. (2018)

For additional information, please contact EDP Sciences – Web of Conferences at the address below.

EDP Sciences – Web of Conferences
17, Avenue du Hoggar
Parc d'Activité de Courtabœuf
BP 112
F-91944 Les Ulis Cedex A
France

Phone: +33 (0) 1 69 18 75 75
Fax: +33 (0) 1 69 28 84 91

contact-edps@webofconferences.org

Additional copies of this publication are available from:

Curran Associates, Inc.
57 Morehouse Lane
Red Hook, NY 12571 USA
Phone: 845-758-0400
Fax: 845-758-2633
Email: curran@proceedings.com
Web: www.proceedings.com

TABLE OF CONTENTS

edp sciences

Web of Conferences

Statement of Peer review

In submitting conference proceedings to *Web of Conferences*, the editors of the proceedings certify to the Publisher that

1. They adhere to its **Policy on Publishing Integrity** in order to safeguard good scientific practice in publishing.
2. All articles have been subjected to peer review administered by the proceedings editors.
3. Reviews have been conducted by expert referees, who have been requested to provide unbiased and constructive comments aimed, whenever possible, at improving the work.
4. Proceedings editors have taken all reasonable steps to ensure the quality of the materials they publish and their decision to accept or reject a paper for publication has been based only on the merits of the work and the relevance to the journal.

Title, date and place of the conference

XVIIth Conference of PhD Students and Young Scientists, May 23-26, 2017, Szklarska Poręba, Poland

Proceedings editor(s):

Tadeusz A. Przylibski, prof. PWr, Wrocław University of Science and Technology, Poland
Damian Kasza, PhD, Wrocław University of Science and Technology, Poland

Date and editor's signature 20.12.2017.

17, avenue du Hoggar - PA de Courtabœuf – BP 112 - 91944 Les Ulis Cedex A (France)
Tél. : 33 (0)1 69 18 75 75 - Fax : 33(0)1 69 07 45 17 – www.edpsciences.org

E3S Web of Conferences **29**, 00001 (2018) https://doi.org/10.1051/e3sconf/20182900001
XVIIth Conference of PhD Students and Young Scientists

Analytical method for establishing indentation rolling resistance

Lech Gładysiewicz[1], and *Martyna* Konieczna[1,*]

[1]Wroclaw University of Science and Technology, Faculty of Geoengineering, Mining and Geology, 27 Wyb. Wyspianskiego St., 50-370 Wroclaw, Poland

Abstract. Belt conveyors are highly reliable machines able to work in special operating conditions. Harsh environment, long distance of transporting and great mass of transported martials are cause of high energy usage. That is why research in the field of belt conveyor transportation nowadays focuses on reducing the power consumption without lowering their efficiency. In this paper, previous methods for testing rolling resistance are described, and new method designed by authors was presented. New method of testing rolling resistance is quite simple and inexpensive. Moreover it allows to conduct the experimental tests of the impact of different parameters on the value of indentation rolling resistance such as core design, cover thickness, ambient temperature, idler travel frequency, or load value as well. Finally results of tests of relationship between rolling resistance and idler travel frequency and between rolling resistance and idler travel speed was presented.

1 Introduction

Belt conveyors are vital elements of transportation systems in both underground and surface mines, as well as in many other branches of industry. They allow the most economical and, frequently, also the fastest transportation of materials. Nevertheless, the maintenance of belt conveyors generates significant costs. Conveyor belt has two important functions in the operation of a belt conveyor – it transports the bulk materials and transmits the driving force; at the same time, however, it generates different types of resistances. Rational belt management must be based on the knowledge of the phenomena that occur during the operation of a belt conveyor. Expanding this knowledge requires increasingly precise tests [1-3] and theoretical analyses, which allow the description and prediction of how a belt will behave in given operating conditions [4- 6]. From the perspective of energy losses, the rolling resistances of the belt on idlers constitute the most significant component of the resistances observed on a belt conveyor. Precise calculation of belt damping factor proves crucial if the theoretically established values of resistances are to correspond to the values established experimentally.

* Corresponding author: martyna.konieczna@pwr.edu.pl

Symbols:

φ_1, φ_2 - entry angle and exit angle of the belt on the idler, radians
σ - compressive stresses acting on the belt, Pa
ξ - measure of belt damping expressed as the ratio of φ_2 and φ_1, -
ψ - belt damping factor expressed through δ, -
t_0 - duration of a single load cycle, s
t_m - time to maximum transverse strain of the belt,
ω_0 - angular velocity of the idler, $\frac{rad}{s}$
v_t - belt speed, $\frac{m}{s}$
δ - phase lag angle, radians
ε - transverse strains, -
ε_1 - particular solution (viscous flow), -
ε_2 - general solution (typical for harmonic load), -
τ_0 - time constant of the belt's model, -
D_k - idler diameter, m
E_c - modulus of elasticity / resultant modulus of the belt under transverse compression (allowing for a number of plies in the belt), Pa
h_0 - active belt thickness undergoing the process of compression, m
F - belt damping function, -
e_1 - unit elastic energy transferred to belt, J
e_2 - recovered energy, J

2 Rolling resistances of the belt on idlers

Various forms of energy conversions occur during the operation of a belt conveyor. These conversions generate a variety of resistances. The motion resistances of a belt conveyor, as classified by their location, include [7]:

- concentrated resistances (on the head pulley, drive pulley, tail pulley and take-up pulley)
- main resistances (distributed along the whole conveyor route),
- lift resistances of the belt and of the bulk material (on the inclined sections of the conveyor route).

For conveyor belts longer than 80 m, main resistances are the dominant component of motion resistance. Depending on how energy is dissipated, main resistances are classified as follows [7]:

- idler rotational resistance W_k,
- rolling resistance of belt on idlers W_e,
- belt flexure resistance W_b,
- flexure resistance of bulk material W_f,
- sliding resistance of belt on idlers W_r.

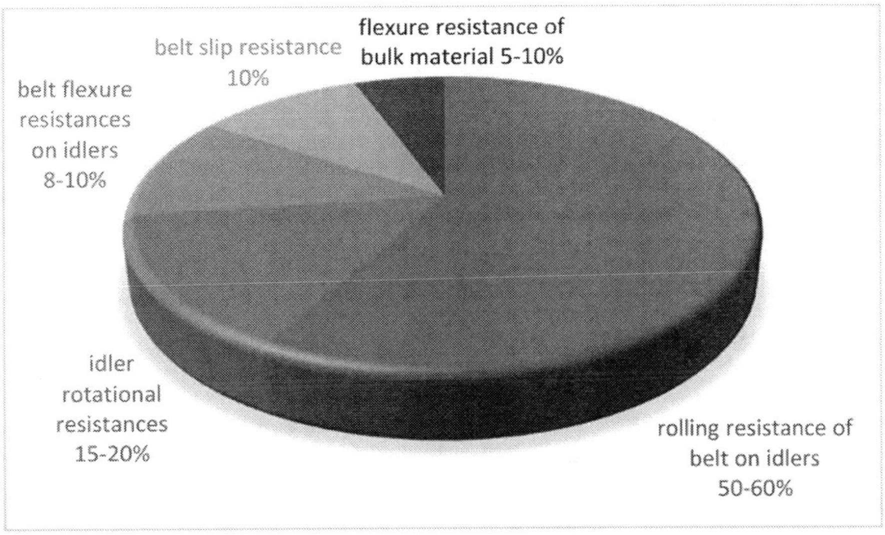

Fig. 1. Approximate values of the components of main resistances on a belt conveyor for the belt's top run [7].

Indentation rolling resistances, also referred to as belt indentation resistances, are the main component of motion resistances which occur on a conveyor. They may account for as much as 60% of the total resistance to motion, and consequently they also generate the greatest energy losses.

Tab. 1. Estimation of savings resultant from limiting belt conveyor resistance to motion in a mine, which operates approx. 100 km of conveyors [8].

Decrease in resistances [%]	**5**	**10**	**15**	**20**	**25**	**30**
Savings $\left[\frac{million\ pln}{year}\right]$	4.69	9.38	14.06	18.75	23.44	28.13

Estimation of savings presented in tab. 1 reveals how even small decrease in resistances of conveyors can be profitable. Resistances reduction of 5% equal more than 4.5 million pln. The precise calculations of indentation rolling resistance will enable to determine the energy consumption more correctly.

3 Theoretical belt damping factor

Belt load on a single idler set is accompanied by temporary strain in belt. Due to the belt's viscoelastic properties, the strain occurs not only in the contact zone of the belt and the idler, but also outside this zone. Therefore, the load-strain relationship is not linear, and the belt resumes its initial state with some delay; this phenomenon is referred to as viscous flow [9, 10] Figure 2 shows the relationship between harmonic load and strain in subsequent cycles.

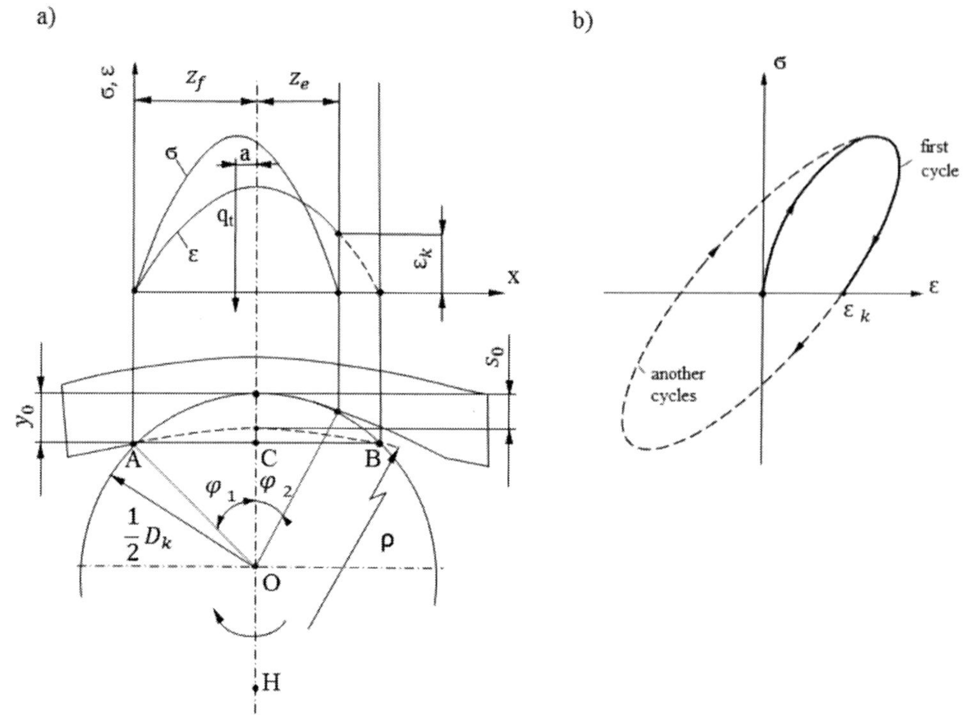

Fig. 2. a) stress and strain distribution in contact zone between belt and idler set, b) deformations in a viscoelastic belt and hysteresis loop for the first and for the consecutive stress cycles [4].

When moving on the supporting idlers, a conveyor belt is subjected to a momentary indentation due to external forces. As the distances between idler sets on an actual belt conveyor are large, before each cycle starts, the belt resumes its initial state. Therefore, the theoretical model was developed on the basis of a hysteresis loop which corresponds to the first stress cycle. The compressive stress curve is a harmonic function, which may be expressed with a simple formula:

$$\sigma = \sigma_0 \cdot \sin(\omega_0 \cdot t) \quad (1)$$

For the belt's model, the formula is as follows:

$$\sigma = E_c \cdot \left(\varepsilon + \tau_0 \cdot \frac{d\varepsilon}{dt} \right) \quad (2)$$

Strain in conveyor belts has two components: a general and a particular solution of the belt equation. The general solution is described by the following equation:

$$\varepsilon_1 = \frac{\sigma_0}{E_c} \cdot \cos\delta \cdot \sin\delta \cdot e^{-\frac{t}{\tau_0}} \quad (3)$$

The particular solution is as follows:

$$\varepsilon_2 = \frac{\sigma_0}{E_c} \cdot \cos\delta \cdot \sin(\omega_0 \cdot t - \delta) \tag{4}$$

The duration of a single load cycle is:

$$t_0 = \frac{D_K}{2 \cdot v_t} \cdot (\varphi_1 + \varphi_2) \tag{5}$$

Whereas the time to maximum transverse strain of the belt is described by the following formula:

$$t_m = \frac{D_K}{2 \cdot v_t} \cdot \varphi_1 = \frac{t_0}{1+\xi} \tag{6}$$

Unit strain energy of the belt may be expressed with the following equation:

$$de = \sigma \cdot h_0 \cdot d\varepsilon = \sigma \cdot h_0 \cdot (d\varepsilon_1 + d\varepsilon_2) \tag{7}$$

$$d\varepsilon_1 = -\frac{\sigma_0}{E_c \cdot \tau_0} \cdot \sin\delta \cdot \cos\delta \cdot e^{-\frac{t}{\tau_0}} \cdot dt \tag{8}$$

$$d\varepsilon_2 = \frac{\sigma_0}{E_c \cdot \tau_0} \cdot \omega_0 \cdot \cos\delta \cdot \cos(\omega_0 \cdot t - \delta) \cdot dt \tag{9}$$

During the time between t=0 and $t=t_m$, energy e_1 is transferred to the belt, while during the time between $t=t_m$ and $t=t_0$, only part of this energy e_2 is recovered. Thus, the belt's damping factor, which depends on the energy lost and on the energy transferred to the belt in a single compression cycle, is described by the following formula:

$$\psi_{th} = \frac{\int_0^{t_0} \sigma(t) \cdot d\varepsilon}{\int_0^{t_m} \sigma(t) \cdot d\varepsilon} \tag{10}$$

Unit elastic energy converted (lost) in a single compression cycle is:

$$e_1 = h_0 \cdot \int_0^{t_0} \sigma(t) \cdot d\varepsilon = \frac{h_0 \cdot \sigma_0^2}{E_c} \cdot \cos\beta_z \cdot \int_0^{t_0} \left[-\frac{\sin\delta}{\tau_0} \cdot \sin(\omega_0 \cdot t) \cdot e^{-\frac{t}{\tau_0}} + \omega_0 \cdot \sin(\omega_0 \cdot t) \cdot \cos(\omega_0 \cdot t - \delta) \right] \cdot dt \tag{11}$$

Elastic energy transferred to the belt in a single compression cycle is:

$$e_2 = h_0 \cdot \int_0^{t_m} \sigma(t) \cdot d\varepsilon = \frac{h_0 \cdot \sigma_0^2}{E_c} \cdot \cos\beta_z \cdot \int_0^{t_m} \left[-\frac{\sin\delta}{\tau_0} \cdot \sin(\omega_0 \cdot t) \cdot e^{-\frac{t}{\tau_0}} + \omega_0 \cdot \sin(\omega_0 \cdot t) \cdot \cos(\omega_0 \cdot t - \delta) \right] \cdot dt \tag{12}$$

The calculations are based on the following assumptions:

$$\omega_0 \cdot \tau_0 = tg\delta \tag{13}$$

$$\frac{\sin\delta}{\omega_0^2 \cdot \tau_0^2 + 1} = \cos^2\delta \cdot \sin\delta \tag{14}$$

$$\omega_0 \cdot t_0 = \pi \tag{15}$$

$$\omega_0 \cdot t_m = \frac{\pi}{1+\xi} \tag{16}$$

$$-\frac{t_m}{\tau_0} = -\frac{\pi}{(1+\xi)\cdot tg\delta} \tag{17}$$

$$-\frac{t_0}{\tau_0} = -\frac{\pi}{tg\delta} \tag{18}$$

$$\delta_z = \frac{\pi}{2} \cdot \frac{1-\xi}{1+\xi} \tag{19}$$

$$\omega_0 = \frac{2\cdot\pi\cdot v_t}{D_K \cdot (\varphi_1 + \varphi_2)} = \frac{2\cdot\pi\cdot v_t}{D_K \cdot \varphi_1 \cdot (1+\xi)} \tag{20}$$

The final shapes for the e_1 and e_2 equations are as follows:

$$e_1 = \frac{h_0 \cdot \sigma_0^2}{E_c} \cdot cos\left(\frac{\pi}{2}\cdot\frac{1-\xi}{1+\xi}\right) \cdot \left\{ e^{-\frac{\pi}{tg\left(\frac{\pi}{2}\frac{1-\xi}{1+\xi}\right)}} \cdot cos^2 \cdot \left(\frac{\pi}{2}\cdot\frac{1-\xi}{1+\xi}\right) \cdot \sin\left(\frac{\pi}{2}\cdot\frac{1-\xi}{1+\xi}\right) \cdot (-\pi) + \frac{\pi \cdot sin\left(\frac{\pi}{2}\frac{1-\xi}{1+\xi}\right)}{2} \right\} \tag{11'}$$

$$e_2 = \frac{h_0 \cdot \sigma_0^2}{E_c} \cdot cos\left(\frac{\pi}{2}\cdot\frac{1-\xi}{1+\xi}\right) \cdot \left\{ e^{-\frac{\pi}{(1+\xi)\cdot tg\left(\frac{\pi}{2}\frac{1-\xi}{1+\xi}\right)}} \cdot cos^2 \cdot \left(\frac{\pi}{2}\cdot\frac{1-\xi}{1+\xi}\right) \cdot \sin\left(\frac{\pi}{2}\cdot\frac{1-\xi}{1+\xi}\right) \cdot \left[\frac{\pi}{1+\xi} \cdot cos\left(\frac{\pi}{1+\xi}\right) + sin\left(\frac{\pi}{1+\xi}\right)\right] + \frac{\frac{2\pi}{1+\xi}\sin\left(\frac{\pi}{1+\xi}\right) - cos\left(\frac{2\pi}{1+\xi} - \frac{\pi}{2}\frac{1-\xi}{1+\xi}\right)}{4} + \frac{cos\left(-\frac{\pi}{2}\frac{1-\xi}{1+\xi}\right)}{4} \right\} \tag{12'}$$

After reduction, the measure for belt damping is expressed as the relationship between the energy lost and the energy transferred to the belt:

$$\psi_{th} = \frac{e_1}{e_2} \tag{21}$$

Thus, the formula for damping factor in function of phase lag angle is as follows:

$$\psi(\delta)_{th} = \frac{e^{-\frac{\pi}{tg\delta}} \cdot cos^2\delta \cdot (-\pi) + \frac{\pi}{2}}{e^{\frac{-\frac{\pi}{2}+\delta}{tg\delta}} \cdot cos^2 \cdot \left[-\left(\frac{\pi}{2}+\delta\right) \cdot sin\delta + cos\delta\right] + \left(\frac{\frac{\pi}{2}+\delta+ctg\delta}{2}\right)} \tag{22}$$

4 Damping factor determined in laboratory tests

The hysteresis loop, which is the result of cyclic compression, shows the belt's load-strain relationship. The damping factor may be expressed as the relationship between the area inside the hysteresis loop (converted energy) and the area below the load curve (delivered energy). The belt's modulus of elasticity may be calculated as the inclination angle of the straight line which connects the vertices of the loop. These values are calculated on the basis of the hysteresis loop established in laboratory tests. The belt's hysteresis loop which would be adequate to the belt's cyclic indentation process is not obtainable in laboratory conditions. Therefore, a loop recorded during continuous harmonic load is used (Figure 2). In the case of laboratory tests, the course of compressive stresses may be expressed with the following formula:

$$\sigma = \sigma_0 + \sigma_0 \cdot \sin(\omega \cdot t) \tag{23}$$

Solution of the belt's differential equation:

$$\varepsilon = \varepsilon_0 + \frac{\sigma_0}{E_C} \cdot cos\delta \cdot \sin(\omega_0 \cdot t - \delta) \tag{24}$$

$$d\varepsilon = \frac{\sigma_0}{E_C} \cdot cos\delta \cdot \omega_0 \cdot \cos(\omega_0 t - \delta)dt \tag{25}$$

$$\sigma = \sigma_0 + \sigma_0 \cdot \sin\alpha \tag{26}$$

$$\varepsilon = \varepsilon_0 + \frac{\sigma_0}{E_C} \cdot cos\delta \cdot \sin(\alpha - \delta) \tag{27}$$

$$d\varepsilon = \frac{\sigma_0}{E_C} \cdot cos\delta \cdot \omega_0 \cdot \cos(\alpha - \delta)d\alpha \tag{28}$$

$$de = \sigma \cdot h_0 \cdot d\varepsilon \tag{29}$$

$$de = h_0 \cdot (\sigma_0 + \sigma_0 \cdot sin\alpha) \cdot \frac{\sigma_0}{E_C} \cdot cos\delta \cdot \cos(\alpha - \delta)\, d\alpha \tag{30}$$

$$de = \frac{h_0 \cdot \sigma_0^2}{E_C} \cdot cos\delta \cdot [cos(\alpha - \delta) + sin\alpha \cdot cos(\alpha - \delta)]d\alpha \tag{31}$$

Unit elastic energy lost in a single compression cycle in laboratory conditions is:

$$e_1 = \frac{h_0 \cdot \sigma_0^2}{E_C} \cdot cos\delta \cdot \left[\int_0^{2\pi} \cos(\alpha - \delta)\, d\alpha + \int_0^{2\pi} sin\alpha \cdot \cos(\alpha - \delta) d\alpha\right] \tag{31}$$

Where:

$$\int_0^{2\pi} \cos(\alpha - \delta)\, d\alpha = 0, \qquad and \int_0^{2\pi} sin\alpha \cdot \cos(\alpha - \delta) d\alpha = \pi \cdot sin\delta$$

$$e_1 = \frac{h_0 \cdot \sigma_0^2}{E_C} \cdot cos\delta \cdot \pi \cdot sin\delta \tag{32}$$

Elastic energy transferred to the belt in a single compression cycle is:

$$e_2 = \frac{h_0 \cdot \sigma_0^2}{E_C} \cdot cos\delta \cdot \left[\int_{\frac{3}{2}\pi+\delta}^{\frac{5}{2}\pi+\delta} cos(\alpha - \delta)d\alpha + \int_{\frac{3}{2}\pi+\delta}^{\frac{5}{2}\pi+\delta} sin\delta \cdot \cos(\alpha - \delta) d\alpha\right] \tag{33}$$

$$e_2 = \frac{h_0 \cdot \sigma_0^2}{E_C} \cdot cos\delta \cdot \left[\frac{1}{2} \cdot \pi \cdot sin\delta + 2\right] \tag{34}$$

The above served to describe the damping factor established in laboratory tests:

$$\psi_{lab} = \frac{e_1}{e_2} \tag{35}$$

$$\psi_{lab} = \frac{\pi \cdot sin\delta}{\frac{\pi}{2} sin\delta + 2} \tag{36}$$

5 Comparison of the theoretical model with the results of laboratory tests

Below are presented two hysteresis loop. First (fig. 3a) in adequate for the theoretical model and describes only the first cycle of loading. The next one has schematic shape of loop established in laboratory tests. To obtain the regular loop it is necessary to start the measurement with load. That is why the laboratory loop is transferred in comparison to the theoretical loop.

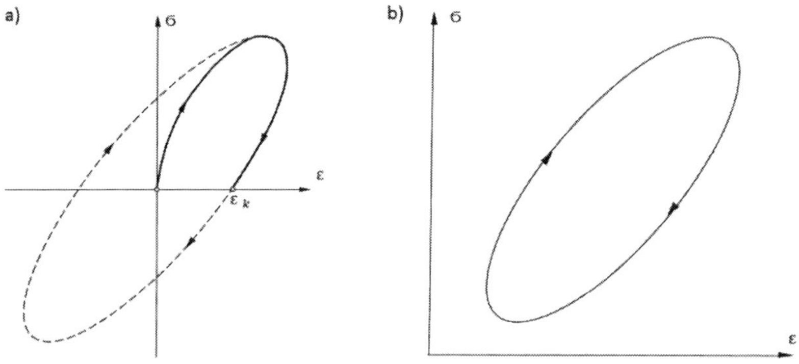

Fig. 3. Hysteresis loop: a) for the theoretical model b) schematic shape of loop adequate for laboratory tests [11].

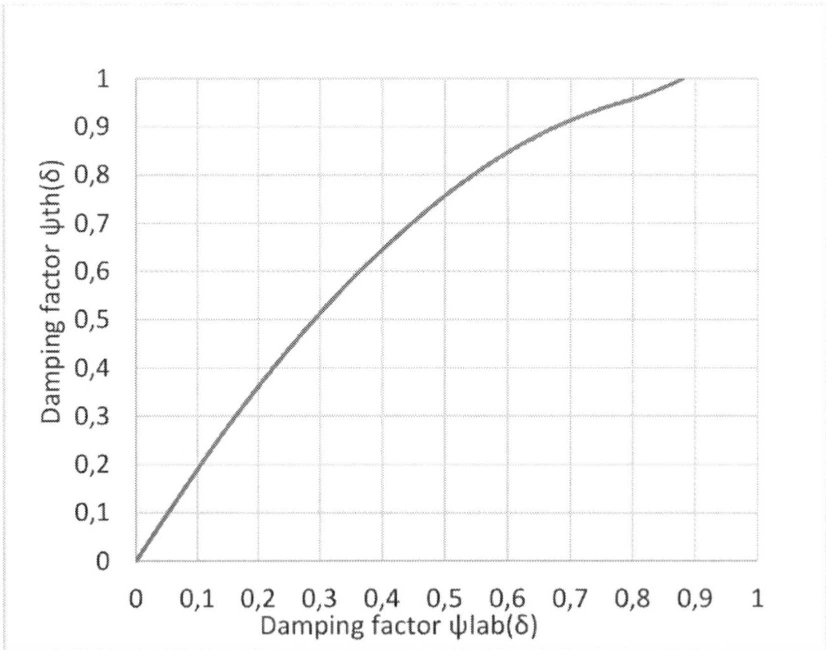

Fig. 4. Relationship between the damping factor established theoretically and in laboratory tests [11].

An equation describing the relationship between the damping factor as calculated theoretically and as found in laboratory tests:

$$\psi_{th} = -1.0385\psi_{lab}^2 + 2.0334\psi_{lab} - 0.0021 \tag{37}$$

The origin of the above diagram (in point 0,0) is purely hypothetical, since damping does not occur then, and this is an abstract situation in the case of a conveyor belt. Equation (37) describes the curve until damping factor 0.8 is reached. The part of the curve which is designated with blue color was not included in the following analyses, since values above 0.8 are practically unobtainable and only introduce disturbance to the characteristic curve. Equation (37) may serve to draw an analytical relationship between the rolling resistance based on the laboratory-determined damping factor and the rolling resistance determined for a conveyor belt operated in actual mining conditions.

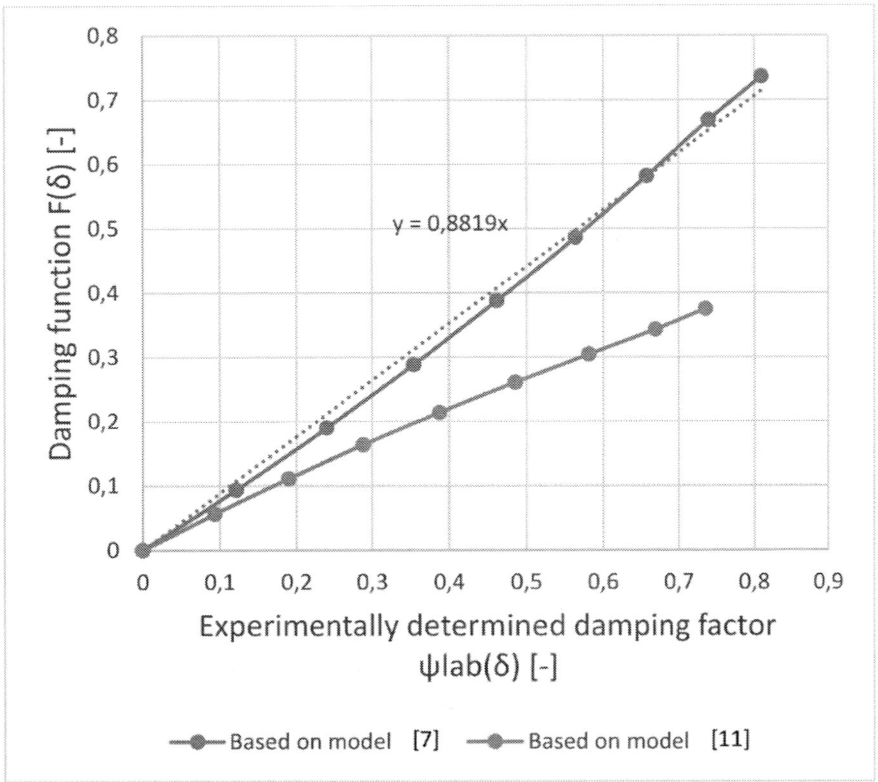

Fig. 5. Relationship between the damping function and the experimentally determined damping factor.

Regardless of the model used [7, 11], if it is bi-parameter, the rolling resistances are calculated from the relationship between the product of the damping factor and the structural parameters [7]. Comparison of different conveyor belts is based on an assumption that they are used similarly and that they work in identical conditions. In such case, damping function F is sufficient for analysis. In the diagram above, damping function is compared with laboratory damping factor. As can be observed, the convergence between the damping function and the experimental damping factor approaches 90% (0.8819). Moreover, the values are convergent regardless of the value of rolling resistance. The convergence is fixed for both low and high rolling resistances. The previously determined coefficient of convergence between the two values was lower: 0.463 [11]. Defining more precise assumptions for the model allowed improving the coefficient almost twice [7], and thus allowed approximating the theoretically calculated values of rolling resistances to the values measured in the laboratory.

$$w_e = F \cdot \sqrt[3]{\frac{q_T^4}{D_K^2 \cdot \lambda \cdot c_e}} \tag{38}$$

$$F(\xi) = \sqrt[3]{\frac{\pi \cdot (1-\xi)^3}{2 \cdot (\xi + 1)} \cdot \left\{ \frac{1}{2} \cdot \sin\left(\pi \cdot \frac{1-\xi}{1+\xi}\right) \cdot e^{-\frac{\pi}{(1+\xi) \cdot tg\left(\frac{\pi}{2} \frac{1-\xi}{1+\xi}\right)}} + \cos\left(\frac{\pi}{2} \cdot \frac{1-\xi}{1+\xi}\right) \right\}} \tag{39}$$

$$w_e = 0.882 \cdot \psi_{lab} \cdot \sqrt[3]{\frac{q_T^4}{D_K^2 \cdot \lambda \cdot c_e}} \tag{40}$$

Theoretical models are aimed at obtaining a possibly most precise representation of observed phenomena. Nevertheless, such models are only an approximation of real conditions. Figure 6 shows the relationship between the total resistance to motion of an idler set in various operating conditions and the conveyor's mass capacity

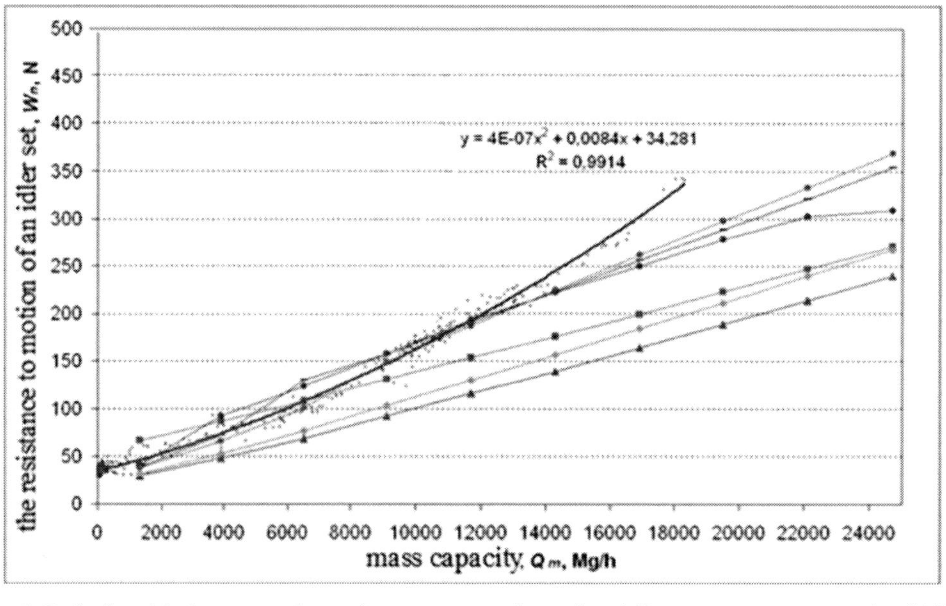

Fig. 6. Relationship between the resistance to motion of an idler set to mass capacity [12].

In the mass capacity range of 1000-12000 $\frac{Mg}{h}$, the resistances calculated for different variants are similar to the measurement results (dark blue line). As the capacity increases, however, the difference between the real values of motion resistances and the theoretical values becomes significant.

Although the above diagram has been plotted for total motion resistances, rolling resistances are notably the most significant component of main resistances. Rolling resistances may therefore, despite a number of additional elements, determine the motion resistance values of the whole conveyor. The diagram in Fig. 4 has been plotted on the basis of the previous theoretical model (5). Had the diagram been based on the assumptions for the new model, the actual, higher values of resistances would have been represented with greater precision.

6 Conclusions

The rolling resistances of the belt on idlers must be analyzed with regard to parameters which accommodate belt properties. In the analytical method, these properties are represented by damping factor and elastic modulus during cyclical compressing. Importantly, these parameters should be determinable in laboratory conditions. The hysteresis loops which are obtained in laboratory tests do not fully correspond to the loops obtained during cyclic indentation processes which occur on actual conveyors operated in mining conditions. The method described in this paper enables converting the damping factor calculated in laboratory tests into values which correspond to the actual operating conditions of a belt conveyor. Moreover, the share of rolling resistances in total motion resistances (for belt conveyors longer than 80 m) has been proven to be greater than previously believed. The new analytical model allows significantly more precise estimations of the resistances occurring on a belt conveyor.

Financial support by the project: Research, diagnostics and optimization of conveyor belt systems No. 0401/0166/16 and Grant no. 0401/0128/17.

References

1. R. Król, W. Kisielewski, Mining Science – Fundamental Problems of Conveyor Transport, Vol. **21**(2) 61–72, Prague (2014)
2. R. Król, WMESS, 11-15 September (2017)
3. M. Bajda, R. Król, Procedia: Earth and Planetary Science, World Multidisciplinary Earth Sciences Symposium WMESS, Vol. **15**, 702-711 (2015)
4. L. Gładysiewicz, M. Konieczna, Mining Science, **Vol. 23**, 105-119 Wrocław (2016)
5. C.O. Jonkers, Forden und Haben, Vol. **30** No. 4 (1980)
6. G. Lodewijks, Bulk Solids Handling **23**(6), 384-391 (2003)
7. L. Gładysiewicz, *Przenośniki taśmowe Teoria i obliczenia,* Oficyna Wydawnictwo Politechniki Wrocławskiej, Wrocław (2003)
8. M. Hardygóra, M. Bajda, L. Gładysiewicz, Transport Przemysłowy, **3** (2007)
9. S. Drenkelford, *Energy-saving potential of Aramid-based conveyor belts*, Delft University of Technology (2015)
10. L. Gładysiewicz, Prace Naukowe CPBP 02.05. Wyd. Politechniki Warszawskiej, Warszawa (1990)
11. L. Gładysiewicz., M. Konieczna, Transport przemysłowy i maszyny robocze, **3** (2017)
12. W. Kisielewski, *Wpływ wybranych parametrów eksploatacyjnych i konstrukcyjnych na opory główne przenośników taśmowych,* PhD Thesis (not published), Wrocław University of Science and Technology (2016)

Experimental tests of the impact of selected parameters on the indentation rolling resistances

Dariusz Woźniak[1], *Lech* Gładysiewicz[1], and *Martyna* Konieczna[1,*]

[1]Wroclaw University of Science and Technology, Faculty of Geoengineering, Mining and Geology, 27 Wyb. Wyspianskiego St., 50-370 Wroclaw, Poland

Abstract. Belt conveyors are main part of transporting systems in mines and in many other branches of industry. During conveyor belt works different types of resistances are generated. Indentation rolling resistance is the most significant component of the resistances from the perspective of energy losses and it cause the biggest costs as well. According to latest state of analyses and measurements it is well known that theoretical rolling resistance were underestimated in comparison with the measured in-situ one. In this paper new method for determination indentation rolling resistance is presented. The authors compared theoretically and experimentally established damping factors. The relation between these two values enabled to obtain more precise equation for damping function. This function is one of the most important component in calculation of the rolling resistance. In new theoretical model value of rolling resistance is nearly twice higher than this used so far.

1 Introduction

Present-day belt conveyors are highly reliable machines capable of transporting materials at distances of several dozen kilometers, passing through curvatures and operating on inclined terrain. Special operating conditions do not affect their efficiency. Therefore, research in the field of belt conveyor transportation increasingly focuses on lowering the power consumption of belt conveyors while maintaining their capability to serve the same transport functions. Lowered power consumption of belt conveyors translates directly into the reductions of maintenance costs [1]. The lowering of power consumption, however, requires performing both theoretical analyses [2-4] and laboratory tests [5, 6]. The reduction of motion resistances, and hence the reduction of the energy absorbed by belt conveyors, remains the object of research in many countries [7, 8], as well as in Poland [5, 9, 10]. This paper offers a simple and inexpensive method of testing belt rolling resistance (also referred to as belt indentation resistance or indentation rolling resistance). The method allows the analysis of such factors as core design, cover thickness, ambient temperature, idler travel frequency, or load on the total value of rolling resistance. The analysis presented in this paper focuses on

* Corresponding author: martyna.konieczna@pwr.edu.pl

how the frequency and speed of rolling the idler on the belt sample influence the belt's rolling resistance.

2 Methods for Testing Belt Rolling Resistance

The translational motion of a belt on a conveyor is accompanied by a number of physical phenomena which entail various forms of energy conversion. In the case of belt conveyors longer than 80 m, main resistances are economically most important. The dominant component of main resistances are belt rolling resistances, which are related to a cyclical process of pressing the coat of an idler into the bottom cover of the belt at each support idler [11].

Indentation rolling resistances are investigated in many research centers and belt manufacturing companies worldwide. The main goal is to develop novel solutions which will lead to the reduction of motion resistances in conveyor belts, while retaining their strength parameters. The known and used methods for determining the bellt-on-idler rolling resistances include: 1) analytical methods based on the parameters of the bottom cover as calculated in laboratory tests, 2) actual measurements of belt rolling resistances in laboratory conditions and 3) actual measurements of belt rolling resistances in belt conveyor operating conditions [12].

The most widespread methods for laboratory measurements of indentation rolling resistances include methods developed at the University of Hannover [7, 13], at the Newcastle University [8] and the inclined plane method developed at Wroclaw University of Science and Technology [11]. The dimensions of the testing rigs used in the above methods necessitate the use of large sample test pieces having nominal length of 7.5 m and also impede tests in sub-zero temperatures. Moreover, the above methods do not allow tests with idler rolling frequencies typical of actual operating conditions of a belt conveyor.

This paper describes a test rig for determining conveyor belt rolling resistances, which allows a small-scale and inexpensive, yet precise measurement of resistance values. Smaller size belt samples allow the belt top cover to be removed with greater ease, and hence the core and the bottom cover can be tested separately. The tests can be performed at temperatures both above and below zero degrees Celsius, as well as at various load values and idler travel frequencies. Thus, the operation of a belt conveyor in actual mining conditions may be simulated in a number of variants. Research to date allowed precise adjustment of both the test method and the result analysis method [14].

3 A new method for testing indentation rolling resistance

The method has been developed at Machine Systems Division, Wroclaw University of Science and Technology, Faculty of Geoengineering, Mining and Geology. It consists in the measurement of rolling resistance during cyclical rolling of an idler on a belt sample approx. 400x60 mm in size. The test, in which the idler travels (passes) on the belt, takes a form reverse to actual conditions, in which it is the belt that passes on a series of idler sets. Notwithstanding, the physical phenomena involved in the process are identical. The main aim of the research was to develop a simple and inexpensive method for testing indentation rolling resistance, which would enable the analysis of how the resistance is influenced by such factors as temperature, load, idler travel frequency (indentation frequency), as well as belt design and bottom cover thickness. Small dimensions of the test rig allow placing it in a thermal chamber and conducting tests in various temperatures above and below zero degrees Celsius; in addition, small dimensions of test samples reduce the cost of tests (no need for long belt sections and splicing) and enable optimal selection of materials and the belt regarding the energy-efficiency of its design.

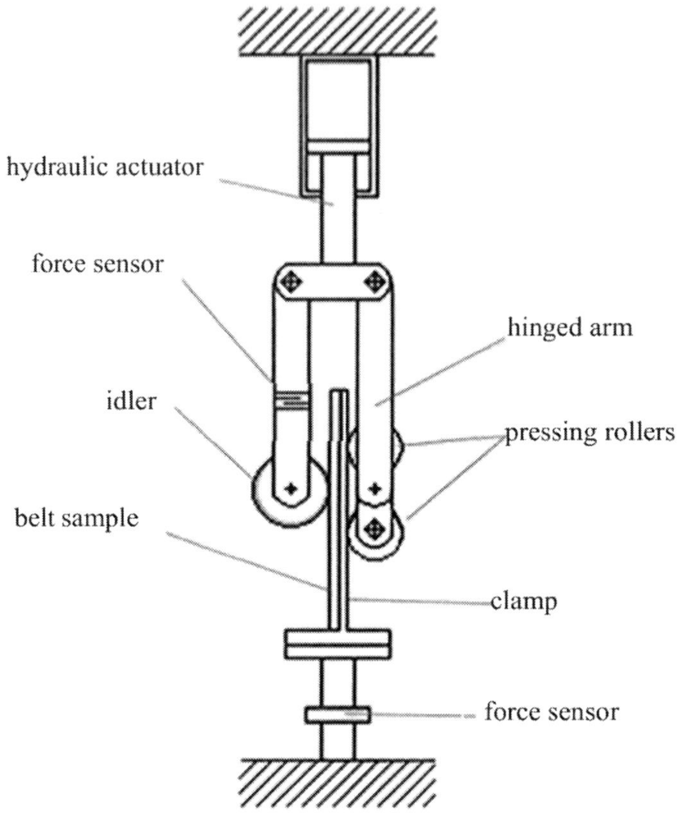

Fig 1. Diagram of a test rig for measuring indentation rolling resistances – small scale tests [4].

Figure 1 is the diagram of the test rig. After the top cover is removed, the belt sample is secured in the clamp (and tensioned). The idler rolls on the test piece. The idler is pressed against the belt using a tensioning system equipped with a force sensor. The rolling resistance is measured with force sensors. To compensate for the transverse forces transmitted to the force sensor and the piston rod of the hydraulic actuator, additional rollers were used, which always roll on the metal part of the clamp. Both the idler and the additional rollers are guided with hinged arms. The idler's reciprocating movement is effected by a hydraulic actuator. Prior to measurements, own resistances of the test rig must be determined. This is done by first assembling the test rig without the belt sample inserted. Own resistances are determined for the idler rolling on the metal clamp, at a set temperature and at a set load. In the next step, the measurements are performed with the belt sample inserted. The difference between motion resistances recorded with the sample and those recorded without the sample correspond to the rolling resistance of idler on belt [5].

Fig. 2. Test rig for measuring indentation rolling resistances (PWr) [12].

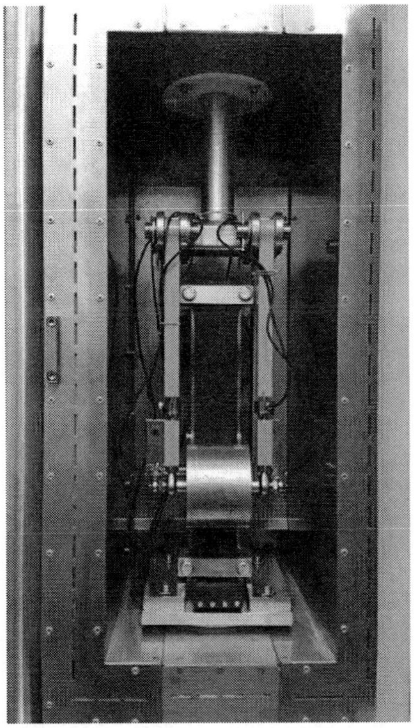

Fig. 3. Test rig for measuring indentation rolling resistances in a thermal chamber (PWr) [12].

Fig. 4. The measurement system of the test rig for measuring indentation rolling resistances (PWr) [12].

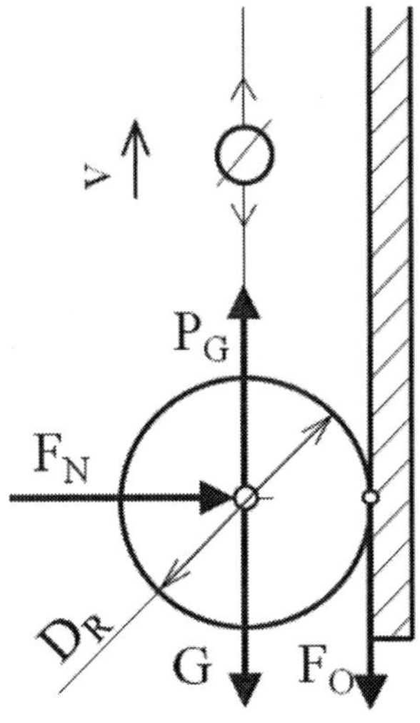

Fig. 5. System of forces during the steady motion of the idler upwards (PWr) [12].

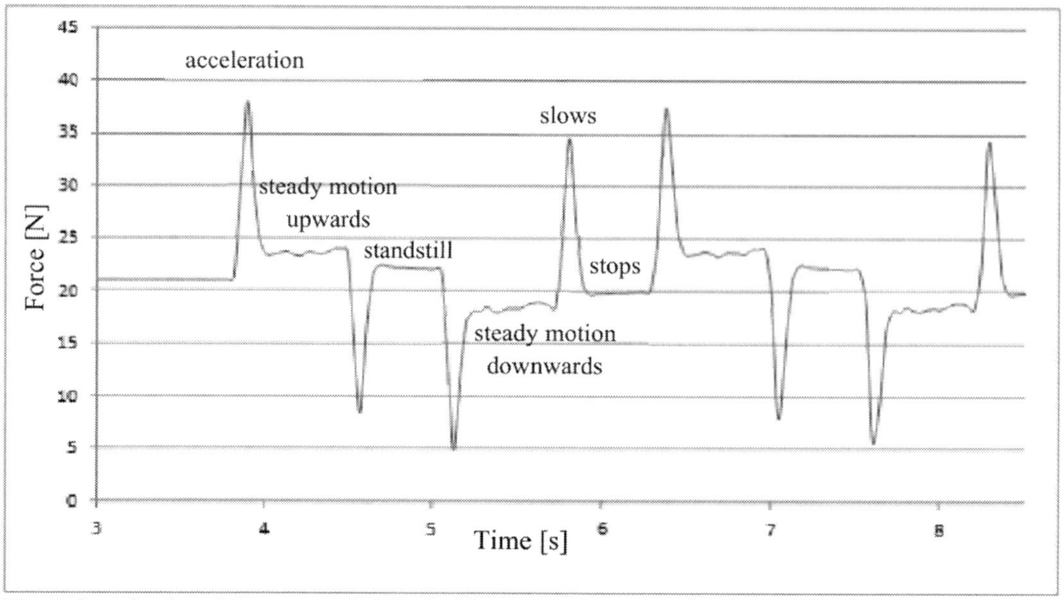

Fig. 6. Sample plot of forces recorded for the idler rod during tests of rolling resistance of idler on belt [12].

The idler travels on the belt sample upwards and downwards with constant speed and stops to a standstill at return points. Drawing force is recorded with force sensors (1) shown in Fig. 4, located in the idler arms. Pressing force of the idler on the belt is recorded with force sensors (2), located in the roller rods. Fig. 6 shows a sample plot of force recorded for the idler rods during the tests. A single motion cycle of the idler (one motion upwards and one motion downwards) can be divided into several stages, which are represented in the plot of force recorded for the idler rods. In the first stage, the idler accelerates to a set speed – here a force peak occurs which is related to the acceleration of a given idler mass and to the setting in rotary motion of the mass of idler's moving part. In the following stage, the idler travels on the belt with a steady motion upwards – here the force is stable. In the third stage the idler slows and stops to a standstill – here the force in the rod decreases rapidly. As the idler remains in standstill for a brief time, the force becomes stable, and in the following stages the idler accelerates again to a set speed, moves with a steady motion downwards, slows and stops at the bottom return point [12].

The calculation of idler motion resistance requires considering the courses of forces during constant motion upwards and downwards. Fig. 5 shows the system of forces during the steady motion of the idler upwards. Steady motion is characterized by a balance of forces, and hence the force recorded for the idler rod during the upward motion (P_G) is balanced by force of gravity G and by motion resistances F_O. In the case of the idler's steady motion downwards, only the vector of force for motion resistances will change its sense. The following equation can be constructed:

$$-P_G + F_O + G = 0 \quad (1)$$

$$-P_D - F_O + G = 0 \quad (2)$$

Thus, a motion resistance equation can be derived:

$$F_O = \frac{P_G - P_D}{2} \quad (3)$$

where

P_G – force in the idler rod during the upward motion, in N
P_D – force in the idler rod during the downward motion, in N
G – force of gravity, in N
F_O – motion resistances, in N

As the idler travels on the belt, motion resistances comprise idler rotational resistance F_R and idler rolling resistances on belt (idler indentation resistances) F_E. In the case when the idler travels on the clamp without a belt, the rolling resistance values for a steel idler traveling on a steel surface are negligible and thus motion resistance F_{OS} may be assumed to be equal to idler rotational resistance F_R. Idler rolling resistance on belt (idler indentation resistance) F_E will be:

$$F_E = F_O - F_{OS} \tag{4}$$

$$F'_E = \frac{F_E}{b_R} \tag{5}$$

where

F_E – idler rolling resistance on belt, in N
F'_E – idler unit rolling resistance on belt, in N/m
F_{OS} – motion resistance in a test without a belt, in N
b_R – the length of the contact line between the belt and the idler coat (sample thickness), in m

Initial tests were performed for an ST 2500 steel-cord conveyor belt with the bottom cover 7 mm in thickness. The idler was pressed against the belt with force F_N=250 N, which results in a unit load per 1 m of the length of the contact line between the belt and the idler coat having value F_N'=4.2 kN/m. The idler had diameter D_R=108 mm. The idler traveled on the belt with speed v=250 mm/s and frequency f=1.2 Hz. The tests were performed in ambient temperature T=20°C. The idler traveled on the belt in 50 cycles (1 cycle understood as a motion upwards and a motion downwards). During the test, the recorded values included time, force in idler rods and pressing force of idler on belt. In the following step, the same conditions and parameters were used to perform measurements without the belt: the idler traveled on the steel clamp. The measurement results for both cases are shown in Fig. 7. Indentation unit rolling resistance for the tested belt is F'_E=27,1 N/m.

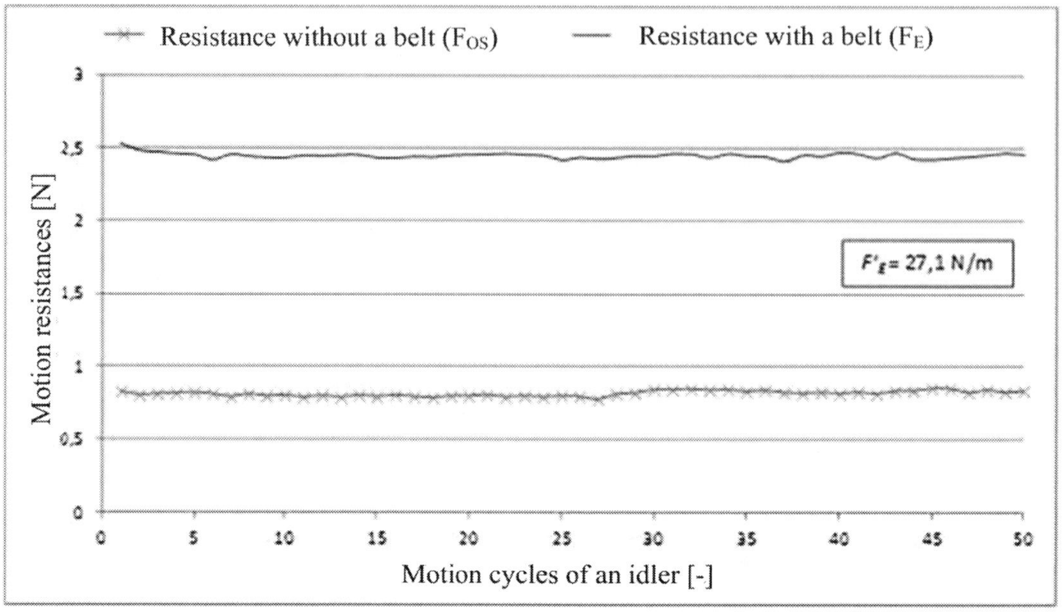

Fig. 7. Measurement results for idler rolling resistance on belt ST 2500 (pulley cover thickness – 7 mm, load – 4.2 kN/m, idler diameter – 108 mm, speed – 250 mm/s, travel frequency – 1.2 Hz, temperature – 20°C) [12]

4 Tests of the relationship between rolling resistance and the selected motion parameters of the idler

4.1 Relationship between rolling resistance and idler travel frequency

In accordance with the methodology described in the previous chapter, a series of rolling resistance tests was performed for two belts: ST 2500 – steel-cord belt described by the manufacturer as a belt with low rolling resistances (energy-efficient) and ST 2000 – a regenerated steel-cord belt. In both cases the tests were performed for various idler travel frequencies, in the range from 0.2 Hz to 1.4 Hz. For each of the samples, a total number of 7 various rolling resistance values was determined, which correspond to various frequencies. The tests were performed at a constant idler travel speed of 250 mm/s, as the standstill time of the idler at the return point can be regulated (Fig. 6).

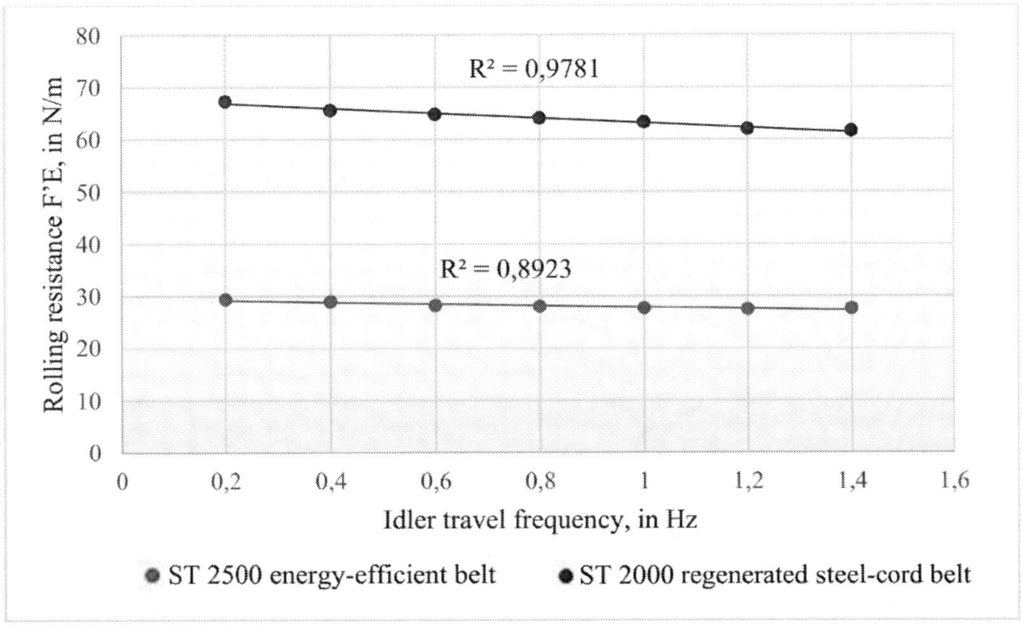

Fig. 7. Relationship between rolling resistance and idler travel frequency

As expected, the energy-efficient belt showed a significantly (over 200%) lower rolling resistance than the regenerated belt. In both cases, the relationship between the measured resistances and idler travel frequencies are linear and have a coefficient of determination equal to 0.8923 for the ST 2500 belt and equal to 0.9781 for the ST 2000 belt. Determination at this level implies that the selected characteristics offer highly precise representation of the investigated relationships. For both the energy-efficient and the regenerated belt, rolling resistances decreased as the travel frequencies increased. Albeit, the decrease for the regenerated belt was faster than for the energy-efficient belt.

4.2 Relationship between rolling resistance and idler travel speed

Another investigated parameter was the speed at which the idler traveled on the belt sample. Here the tests were also performed on the ST 2000 regenerated belt and on the ST 2500 energy-efficient belt. In the case of belt ST 2000, the measurements were performed for four different idler travel speeds, in the range from 100 mm/s to 250 mm/s (speed was incremented at 50 mm/s intervals). In the case of belt ST 2500, 7 measurements were performed for different idler travel speeds, in the range from 50 mm/s to 350 mm/s (speed was also incremented at 50 mm/s intervals). All measurements were performed at a constant idler travel frequency equal to 0.25 Hz.

 https://doi.org/10.1051/e3sconf/20182900002

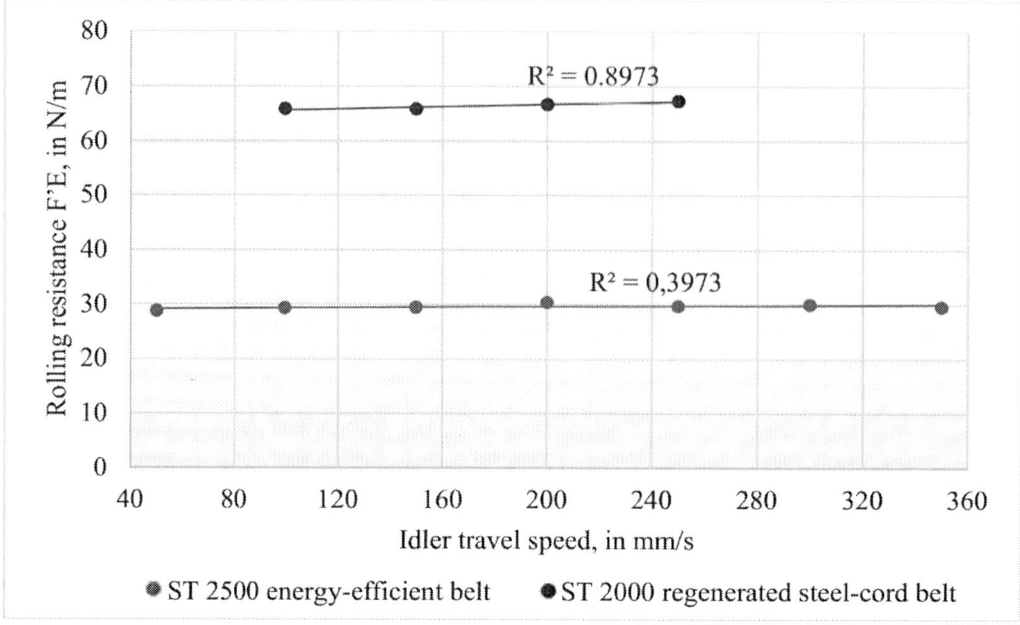

Fig. 8. Relationship between rolling resistance and idler travel speed

As was the case with the previous tests, here rolling resistances for the regenerated belt are also twice higher than for the energy-efficient belt. The values of rolling resistance increase together with increasing speed. These changes, however, have a distinctively smaller scale than the changes effected by the varying idler travel frequencies (which are practically negligible for the energy-efficient belt). This fact correlates with the results of theoretical analyses performed to date by researchers from a number of countries [2, 4], which assume that the influence of speed on rolling resistance is negligible. It also correlates with the results of experiments which show this influence to be limited [8]. However, in the case of the regenerated belt, which has higher rolling resistances, the range of resistance changes is greater than for the energy-efficient belt. Therefore, the influence of speed on the changes in rolling resistances depends on the elastic and damping properties of the rubber cover.

6 Conclusions

The new method for testing indentation rolling resistance enables an uncomplicated and inexpensive analysis of the influence of various belt motion parameters on rolling resistances. Research into the influence of idler travel frequency on rolling resistance was performed for the first time. Such research has not yet been described in literature.

The performed tests (4.1) show a significant influence of idler travel frequency on the rolling resistance measured on the belt. An increase in the frequency causes the resistance to decrease, which can be explained by reduced relaxation time of the rubber cover. In the case of the regenerated belt, the easured resistance was more than twice higher than in the case of the energy-efficient belt.

Tests performed for various idler travel speeds confirm that the influence of speed on rolling resistances is limited/negligible. Significantly, the above changes depend on the elastic and damping properties of the belt, which have the most significant influence on the rolling resistances generated as the idler travels on the belt (i.e. on the energy-efficiency of the belt).

Financial support by the project: Research, diagnostics and optimization of conveyor belt systems No. 0401/0166/16 and Grant no. 0401/0128/17.

References

1. J. Kudelko, Gospodarka Surowcami Mineralnymi - Mineral Resources Management, Volume **32,** Issue 4, 157-179 (2016)
2. C. Spaans, Bulk Solids Handling Vol. **11** No.4, 809-826 (1991)
3. X. Qiu, Journal of Engineering Mechanics, Vol. **132** No.11, 1241-1251 (2006)
4. L. Gładysiewicz, *Przenośniki taśmowe Teoria i obliczenia,* Oficyna Wydawnictwo Politechniki Wrocławskiej, Wrocław (2003)
5. D. Woźniak, L. Gładysiewicz, M. Hardygóra, D. Kaszuba, W. Kisielewski, *Stanowisko pomiarowe do badania oporu toczenia krążników,* Patent PL223986 (2016)
6. R. Król, W. Kisielewski, D. Kaszuba, L. Gładysiewicz, Procedia: Earth and Planetary Science, WMESS Vol. **15**, 712-719 (2015)
7. T. Wennekamp, S. Hotte, S. Von Daacke, L. Schulz, L. Overmeyer, Bulk Solids Handling Vol. **32**, Issue 6 (2012)
8. C. Wheeler, P. Munzenberger, International Materials Handling Conference (Beltcon 16) (2011)
9. M. Bajda, R. Król, Procedia: Earth and Planetary Science, WMESS vol. **15**, 702-711 (2015)
10. R. Król, W. Kisielewski, Mining Science – Fundamental Problems of Conveyor Transport, vol. **21**(2), 61–72 (2014)
11. M. Hardygóra, H. Komander, M. Bajda, Górnictwo i Geoinżynieria, Wydawnictwo AGH, Rok **35**, Zeszyt 3/1, Kraków (2011)
12. L. Gładysiewicz, D. Woźniak, M. Konieczna, Transport Przemysłowy i Maszyny Robocze, no **2**, 6-10 (2017)
13. DIN 22123: 2012-10 *Fördergurte - Gurtbreitenbezogener Eindrückrollwiderstand von Fördergurten - Anforderungen, Prüfung* (2012)
14. D. Woźniak, L. Gładysiewicz, M. Hardygóra, D. Kaszuba, W. Kisielewski, Transport Przemysłowy i Maszyny Robocze, no 3, 6-10 (2015)

The influence of drilling process automation on improvement of blasting works quality in open pit mining

Maciej Bodlak[1], *Dominik* Dmytryk[2], *Piotr* Mertuszka[3,*], *Marcin* Szumny[3], and *Grzegorz* Tomkiewicz[2]

[1]Wroclaw University of Science and Technology, Faculty of Geoengineering, Mining and Geology, 27 Wyb. Wyspianskiego St., 50-370 Wroclaw, Poland
[2]MAXAM Poland Ltd., Duninów 3, Chocianów, Poland
[3]KGHM CUPRUM Ltd. Research and Development Centre, 2-8 Gen. Sikorskiego St., 53-659 Wrocław, Poland

Abstract. The article describes the monitoring system of blasthole drilling process called HNS (Hole Navigation System), which was used in blasting works performed by Maxam Poland Ltd. Developed by Atlas Copco's, the HNS system – using satellite data – allows for a very accurate mapping of the designed grid of blastholes. The article presents the results of several conducted measurements of ground vibrations triggered by blasting, designed and performed using traditional technology and using the HNS system and shows first observations in this matter.

1 Introduction

Currently in the world mining industry, modern mines use technologies that allow for remote control, automation and navigation of heavy machines. Manual control in this type of machines nearly reached excellence thanks to hydraulic and electrohydraulic joysticks application. Machine automation in the mining industry allows for the precise drilling of the blastholes, which results in more uniform fragmentation and consequently for reduction of operating costs [1].

In 2016, Maxam Poland Ltd., which provide the blasting service for opencast mining, have purchased the Atlas Copco SmartROC T45 drill rig (Fig. 1) to raise the competitiveness of the company on the market. The drill rig has been equipped with the total-site Hole Navigation System based on the Global Navigation Satellite System (GNSS) to ensure all holes are drilled precisely parallel to each other. GNSS-guided drilling is accurate to within 10 centimeters and automatically calculate its depth to the given elevation [2].

In spite of the use of high-tech machines, the quality of the blasthole drilling, according to designed drilling and firing patterns, largely depends on experiences and skills of operator. Precisely drilling of the holes permit optimized blasting patterns while maintaining control of fracture size, fly-rock and ground-borne vibrations, contributing to reduced costs [3].

The vibrations induced by blasting operations in opencast mining have been analyzed in a number of research works [3-6]. The results of measurements and observations have

* Corresponding author: pmertuszka@cuprum.wroc.pl

proved, that some factors may affect their intensity, including the distance from the blasting site, geological structure of the rock mass, the amount of explosives used and applied delays.

The results of measurements of ground-borne vibrations induced by blasting have been analyzed within the framework of presented paper. Two different methods of blasthole drilling were considered, i.e. using the HNS system and traditional drilling method. The tests were carried out in selected opencast mine in Poland, in which blasting service is provided by Maxam Poland Ltd.

2 Hole Navigation System

The idea behind the application of the HNS system is to support the process of determining the position of the drill rig and the designed blastholes using correction data, e.g. ASG-EUPOS system (EUPOS Active Geodetic Network). To generate a stream of observations and/or adjustments based on data from the group of ASG-EUPOS stations, the computing system must have approximate user position information, which uses the Leica GNSS receiver fitted to the drill rig (Fig. 1).

Fig. 1. SmartROC T45 drill rig with a GNSS Leica receiver [7].

This receiver sends information on its position to the computation module on the ASG-EUPOS server. In response to the approximate position of the user, the ASG-EUPOS computation module generates a stream of correction network data and sends it to the receiver in the drill rig. Communication between the computation module on the support service server and the receiver module on the drill rig is carried out through the GSM network. In addition to the use of the ASG-EUPOS service, the user must also be able to use data transmission in the GSM network from one of the local mobile network operators that provides network coverage at the place of machine operation. Lack of permanent access to the cellular network will prevent transmission of data between the drill rig and the ASG-EUPOS computation module and the transmission of correction data, thus precluding the use of the HNS system on the drill rig [7]. The drilling pattern is designed in the ROC Manager software, which is part of the HNS system. It allows for analysis of the data obtained during drilling of blastholes using the Measuring While Drilling (MWD) technology (Fig. 2). In addition, the system enables the tracking of drilled blastholes and their deviations.

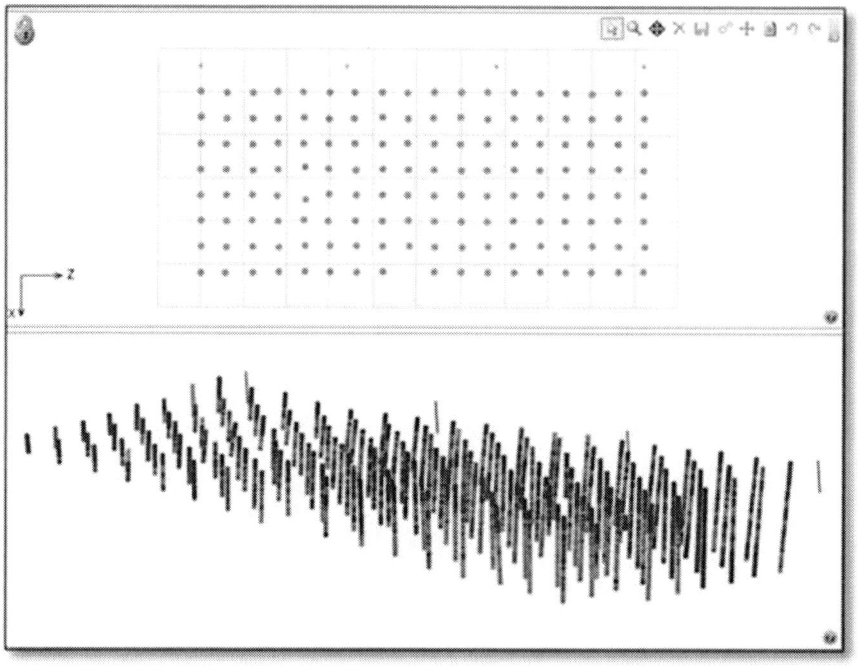

Fig. 2. Location of the drilled holes using the MWD data, generated in the ROC Manager software [7].

The HNS system collects data based on sensors located on the drill rig platform. By assigning colours to numerical values (e.g. rock hardness), the user can trace their variations at each meter of the blasthole. The ROC Manager uses the GPS coordinates of the two reference points that have been set by positioning the drill mast with the GNSS receiver mounted above them. The mapping of the blasthole grid in the 3D space, with no support tools, is both cumbersome and time-consuming. So far, the mapped drilling and firing patterns have shown a view of the holes in two planes. Incorporation of vertically deviated blastholes in the drilling pattern plan have caused problems in maintaining the appropriate spacing between them, especially in the bottom section, which in turn led to uneven distribution of the explosives in the blastholes. Currently, thanks to the Hole Navigation System, the drilling machine operator receives a digital pattern layout, which is exported into the computer memory of the drill rig. The design of drilling pattern takes into account the complex geometry of the slope and the spatial orientation of the blastholes, eliminating errors at the design stage.

The blasthole drilling process is almost fully automated, thus the human factor is reduced, i.e. the errors of the drilling machine operator. Drilling of the individual hole is carried out up to the set level ordinate, taking into account the subdrilling and the frequent denivelations of the floor, on which the drill rig is travelling. So far, during drilling operations, many attempts were made to obtain the exact hole lengths specified in the blasting pattern layout, which, in the case of complicated grids, could lead to incorrect performance of subdrilling. One of the example of blasting works design and control using the HNS system, which was aimed at access pit execution is presented in Figures 3 to 5. The views of the designed and performed blastholes were generated in the ROC Manager software.

In this case, determination of blasthole grid in a conventional way would take approximately 2 hours. Determining of the same grid by the use of HNS system takes not more than 15 minutes. By changing the final ordinate level of each rows of holes, the access pit to the next level can be performed with very high accuracy (each row of holes which is drilled 30 cm deeper, in case of 3-metre spacing between the rows, gives the slope inclination of 1:10, i.e. 6°). Accurate determination of hole lengths with the precise subdrilling results in regular surface of the floor with a simultaneous optimization of fuel consumption (additional unnecessary metres = additional fuel consumed) and finally – reduce the drilling time.

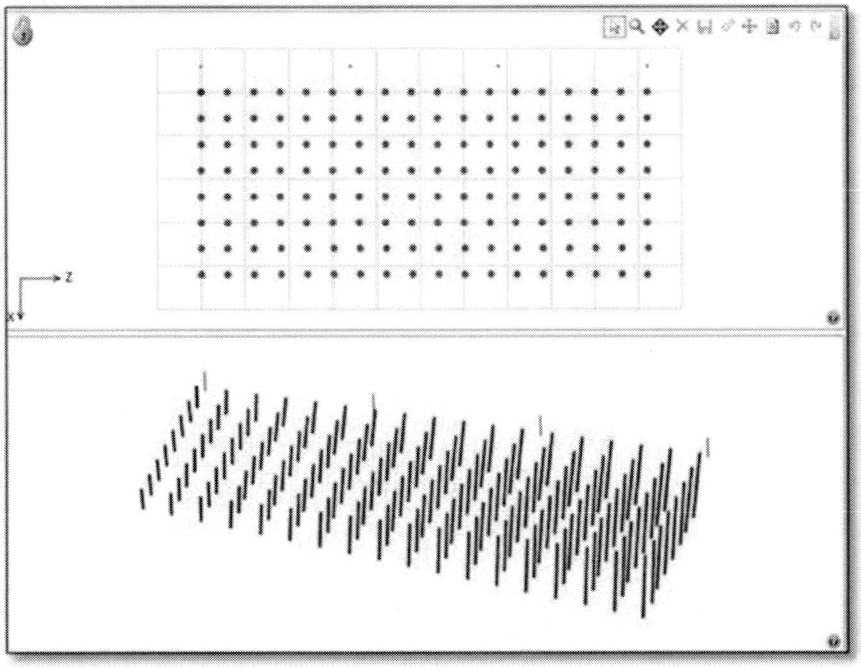

Fig. 3. View of the designed blasthole grid for an access pit [7].

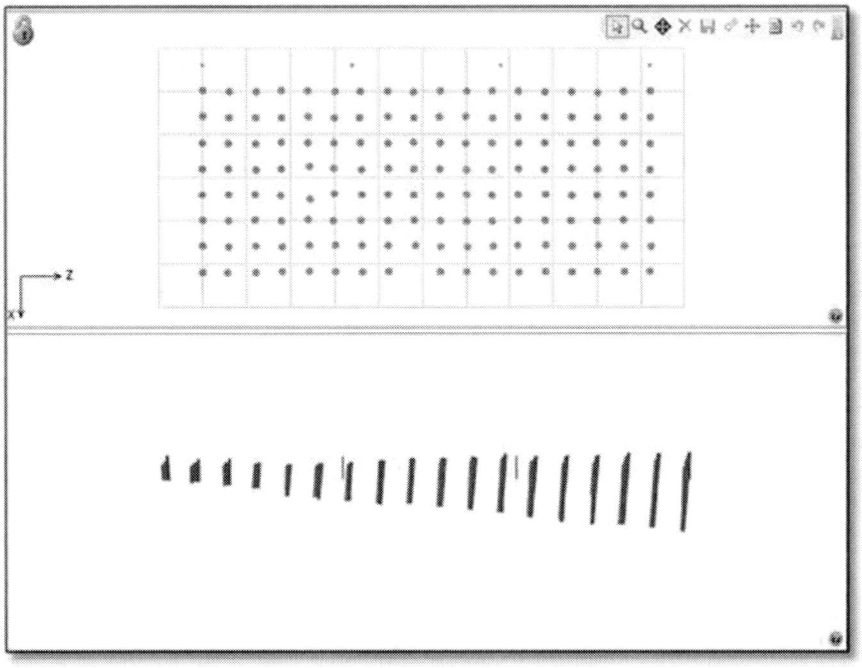

Fig. 4. Locations of drilled blastholes for an access pit [7].

Fig. 5. View of the access pit at the mine site prior to firing [7]

The length of the subdrilling is determined by the experts during the measurements of the range of paraseismic vibrations and the determination of the maximum co-operating explosives charges [8]. One of the reasons of increased vibration is incorrect performance of subdrilling. Up until now, inaccurate drilling of the holes, mainly in the bottom section, has generated the stumps above the theoretical grade and has increased levels of ground vibrations. By selecting the right mining method and appropriate drilling and firing patterns, the size of the ground vibrations can be controlled and finally reduced. After the first tests of the HNS system, the decrease of ground vibrations caused by rock blasting was observed. For that reason, an analysis of the influence of the correct drilling of the holes on the level of paraseismic activity has been developed.

3. Seismic analysis

So far, a number of research works were made to determine the impact of the drilling accuracy on the quality and effectiveness of blasting works in both open-cast and underground mining. In the case of underground mining, an increase of the round during the attempts of the Feeder Guiding System (FGS) implementation was observed. This was possible by controlling the correctness of the work with respect to the internal procedures, i.e. drilling of the blastholes according to the applied drilling and firing pattern and obtaining the required quantity and length of the holes. The trials have also showed the improvement of blasting works efficiency in terms of the rock fragmentation [9].

The technology of designing and performance of blasting operations in opencast mining is significantly different from blasting in underground mining. One of the aim of underground blasting, apart from throwing rocks away, is to distress the rock mass by amplifying the elastic wave caused by detonation of explosives, which is one of the main elements of active rockburst prevention method. In the opencast mining, the ground vibration induced by detonation of explosives is undesirable because it is energy loss which could have negatively influences the technical condition of the buildings and other constructions located in the vicinity of blasting works area. That means that the drilling and firing pattern should be designed according to the expert's guidelines. The analyses on the ground vibrations generated by blasting have been developed and examined by many researchers. It follows that the observed paraseismic activity is influenced by many factors of geological, mining and technological nature. The most important include [3, 4]:

- quantity of explosives and number of co-operating charges,

- type of explosive used,
- blastholes geometry (including the length and spacing of holes),
- type of initiation (instantaneous or milli-second detonators),
- applied delay time,
- physical and mechanical properties of the rock mass,
- the distance from the blasting site.

In order to preserve the admissible values of ground vibrations, appropriate performance of the drilling grid, in accordance with the designed drilling and firing pattern, which consider the recommendations of experts is required. Previous mining experiences and earlier research done in this field have proved, that insufficient accuracy of the drilling grid performance will be reflected in the level of ground vibrations [3, 4, 10]. And thus, it will be possible to determine the impact of the automation of the drilling process on the final effect of blasting operation. Considering the above, in order to draw reliable conclusions from the comparative analysis, the number of variables should be reduced and investigated in similar geological and mining conditions, which determines the selection of given explosive type and the geometric parameters of the blasting works (e.g. size of burden, subdrilling and hole spacing). These parameters are related in turn with selection of the column charge construction and its size. It is commonly assumed that these parameters are constant during the study in a particular location. Therefore, two following variables should be determined: the weight of the co-operating charge (Q) and the distance of the measuring point from the blasting site (r). Vibration intensity (velocity of vibration [mm/s]) at the measuring point can be calculated from the following relationship:

$$v = k \cdot Q^{\alpha} \cdot r^{-\beta} \tag{1}$$

where:
k – coefficient characterizing geological conditions and parameters of blasting works,
r – distance of the measuring point from the blasting site,
Q – weight of explosives per co-operating charges.

During the control tests carried out by experts, the specific charge should be obtained. It may be calculated from the following relationship:

$$\rho = \frac{Q^{n}}{r} \tag{2}$$

where: $n = \alpha / \beta$.

Unknown parameters of the equation (1) are: coefficient k and exponents α and β. They are obtained during the control tests aimed at determining the maximum explosive charge per one delay (co-opearing charge) and permitted levels of ground vibrations.

Measurements of the ground vibrations levels induced by detonation of explosive charges were carried out in one of the Polish granite mines. Conducting the measurements in a specific zone allows to investigate the seismic activity triggered by blasting works of similar parameters, i.e. the maximum co-operating charge weight, total charge (Q), the initiation method and the blastholes geometry. Monitoring of the vibration velocity was made using the pocket size multi-channel vibration instrument called Vibraloc (Fig. 6). It is tri-axial geophone integrated in one device. Accuracy is equal +/- 5 % for vibration velocity from 1 to 240 mm/s in the frequency range 15÷250 Hz [11]. The following set up was used during survey:

- Sampling frequency - 4000 Hz
- Trigger level - 0,5 mm/s (for all channels L, T, V).

Device was installed on the ground (from 200 to 600 m from blasting point) by means of three anchors for good connection between device and soil. Figure 7 shows an example plot of the velocity of ground vibrations generated by the Vibraloc system.

Fig. 6. Vibraloc instrument

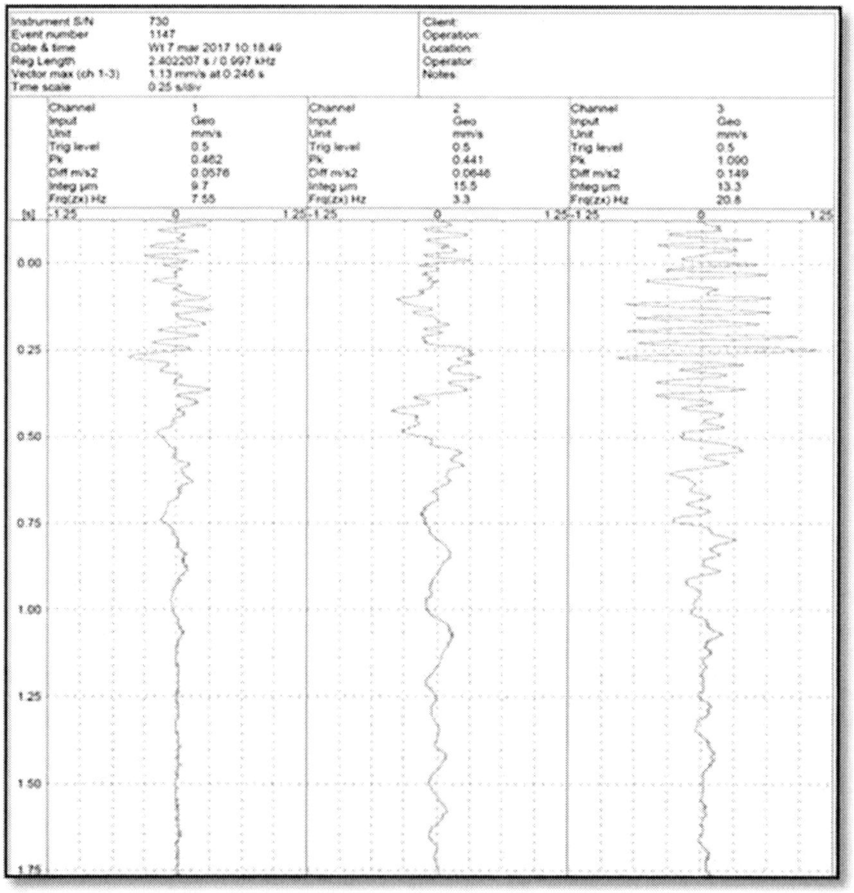

Fig. 7. An example plot of the velocity of ground vibrations from the measurement of 7 May 2017

The velocities of vibrations generated by blasting works, which were designed and performed using the HNS technology (variant I) and using the conventional method (variant II) were analyzed. Each case consisted of 7 series of blasting. Drilling and firing patterns for individual blasting works were similar. It was determined by the geological and mining conditions of the mine site. The blasting parameters and the total volume of the explosives were consistent with the mining technology currently used in that mine. The selected parameters of the blasting works carried out within the framework of presented analysis for both variants are presented in Table 1. The schematic diagram of the charge construction in an explosive column, as well as the blasthole spacing are shown in Figure 8.

Table 1. Selected blasting parameters considered in analysis.

Parameter	Symbol	Unit	Blasting with HNS system (variant I)	Conventional blasting (variant II)
Bench height	H	m	16.3÷15.8	14.0÷15.5
Hole depth	l	m	17.3÷16.8	15.0÷16.5
Subdrilling	p	m	1.0	1.0
Stemming	b_p	m	2.5	2.5
Intermediate stemming	l_{pp}	m	-	1.0
Burden	z	m	3.3	3.0
Distance between holes	a	m	3.8	3.8
Distance between rows	b	m	3.0	3.0
Hole inclicantion	α	degrees	83÷88	83÷88
Diameter of the blasthole	Φ	mm	89	89
Number of blastholes	n	szt.	42	30
Number of rows	i	szt.	4	2
Charge weight per delay	Q_z	kg	120	110
Charge weight per hole	Q_o	kg	120	110
Total charge	Q_c	kg	3991	3120
Connector delay	-	ms	25, 42, 67	25, 42, 67

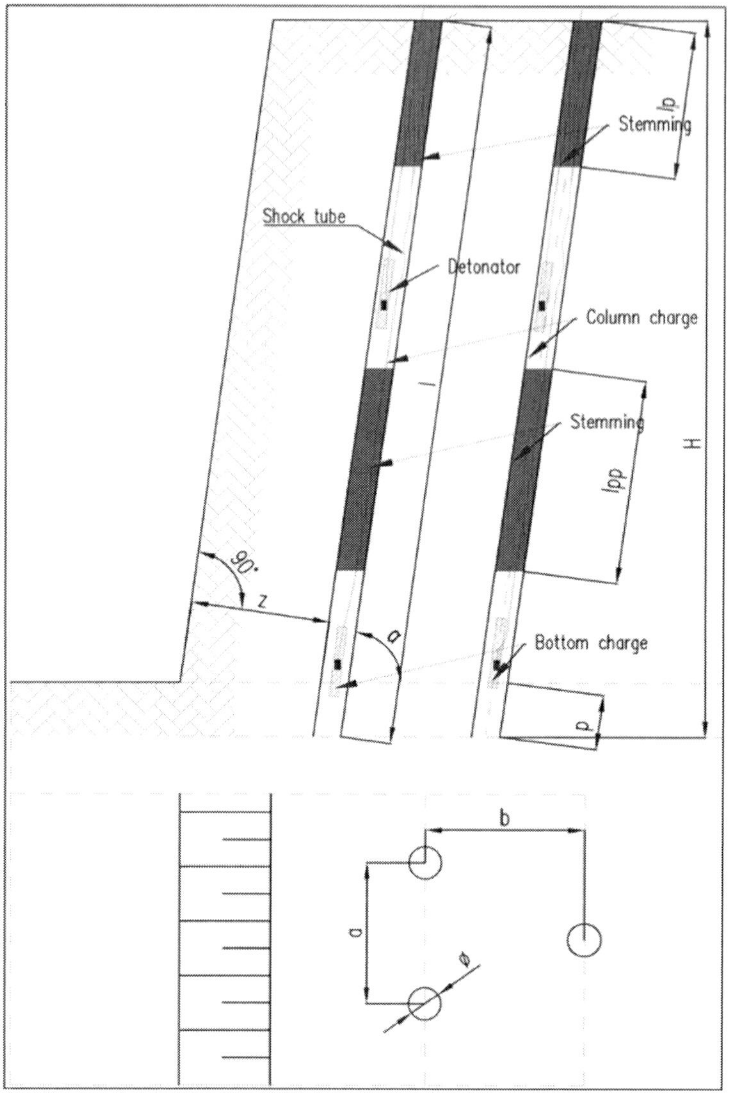

Fig. 8. Geometrical parameters of the blasting grid and construction of an explosive charge

The mathematical propagation model of ground vibration (1) was determined based on the studies of the range of the vibration impact on the surrounding environment in the vicinity of the analyzed granite mine. This research was carried out by the Poltegor Opencast Mining Institute. It has allowed to calculate the maximum specific charge (6), the predicted charge (4) and the charge calculated on the basis of the values of vibration velocity (5). Specific charge allows for the comparative seismic analysis (7) for selected blasting configurations differing in the volume of the charge Q and the distance of the measuring point from the vibration epicentre. For analysis of the vibration propagation, the following formulas were used [3]:

- propagation equation

$$v = k \cdot \rho^{\beta} \; [mm/s] \tag{3}$$

k ,β – coefficient and exponent, which are dependent on the geological and mining conditions and are determined experimentally,

ρ – specific charge.

The predicted specific charge, calculated from the blasting works parameters (quantity of explosive charge and the distance of the measuring point from the vibration epicentre) according to the following relationship:

$$\rho_{rk/rhns} = \frac{Q^{1/\beta}}{r} \; [kg/m] \tag{4}$$

where:
β – exponent, equal 2 for local geological and mining conditions [6][7],

$\rho_{rk/rhns}$ – predicted value for the variant with conventional blasting ρ_{rk} and designed and performed using the HNS system ρ_{rhns} determined on the basis of the blasting works parameters (e.g. explosive charge) and the distance of the measuring point from the blasting site. This value should be lower than admissible charge ρ_{dop}.

Specific charge calculated from the results of measurements:

$$\rho_{k/hns} = \left(\frac{v_{zxy}}{k}\right)^{1/\beta} [kg/m] \tag{5}$$

$\rho_{k/hns}$ – the value calculated from the measurements, for the variant with conventional blasting ρ_k and variant designed and performed using the HNS system ρ_{hns} determined from the relationship (5) by applying propagation equation parameters (3) and measured vibration velocity. The value of this parameter should be less than or equal $\rho_{rk/rhns}$ and consequently also less than ρ_{dop}.

The admissible specific charge, calculated on the basis on the propagation equation and acceptable vibration velocity:

$$\rho_{dop} = \left(\frac{v_{dop}}{k}\right)^{1/\beta} [kg/m] \tag{6}$$

Finally, the volume of specific charge can be represented in the form of following inequality:

$$\rho_{dop} \geq \rho_{rk/rhns} \geq \rho_{k/khns} \tag{7}$$

Afterwards, all calculated values of specific charge ρ have been averaged and compared to each other. Direct comparison of the specific charge calculated for the variant of the HNS system (variant I) ρ_{khns} with the specific charge calculated for the variant of the conventional blasting (variant II) ρ_k would be unreliable, because the individual blasting operations differed in the volume of the total charge (Q) and the measuring points were located at different distances from the blasting site (r). Figure 9 shows, that the predicted value of specific charge for the blasting works performed using the HNS system ρ_{rhns} is equal to the predicted value for the second variant ρ_k. Thus, the differences between the measured values $\rho_{k/khns}$ and predicted values $\rho_{rk/rhns}$ should be taken into account. Increasing the difference between the triggered value ρ_{khns} and the predicted value ρ_{rhns} compared to the difference for variant II, provides information on the reduction of seismic activity. The results of the measurements are presented in Table 2.

Table 2. Results of vibration velocity measurements with corresponding specific charges for both analyzed variants.

Type of blasting and number of test		Q [kg]	r [m]	v_{zxy} [cm/s]	$\rho_{rk/rhns}$ $[kg/m]$	$\rho_{k/khns}$ $[kg/m]$
Blasting with HNS system (variant I)	1	130	300	0.113	0.0380	0.0221
	2	120	300	0.103	0.0365	0.0214
	3	111	400	0.056	0.0263	0.0173
	4	125	600	0.124	0.0186	0.0229
	5	120	300	0.096	0.0365	0.0209
	6	110	300	0.083	0.0350	0.0199
	7	130	300	0.088	0.0380	0.0203
Conventional blasting (variant II)	1	110	260	0.227	0.0403	0.0283
	2	120	300	0.136	0.0365	0.0236
	3	115	200	0.47	0.0536	0.0365
	4	120	300	0.385	0.0365	0.0340
	5	130	280	0.641	0.0407	0.0407
	6	125	370	0.31	0.0302	0.0315
	7	110	200	0.626	0.0524	0.0404

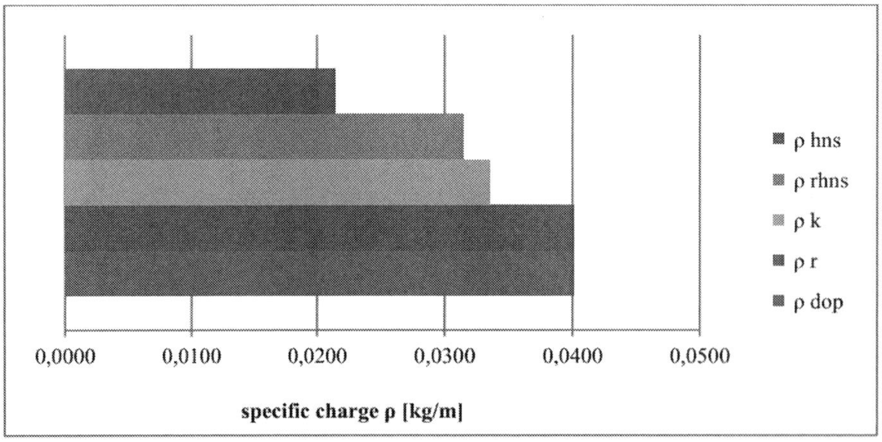

Fig. 9. Graphic analysis of the results of comparative tests of the blasting using the HNS system (ρ_{hns}, ρ_{rhns}) with conventional blasting (ρ_k, ρ_r).

4. Summary

The results of the measurements of ground vibrations triggered by blasting works, despite a small amount of obtained data, show a significant difference in level of specific charge and measured vibration velocities depending on the technology used. Individual vibration measurements were carried out in almost identical mining and geological conditions and at a similar distance from the source of vibration, so that the results could be considered as reliable.

The average value of the specific charge, depending on the measured vibration velocity, in this specified place and situation is approximately 30% lower than the predicted value for variant I. In contrast, the level of specific charge in conventional blasting (variant II) is approximately 20% lower than the predicted value, so the drop in the triggered seismic activity when using the HNS system is noticeable. Figure 10 shows the relationships between the vibration velocity and the predicted specific charge value for each of the analyzed variants.

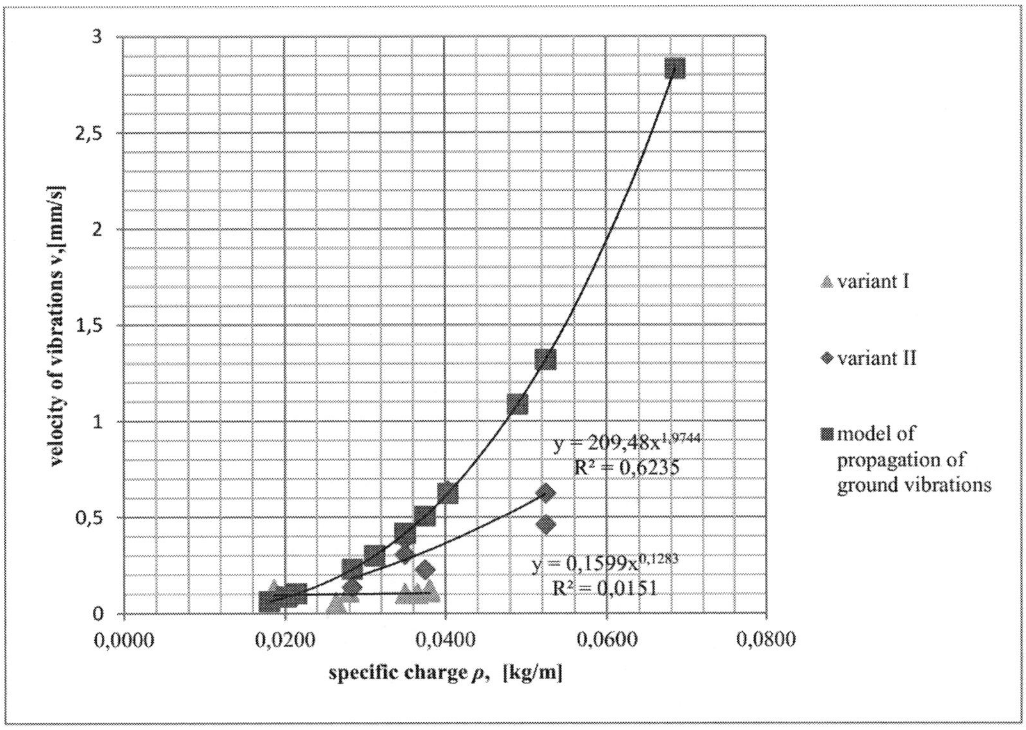

Fig. 10. Graphic analysis of the comparative studies results of vibration velocity as a function of specific charge.

Research work in the aspect of the impact of the blasthole drilling process quality on the final result and effectiveness of blasting works should be continued by expansion of the scope of measurements. Determination of the degree of rock fragmentation from a blast using the HNS system and methods based on conventional drilling seems to be crucial from the point of view of the blasting works effectiveness.

The observed drop in seismic activity when using the HNS system is associated with the maintenance of suitable subdrillings and the uniform distribution of explosives in the blastholes, particularly in the bottom section, which was problematic in the conventional drilling method. Reducing the paraseismic activity in the near future may contribute to a change in the vibration propagation pattern and, as a result, to the increase of the admissible charge weight per delay and total charge, which will directly translate into a reduction on operation costs of mining works. This paper shows preliminary observation of influence of

HNS system usage on inducted vibration and more research is needed to obtain more accurate data in this subject.

This work was financed by the Grant no. 0401/0128/17.

References

1. J. Marianowski, M. Palion, Chippings: production - transport - application, **E. 4**, 10-14 (2013)
2. Promotional materials of Atlas Copco
3. Z. Onderka, J. Sieradzki, Mining and Geoengineering **E. 3/1** Science and Education Publishing House AGH (2004)
4. S. Olofsson, *Applied explosives technology for construction and mining* Applex, Arla (1995)
5. G.R. Tripathy, R.R. Shirke, M.D. Kudale, Journal of Rock Mechanics and Geotechnical Engineering, Volume **8**, Issue 2, 248-255 (2016)
6. A. Kumar, D. Deb, Journal of Geological Resource and Engineering **4,** 203-212 (2015)
7. Material from Maxam Poland Ltd.
8. Polska Norma PN-85/B-02170 „*Ocena szkodliwości drgań przekazywanych przez podłoże na budynki*"
9. K. Ornet, *Research of the influence the drilling process on the durability cutting tools,* PhD thesis AGH (2012)
10. Z. Onderka, Blasting works in mining, 175-208 (2003)
11. Vibraloc, User's Manual, Version 1.2.9 and later 2010-03-24

E3S Web of Conferences **29**, 00004 (2018) https://doi.org/10.1051/e3sconf/20182900004
*XVII*th *Conference of PhD Students and Young Scientists*

Laboratory testing of a long expansion rock bolt support for energy-absorbing applications

Krzysztof Skrzypkowski[1,*]

[1]AGH University of Science and Technology, Faculty of Mining and Geoengineering, 30 A. Mickiewicz Av., 30-059 Kraków, Poland

Abstract. The main purpose of rock support and reinforcement in underground mining is to maintain excavations safe and open for their intended lifespan. The basic type of rock mass reinforcement method both in ore and hard coal mining is rock bolt support. Very often, existing bolt support systems are not always capable of providing a reliable controlled performance. Therefore, in recent years energy-absorbing bolts which are exposed to dynamic loading, for example from rock burst caused by high rock stresses, earthquakes, or blasting have appeared. In this article particular attention was paid to short and long expansion bolts. Quasi-static tests of expansion bolts were carried out at the laboratory test facility in simulated mining conditions, especially for the KGHM Polska Miedź S.A. mines. In the underground mines of the Legnica-Głogów Copper District (LGOM) the main way to protect the room excavation is rock bolt support with a length from 1.2 m to 2.6 m. Rock bolt support longer than 2.6 m is considered as additional support of excavations and is increasingly being used to reinforce the roofs. The comparisons of energy-absorbing short and long expansion bolts with a length of 1.8m, 3.6m and 5.2m were presented. In addition, for elastic and plastic range of each bolts were determined.

1 Introduction

Many mines around the word, for example those in Poland, Chile, Canada, Australia and South Africa, are currently being operated at depths greater than 1000m. At such levels, the rock mass can be highly stressed causing an increase in the severity and frequency of rock bursts. The rejected rock mass can reach velocities between 3 and 10m/s with corresponding energy levels varying from 10 to 50kJ/m^2 [1]. Mining supports must be adapted to the natural hazards, especially rock bursts, whose intensification will be increasing with the increase in depth of exploitation [2, 3]. The consequence of the execution of any underground excavation is the formation of a new equilibrium state defined by the difference between the original state of stress and displacement. This is accompanied by specific geotechnical processes, their type and intensity being a function of the stress state and geomechanical properties of the rocks surrounding the excavated area. In general, the magnitude of the so-called rock mass pressure is determined by the energy

* Corresponding author: skrzypko@agh.edu.pl

changes that characterize the energy balance equation for any point in the rock medium by the following formula:

$$A_g + A_t + A_e + A_d = A_{os} + A_{or} + L_z + A_k + A_r \qquad (1)$$

where:
A_g – density of energy caused by gravitational stresses,
A_t – density of energy caused by tectonic stresses,
A_e – density of energy caused by the effects of stresses from the past influences of exploitation,
A_d – density of kinetic energy resulting from the impact of a possible bumps,
A_{os} – density of energy of elastic deformation,
A_{or} – density of energy used for rheological deformation,
L_z – work worn for destruction - fracturing (for formation a new surfaces),
A_k – density of kinetic energy used for the possible movement of the fractured rock area,
A_r – density of dissipated energy used for thermal, friction, acoustic, seismic effects.
Based on the above energy balance, can be made the following classification [4].

Table 1. Classification of energy balance.

No	Energy balance	Support of the excavation
1.	$A_{os} > 0$ (2) $A_{or} = L_z = A_k = A_r = 0$ (3) $A_g + A_t + A_e + A_d = A_{os}$ (4)	The excavation can be carried out without a support and will retain the required stability and functionality throughout its life, as no fractures are formed.
2.	$A_{os} > 0$ (5) $L_z > 0$ (6) $A_{or} = A_k = A_r = 0$ (7) $A_g + A_t + A_e + A_d = A_{os} + L_z$ (8)	Around the excavation a fracture zone is formed. The excavation must be supported because will be exposed to static loads.
3.	$A_{os} > 0$ (9) $A_{or} > 0$ (10) $A_{or} = A_k = A_r = 0$ (11) $A_g + A_t + A_e + A_d = A_{os} + L_z$ (12)	The excavation must be supported, but due to the lack of fractures, the support will be exposed to deformation loads and the excavation itself will be under convergence successively, although no stress changes in its vicinity will occur. The intensity of convergence will depend on the rheological properties of the rock medium and the stresses in the rock mass.
4.	$A_{os} > 0$ (13) $A_{or} > 0$ (14) $L_z > 0$ (15) $A_{or} = A_r = 0$ (16) $A_g + A_t + A_e + A_d =$ $= A_{os} + A_{or} + L_z$ (17)	The excavation must be supported and the support will be subjected to both static and deformation loads, while the excavation itself will be successively under convergence even though there will be changes in the stress state in its vicinity.
5.	$A_{os} \approx 0$ (18) $A_{or} \approx 0$ (19) $L_z > 0$ (20) $A_k > 0$ (21) $A_r \approx 0$ (22) $A_g + A_t + A_e + A_d = L_z + A_k$ (23)	In view of the existence of a fractures zone and excess energy in the system around the excavation (the latter converts into kinetic energy, specifically the energy of the fractured rock elements within the fractures zone. We will deal with the phenomenon of rock burst.

Unfortunately, despite the great simplicity of the presented balance, the creation of dynamic assessments characterizing the possibility of occurrence of certain phenomena is, unfortunately, much more complicated, since the complexity of the rock formation, expressed in its form: discontinuity, heterogeneity, anisotropy or nonlinearity, is further enhanced by the spatial structure of the excavations. Faithful description of phenomena occurring in a rock mass outcrop are prone to failure.

The efficiency of the rock bolting is determined by two groups of factors:

a) geotechnical conditions understood as the effect of stress interactions and geotechnical properties of rocks,

b) technical and organizational factors, which are expressed primarily in the quality of workmanship and in the current control of the support.

In view of such conditions, it will be apparent that the rock bolting is one of many types of support and only in certain situations its use is justified.

Rock bolt support fulfills three functions as:

a) reinforce the rock mass to strengthen it and to control bulking,

b) retain broken rock to prevent fractured block failure and unraveling,

c) hold fractured blocks and securely tie back the retaining elements to stable ground.

The mechanics of rock support is very complex, and no models exist that can fully explain the interaction of various support components in a rock support systems [5, 6].

2 Energy-absorbing bolts

Rock bolt support is used in underground mining primarily in ore mining, although increasingly this method of rock reinforcement can also be found in hard coal mines. In mining practice there are different construction solutions of bolts co-operating in different systems with rock mass formations and other types of supports.

The variety of conditions in which the rock bolt support is used makes it suitable for securing the roof and the excavation work as a stand alone or reinforcing rock bolt support. It also serves to suspend various items of mining equipment. Rock bolt can be divided into five categories [7, 8]:

Table 2. Division of rock bolt support.

Criterion	Type of rock bolts		
Type of material	Steel – in the form of rods, pipes, sheets, strings and cables	Wooden rods	Composite materials in the form of rods, pipes and strings
Mechanism of action	Grouted - mounted in the rock mass by means of various types of cement binders or resin cartridges	Mechanical – mounted in the rock mass by means of shell head spread out inside a borehole	Friction – acting on the principle of friction occurring between bolt and borehole
Method of mounting	Continuous - the bolt is fixed in the hole along its entire length by applying adhesive resin catridges or a suitable cement binder and on the friction principle.	Point - this type is characteristic for wedge, jaw or short encapsulated bolts, it allows the bolt be pre-tensioning	
Length	Short – up to 2.7 m in (in hard coal and ore	Long – up to 12 m (in hard coal mining in	

	mining in Poland)	Poland) and up to 7m in copper ore mining in Poland	
Taking over a load	Rigid - with satisfactory strength, breaking in when taking over the load near the limit of strength of bolt rod, with relatively small deformation capacity	Yielding - capable of carrying significant deformation, with relatively low carrying capacity	Energy-absorbing of dynamic load - characterized by high load carrying capacity and the ability to carry relatively high displacement and deformation

A modern, efficient rock bolt support provided for dynamic loading conditions must be made of a material of higher tensile strength and at the same time maintain the required load carrying capacity at its axial displacement. In different scientific centers in the world, in mines and other companies, bolt support is constructed with different stress-strain characteristics and dynamic load parameters. A rock bolt capable of absorbing kinetic energy from loads must be able to yield with the ground movements and also deform plastically over large distance, at high displacement rates [9]. Energy-absorbing bolts are used as part of rock support systems in underground mines that are exposed to detonating explosives and rock burst. In addition, most importantly a special bolt construction are characterized by their high load capacity and also their large deformation capacity. The energy-absorbing bolts have been studied almost 30 years the world. The first concept of such a bolt, the so-called „cone bolt" was proposed in South Africa in the early 1990s [10]. The „cone bolt" was modified later for resin grout in Canada [1]. Another bolts which have variable geometry along entire length are: „Durabar" [11], „D" 5], „CTw" [12]. „NMX" [13]. There are also bolts that have an additional absorbing element added from the roof side of the excavation, e.g. „Deformable component" [14] or on the bottom of the borehole, e.g. „Dynatork" [15], „Yiel-Lok" [16] or inside the borehole (the so-called absorber), e.g. „Roofex" [17], „Garford" [18], „He" [19], „DAP" [20]. A few of those bolts schematics which are available so far are briefly presented on the Fig. 1. The capacity of rock bolt reaches over 300 kN, while the value of the absorbed energy reaches $6{,}6 \cdot 10^4$ J. The authors of the research study report the values of acceptable displacements up to 0.6 m. The disclosed displacements are the result of deformation of the bolt material in the elastic and plastic range as well as the properties of the bonding materials used (cement, resin) and the specific construction and applied mechanisms, the so-called energy-absorbers. It should be emphasized here that too high limits of strain or elongation of the bolt from the borehole, even with the high bolting capacity itself, can cause premature loss of stability of the excavation, for example due to exceeding the limit value of deflection in the bolted roof. This part of the problem should always be considered when designing a rock bolt support intended for use in conditions of dynamic loads of rock mass.

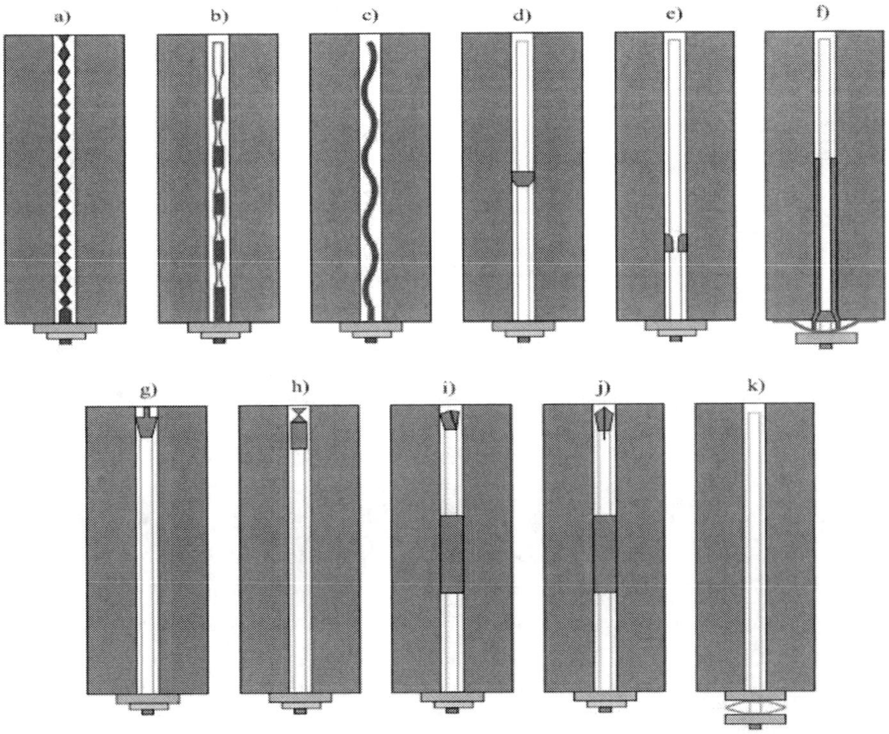

Fig. 1. Simplified mechanisms for absorbing of energy by rock bolts; a) "CTw", b) "D", c) Durabar", d) "Roofex", e) "Garford", f) "He", g) "Modified Cone bolt", h) "Dynatork" i) "DAP", j) "NMX", k) "Deformable component".

3 Laboratory test facility

Quasi-static tests of expansion rock bolts were carried out at the laboratory test facility in simulated mining conditions, in particular for the KGHM Polska Miedź S.A. mine. Studies were carried out using a consistent methodology. In the quasi-static load mode, the rock bolts were stretched and broken by the use of maximum force in the shortest possible time (approximately 4 s) resulting from the power of the pump. Displacement and force sensors were connected to a QuantumX MX840 universal measuring amplifier via 15-pin plugs. During the process of rock bolt expansion, the results of force measurement of displacement and deformation were recorded continuously by a program specialized in the field of measurement technology - known as "CATMAN - EASY" [21]. Expansion rock bolts used in the study form the basis of mining excavation supports at underground mines of the Legnica-Głogów Copper District (LGOM). Rod bolts with a length 1.8 m, 3.6 m and 5.2 m were tested. Rock bolt with a length 1.8 m was equipped with expansion shell with a length 0.095 m whereas bolts with a length 3.6 m and 5.2 m were equipped with expansion shell with a length 0.24 m (Figs. 2, 3). For all free bolts, the diameter of smooth rod was 0.0183 m. A detailed description of used rock bolts can be found in the literature [22, 23]. In order for laboratory studies to best satisfy mining conditions, dual bushings filled with concrete with a compressive strength of at least 75 MPa were used at the laboratory test facility. Openings with a diameter of 0.0375 m were made in the 0.22 m diameter cylindrical concrete blocks. Additionally, for bolts with a length 3.6 m and 5.2 m rock cores obtained from underground excavations were used. Rock cores had holes with a diameter 0.026 m and characterized compressive strength up to 100 MPa.

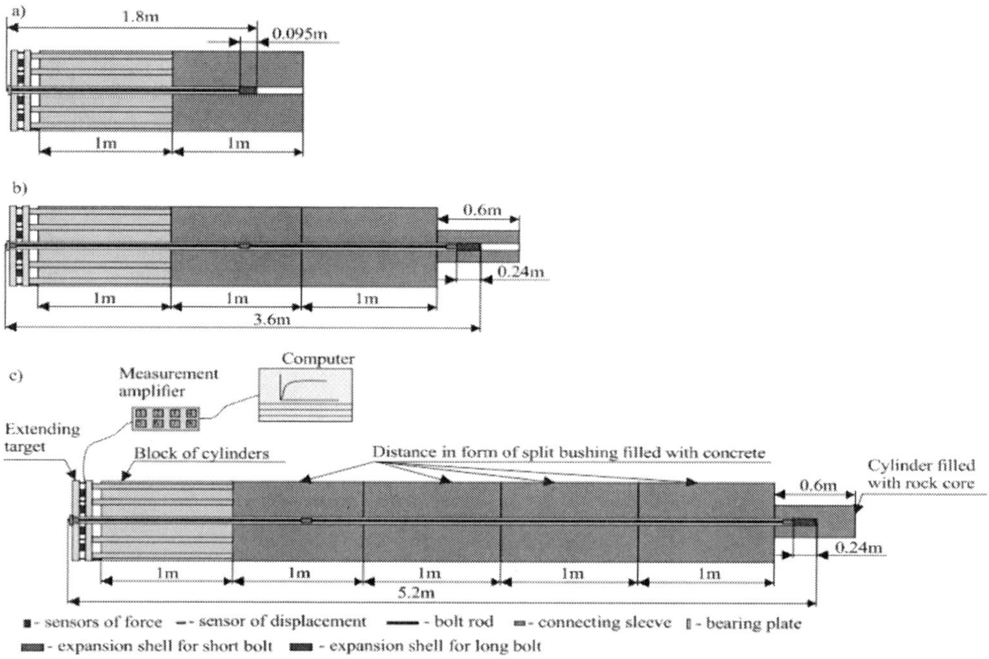

Fig. 2. Arrangement of sensors and elements of rock bolt support on the laboratory test facility; a) rock bolt support with a length 1.8 m, b) rock bolt support with a length 3.6 m, c) rock bolt support with a length 5.2 m.

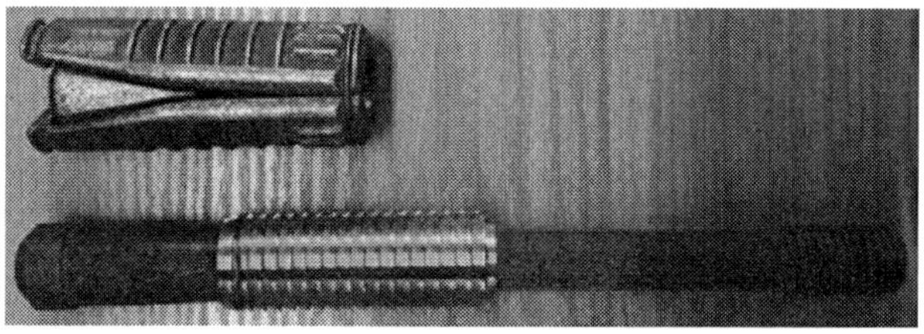

Fig. 3. Expansion shell; upper with diameter 0.036 m equipped with three jaws for bolt with a length 1.8 m, lower with diameter 0.0254 m equipped with four jaws for bolt with a length 3.6 m and 5.2 m (photo: K. Skrzypkowski).

3.1 Work done by short and long expansion rock bolts

The aim of the study was to obtain load-displacement characteristics on the basis of which it will be possible to determine the amount of absorbed energy. During tests the results of force and displacement characteristics were recorded continuously and visualization and evaluation of the measurement were tracked on-going. In addition, after the tests were completed, reports documenting the results of the measurements were created and then stored as an ASCII file extension. Then the field under the curve by means of Simpson method was integrated. Furthemore for elastic and plastic ranges a amount of absorbed energy were determined. Examples of the characteristic parameters obtained in the tests are shown in Figure 4.

a)

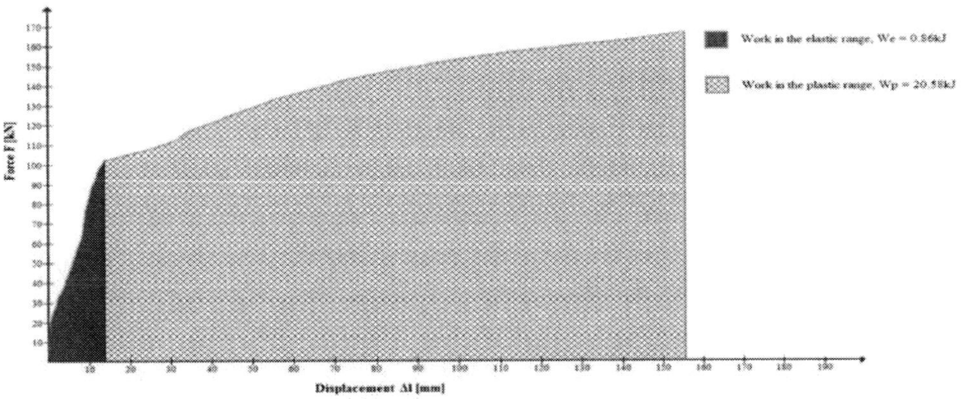

b)

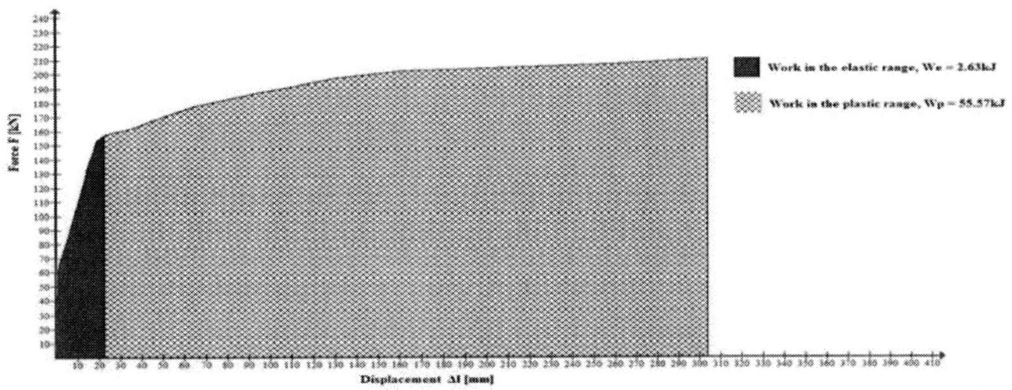

c)

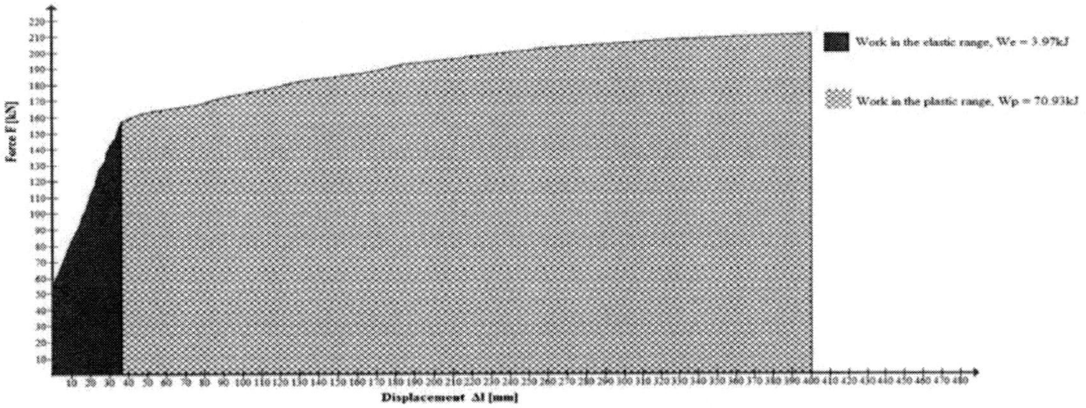

Fig. 4. Energy-absorbing by rock bolt support; a) with a length 1.8 m, b) with a length 3.6 m, c) with a length 5.2 m.

Based on Figure 4, it can be concluded that the most energy is absorbed by the support, which is the longest. This is due to the structural features of the bolt and the connecting rods using the connecting sleeves. The 5.2 m long bolt consists of three rods, which, together with the connecting sleeves and the bearing plate and the slide out from the hole, constitute a deformation of 0.4 m. This corresponds to a maximum breaking load of approximately 213 kN. On the Figure 4 the axes have units kN and mm in order to determine the work in kJ. A 1.8 m long bolt is capable of performing work (absorbing energy) of about 21 kJ. On the other hand, bolts with a length of 3.6 m consume more than 58 kJ. The greatest amount of energy was absorbed by bolts of 5.2 m in length, almost 75 kJ. In the elastic range of 3.6 m and 5.2 m, the bolt absorbs 305% and 461% more energy, respectively, than the bolt of 1.8 m. On the other hand, in terms of plasticity, the bolts of 3.6 m and 5.2 m are 270% and 344% respectively, compared to the 1.8 m bolt.

4 Conclusions

In recent years, the concept of bearing capacity of a rock bolt support, previously identified primarily with the ability to carry a specific load that grows over a long period of time, is related to the load that has grown rapidly - to the dynamic load. With this situation we are increasingly dealing with the successive increase in depth of exploitation resulting in the increase of stresses in the rock mass and to generate seismic events, rock masses and bumps. Dynamic dynamics induces the study and determination of the amount of energy absorbed by commonly used bolt support and to seek more effective ways of rock mass reinforcement. The absorbtion of energy through the bolt support is subject to ongoing research in many scientific research centers around the world. There are more and more new concepts, the common feature of which is the ability to maintain a high carrying capacity while accepting much larger displacement and deformation of the support. Rock bolts support used in dynamic hazards are characterized by varying geometry of structural elements, different grades of steel and addition of special energy absorbers.

Bolt supports are able to absorb the energy of several dozen kJ.

In Polish mining, most of the rock bolt support are used in the underground exploitation of copper ore deposits in the Legnica-Głogów Copper District (LGOM) (about 2 ÷ 3 million units). Bolt support with a length from 1.2 m to 2.6 m is characterized as short. On the other hand, 3.6 m, 5 m and 7 m bolts are used as additional, most commonly used for rebuildings and on the crossings of excavations. Expansion Rock bolt support with is able to absorb energy:

a) 21 kJ for a length 1.8 m and with a shell head with a 0.036 m diameter,
b) 58.2 kJ for a length of 3.6 m and with a shell head with a 0.0254 m diameter,
c) 74.9 kJ for a length of 5.2 m and with a shell head with a 0.0254 m diameter.

The amount of absorbed energy by the rock bolt support constitutes only a few percent of the total energy generated by the rock mass, so it is so important to create new construction solutions dedicated to ever deeper exploitation levels.

The study was carried out within the framework of dean's grant for young scientists no. 15.11.100.075

References

1. L.St-Pierre, F.P. Hassani, P.H. Radziszewski, J. Ouellet, Int. J. Rock Mech. Mining Sci. **46**, 1 (2009)
2. K. Skrzypkowski, Utility PL 68412 Y1 (2016)

3. S. Prusek, M. Rotkegel, Malecki Ł, Eng. Struct. **75** (2014)
4. A. Zorychta, Monthly of State Mining Authority **2** (1996)
5. C.C. Li, Int. J. Rock Mech. Min. Sci. **47**, 3 (2010)
6. P.P. Kaiser, M. Cai, JRMGE. **4**, 3 (2012)
7. K. Skrzypkowski, Mining Review. **68**, 4 (2012)
8. W. Korzeniowski, K. Skrzypkowski, Mining Review, **66**, 3-4 (2011)
9. A. Ansell, Tunn. Undergr. Sp. Technol. **20**, 4 (2005)
10. A.J. Jager, Rock support in mining and underground construction, Balkema (1992)
11. W.D. Ortlepp, J.J. Borman, N. Erasmus, Rockbursts and Seismisicity in Mines, South African Institute of Mining and Metallurgy (2001)
12. W. Pytel, P. Mertuszka, K. Szeptun, Ground Support, Lulea (2016)
13. NMX Dynamics, www.nmvgroup.se (2017)
14. W. Korzeniowski, K. Skrzypkowski, K. Zagórski, SGeM, **39**, 1 (2017)
15. DYNATORK, www.dsiunderground.com (2017)
16. Yield-Lok, www.jennmar.com (2017)
17. B. Neugebauer, Mining&Construction, **3** (2008)
18. Garfordcable, www.garfordcablebolts.com (2017)
19. M. He, W. Gong, J. Wang, P. Qi, Z. Tao, S. Du, Y. Peng, Int. J. Rock Mech. Mining Sci. **67**, (2014)
20. A. Nierobisz, International Mining Forum, London (2007)
21. W. Korzeniowski, K. Skrzypkowski, Ł. Herezy, Arch. Min. Sci. **60**, 1 (2015)
22. K. Skrzypkowski, W. Korzeniowski, K. Zagórski, P. Dudek, SGeM, **39**, 3 (2017)
23. K. Skrzypkowski, XXVI School of Underground Mining. Kraków (2017)

Impact of type of the roof rocks on location and range of endogenous fires particular hazard zone by in goaf with caving

Magdalena Tutak[1,*], and *Jarosław* Brodny[2]

[1]Silesian University of Technology, 2 Akademicka St., 44-100 Gliwice, Poland
[2]Silesian University of Technology, 26-28 Roosevelta St., 41-800 Zabrze, Poland

Abstract. Hazard of endogenous fires is one of the basic and common presented occupational safety hazards in coal mine in Poland and in the world. This hazard means possibility of coal self-ignition as the result of its self-heating process in mining heading or its surrounding. In underground coal-mining during ventilating of operating longwalls takes place migration of parts of airflow to goaf with caving. In a case when in these goaf a coal susceptible to self-ignition occurs, then the airflow through these goaf may influence on formation of favorable conditions for coal oxidation and subsequently to its self-heating and self-ignition. Endogenous fire formed in such conditions can pose a serious hazards for the crew and for continuity of operation of mining plant. From the practical point of view a very significant meaning has determination of the zone in the goaf with caving, in which necessary conditions for occurence of endogenous fire are fulfilled. In the real conditions determination of such a zone is practically impossible. The main aim of the analysis was to determine the impact of type of the roof rocks forming the goaf on the location and range of endogenous fires particular hazard zone by in these goaf. For determined mining-geological conditions, the critical value of velocity of airflow and oxygen concentration in goaf, conditioning initiation of coal oxidation process were determined.

1 Introduction

In mined seams of the Upper Silesian Coal Basin there are five basic lithological types of rocks: sandstones, mudstones, coal shales and hard coal with different type of strength properties.

Up to now strength properties of rocks surrounding the mining heading were considered the most often in the aspect of their influence on the maintenance of the stability of mining headings and yielding capacity for forming goaf with caving filling the space after mined usable mineral.

Practice shows that this influence should be considered also in the aspect of formation of ventilation hazards occurring in the underground mining headings, including the goafs of

* Corresponding author: magdalena.tutak@polsl.pl

longwalls. One of the ventilation hazards commonly present in goafs with caving is hazard of endogenous fires.

Goaf is a space in mined seam rock debris, filled with roof rocks, after the transition of mining front. This filling takes place as a result of spontaneous detachment of these rocks from the rock mass. Degree of filling of the selected surface depends on many factors, to which among the others one can include height from which rock block are falling down (thickness of the layer being exploited or thickness of the seam), type of the roof rocks and their strength properties, as well as the distance from the front longwall and depth of mining exploitation carried out. One of the most important factors deciding about the degree of filling of mined area is a type of roof rock forming goaf with caving and their stratification resistance. This resistance decides about their tendency to transition into caving state and about the permeability of caving debris [1-3]. Robust and compact roof rocks make the packing of cracked rock blocks looser and make them harder to transit into fall state, and weak and brittle rocks transit more easily into fall state, thus the fall is characterized by higher sealing and lower permeability. Independently of goaf with caving being formed by robust and compact rocks or weak and fragile rocks, empty spaces, called gaps, are always presented in this debris. These gaps together with the rock blocks create porous medium through which flow of ventilated air and mining gases is possible.

Flow of gases through the goaf with caving is a result of migration into them a part of ventilation air stream flowing through the wall heading in the process of its ventilation. Amount of migrating stream depends on many factors among the others, also on a type of applied support in the dog and longwall headings. [4-10]. The air stream flowing into the goafs creates dangerous initialization of low temperature reaction of coal oxidation which can lead to self-ignition of remained coal in the goaf with caving. It also can cause the outflow of mining gases (for instance methane) to others headings.

Necessary conditions to initialize oxidation process of coal, besides its presence is a flow through the air goafs with determined velocity and proper oxygen concentration. Fulfilment of these conditions creates possibilities to initialize low temperature reaction of carbon oxidation during which heat is emitted, which is accumulated by coal and causes increase of its temperature. If this state is kept through appropriately long time, it may come to self-ignition of coal, that is occurrence of endogenous fire. The most important factor influencing on the process of heat accumulation is the velocity of air stream flowing through the goaf with caving of exploitation longwall. This velocity is dependent among the others on a type of roof rocks forming the goaf. Type of these rocks influences on the degree of sealing of goafs, which subsequently decides about their permeability.

Results of tests indicate that critical value of air stream, causing self-heating of coal is in the range from 0.0015 to 0.02 m/s [9-14]. Laboratory tests of coal oxidation indicated that boundary value of oxygen concentration below which self-heating did not occur is equal to 8%. [13-15]. Oxidation process occurs in area where above specified necessary conditions for its initiation are satisfied. This area can be defined as a particular hazardous zone of endogenous fires.

Therefore, it is reasonable to conduct studies in order to determine the range of this zone. Determination of this zone in underground conditions in practice is impossible to perform, because it is formed in inaccessible places, what precludes to carry out direct tests.

It is necessary to use also other, more safe methods to determine location of this zone. Such possibilities create modelling tests using the numerical methods of fluid mechanics.

These methods in recent years more and more often are used to analyse ventilation problems. They enable variant analyses of processes connected with ventilation of underground mine heading, and also in analyses of states of emergency [7-15]. In analyzed case, the Finite Volume Method for determination of location and range of endogenous fires particular hazard zone in goaf with caving was used. Analyses were performed taking into account a type of roof rocks forming caving, having significant impact on location and range of this zone. Real longwall heading was submitted to the tests, for which ventilation parameters of air and geological structure of surrounding rock mass were known. The aim of test was to determine location and range of endogenous fires particular hazard zone in goaf with caving treated as porous medium.

Obtained results clearly indicate rightness of application of fluid mechanics for analysis of ventilation phenomena in porous media. This method in connection with geological-mining tests enables to determine the physical and chemical parameters of air and gases mixture in goafs. This subsequently enables to determine location and range of endogenous fires particular hazard zone. Based on these information, one can make with proper advance preventive actions in aim to limit its hazard.

2 Division of the roofs due to the value of destratification resistance of roof rock

The most significant property of roof rocks, due to formation of a caving zone in mined layer is stratification resistance is defined as the strength to loosening under the gravity force. The value of this resistance decides about the rock tendens to pass in a heap and about the permeability of forming goaf with caving in this process [1-3].

Stratification resistance of rocks is determined experimentally using downhole penetrometers or by an indirect method by disruption segments of vertical drilling of bore in the direction of the longitudinal axis of the opening. In a second case its value is calculated from dependence:

$$R_{rrs} = 0{,}8\frac{F}{d^2} \tag{1}$$

In a Table 1 there is presented division of roofs based on rock type littering over exploited seam. Additionally, also a value of stratification resistance and description of behavior of these roofs after selection of layer were given.

Table 1. Division of roofs [3].

Resistance of roof rock, R_{rrs}, MPa	**Description of roof**	**Example of rock**
0-0.5	roof falling immediately after unveiling	clay and sandy slates,
0.5-1.5	falling roof and weakly self-supporting	clay and sandy slates, coals
1.5-3.0	cracked roof, partially self-supporting and bearing, easily passing into a caving state	shales, sandy shale, coals
3.0-4.5	self-supporting roof, it goes automatically into a caving state without sagging into goaf	sandstones
4.5-6.0	supporting roof, without roof falls, hardly passing into caving state, sagging into goaf	sandstones
> 6.0	strongly compacted roof, very difficult for passing into caving state	sandstones

3 The goaf permeability

The coefficient value of gobs permeability is determined based on the value of resistance of rock roof stratification which form the caving. The resistance of roof rock stratification is their proper loosening resistance acting against the gravity force, that is natural capability of the rock mass to oppose stratification and caving of roof rock to the heading surface. This resistance determines their tendency to going into caving state.

The resistance of roof rock stratification forming caving is calculating from dependence [3]:

$$R_{rrs} = \frac{\sum_{i=1}^{n} R_{rri} \cdot m_i}{\sum_{i=1}^{n} m_i} \tag{2}$$

where: R_{rrs} is resistance of roof rock stratification (Pa), R_{rri} tensile strength of the rock layers (Pa), m_i the thickness of layers (m).

This dependence takes into account an influence of the thickness of particular layers of roof rocks on their tendency to going into caving state.

Resistance of roof rock stratification calculated on the basis of Equation 2, enables to determine permeability coefficient of goaf with caving from Equation 3 [3]:

$$k(x) = \frac{\mu_g}{r_0 + ax^2} \quad \text{for } 0 \le x \le 2/3 \cdot l \tag{3}$$

and from dependence [3]:

$$k(x) = \frac{\mu_g}{r_0 + a\left(\frac{4}{3}l - x\right)^2} \quad \text{for } 2/3 \cdot l \le x \le l \quad 4)$$

where: μ_g is the coefficient of dynamic viscosity of air (Nsm^{-2}), l the total length of the longitudinal longwalls (m), r_0 we determine from dependence $r_0 = \frac{\mu}{k_0}$ and a we determine from dependence $a = 6 \cdot 10^{9} R_{rrs}^{-1,74}$.

Value of coefficient k_0 we determine from dependence [3]:

$$k_0 = \frac{\mu_g}{6} \cdot 10^{-10} R_{rrs}^{1,44} \tag{5}$$

Presented dependences were used during determination of permeability coefficient value of goafs in modeling tests.

4 Mathematical model of gas flow

In order to determine particular hazardous zone of endogenous fires in goaf with caving, mathematical model of air stream flowing through porous medium based on below discussed dependences was developed. A turbulent flow of viscous incompressible fluid, was described by the Navier-Stokes system of equations, which together with continuity equation makes a complete system of relationships, allowing to determine pressure and field of flow velocity [16]:

– The continuity equation:

$$\frac{\partial\rho}{\partial t}+\frac{\partial(\rho u)}{\partial x}+\frac{\partial(\rho v)}{\partial y}+\frac{\partial(\rho w)}{\partial z}=0 \tag{6}$$

where: u, v, w are directions velocity (m/s), ρ is density (kg/m^3), and t is time (s).

– The momentum equation:

$$\frac{\partial}{\partial t}(\rho v)+\nabla\cdot(vv)=-\nabla p+\nabla\tau+\rho g+F \tag{7}$$

where: p is static pressure (Pa), τ is the stress tensor (Pa), g is the gravitational body force (m/s^2) and F is the external body force (N).

The basis of the mathematical description of the transport process of the methane emission to the headings is a mass conservation principle related to this gas. Mathematical model of the transport, being a system of equations of advection-diffusion, which for i-th substance it takes the following form:

– The species transport equation

$$\frac{\partial}{\partial t}(\rho Y_i)+\nabla\cdot(\rho v Y_i)=-\nabla\cdot J_i+R_i+S_i \tag{8}$$

where: v is velocity (m/s), Y_i is the local mass fraction of each species, J_i is the diffusion flux of species i (kg/(m^{2}s)), R_i is the net rate of production of species i by chemical reaction and S_i is the rate of creation by addition from the dispersed phase plus any user-defined sources.

– The mass duffusion in turbulent flows:

$$J_i=-(\rho D_{i,m}+\frac{\mu_t}{Sc_t})\nabla Y_i \tag{9}$$

where: $D_{i,m}$ is the mass diffusion coefficient for species i in the mixture (m^2/s), μ *is* the viscosity (Pa·s) and Sc_t is the turbulent Schmidt number, 0.7.

Airflow through the goaf with caving is a flow through the porous medium. Therefore, to the equation of conservation of momentum, an additional source member Si describing this flow was introduced:

$$S_i=-\left(\frac{\mu_g}{k}+C_2\frac{1}{2}\rho_g|v|v\right) \tag{10}$$

where: Si is the pressure loss items defined by Darcy's law and C_2 is the inertial resistance factor.

Flow of air stream through the exploitation longwall and segment of goaf with caving located just behind the sections of mechanical support, in which there was no full caving (determined by so-called caving step) does not correspond to laminar flow (Fig. 1).
Flow of gas streams through the excavations has turbulent character, that is, in which irregular movement of air particles occurs, and its flow parameters undergoes to unpredictable random changes in space and time. In the presented flow analysis the „*k-ε standard*" turbulence model was applied. This model describes components of the Reynolds turbulent stress tensors according to the Boussinesq hypothesis.

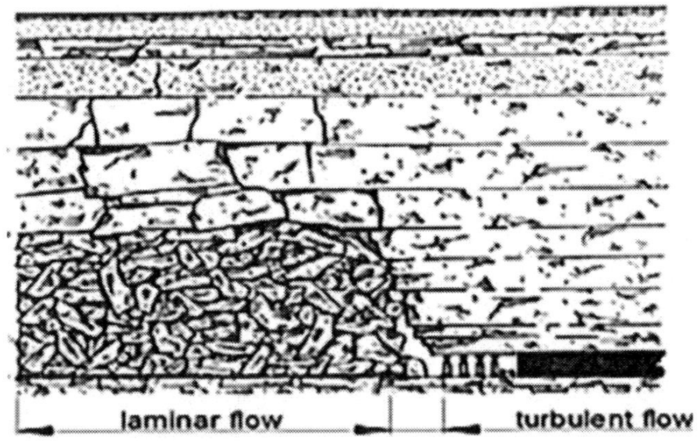

Fig. 1. The character of flow in goaf [1].

In the formula for the components of stress tensor there occur k and ε quantities. These quantities requires additional two transport equations for single phase flows in a form [15]:

$$\rho\frac{\partial k}{\partial t}+\frac{\partial}{\partial x_i}(\rho k u_i)=\frac{\partial}{\partial x_j}[(\mu+\frac{\mu_t}{\sigma_k})\frac{\partial k}{\partial x_j}]+G_k+G_b-\rho\varepsilon-Y_M+S_k \tag{11}$$

$$\rho\frac{\partial \varepsilon}{\partial t}+\frac{\partial}{\partial xi}(\rho\varepsilon u_i)=\frac{\partial}{\partial x_j}[(\mu+\frac{\mu_t}{\sigma_\varepsilon})\frac{\partial \varepsilon}{\partial x_j}]+C_{1\varepsilon}\frac{\varepsilon}{k}(G_k+C_{3\varepsilon}G_b)-C_{2\varepsilon\rho}\frac{\varepsilon^2}{k}+S_\varepsilon \tag{12}$$

where: $C_{1\varepsilon}$, $C_{2\varepsilon}$, $C_{3\varepsilon}$ are constans, σ_k, σ_ε turbulent Prandtl numbers for k and ε, G_b the generation of turbulence kinetic energy due to buoyancy, G_k the generation of turbulence kinetic energy due to the mean velocity gradients, Y_M contribution of the fluctuating dilatation in compressible turbulence to the overall dissipation rate, S_k, S_ε user-defined source terms.

5 Characteristic of investigated system

In order to perform the analysis, a geometrical model of goaf with caving and exploitation longwall and longwall headings was developed. The vertical extent of airflow in goaf with caving amounted to 3.5 times of the height of exploiting seam (10.5 m). InFigure 2 there is presented scheme of longwall ventilation with indicated line of air distribution.

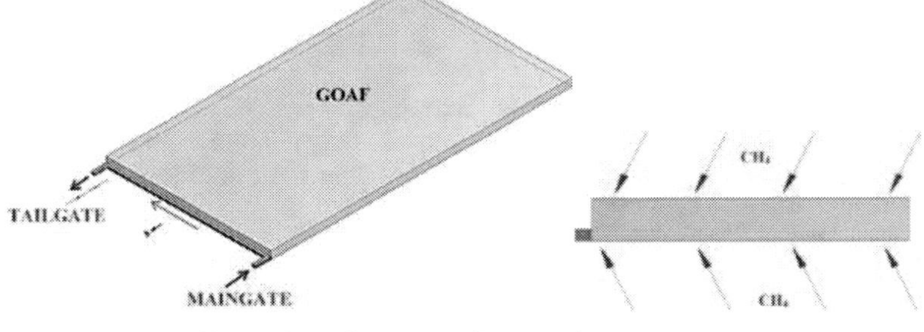

Fig. 2. Geometrical model of investigated system of excavations.

The oxygen part in the air stream supplied to the longwall equals to 21%. Release of methane to the exploitation longwall amounted average to 8,0 m^3/min. The temperature of the air stream supplied to the longwall equals 20 °C.

Modelling tests order to determination of the impact of roof rock type on location and range of endogenous fires particular hazard zone in goafs were performed for six types of roof rocks forming the goafs. A parameter characterizing type of these rocks was destratification resistance value which decides about goafs permeability. In studies there was assumed that in goafs is located roof rocks with weighted means values appropriately equal to 2, 3, 4, 5, 6 and 7 MPa.

6 Results

Based on performed calculations distributions of velocity of mixture of air and gases flowing through the goaf with caving and oxygen concentration in this mixture were determined. These values for tested region including goaf with caving with adjacent headings were determined.

In order to clear imaging of obtained results in Figures 3, 4, 5 and 6 there are presented distributions of tested values in parallel planes to slough of exploited seam located at different heights. In Figure 3 there are presented distributions of velocity of air and gases mixture and dangerous velocity because of hazard of endogenous fires in goaf with caving in a distance of 0.5 m from slough of exploited seam. Whereas, in Figure 4 there is presented oxygen concentration distribution for the same layer. Presented distributions in Figures 3 and 4 were made for remaining rocks in goafs with weighted means values of destratification resistance amounted to 3 MPa and 6 MPa.

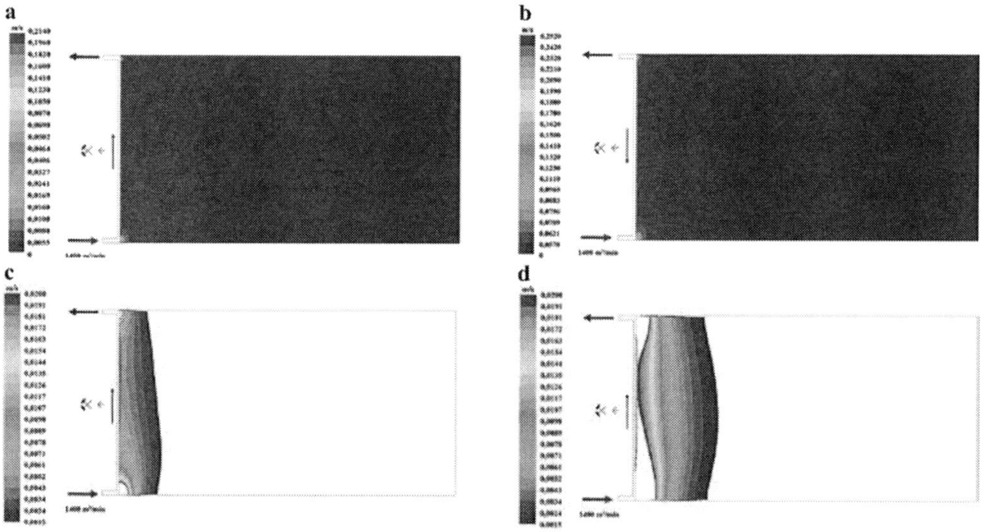

Fig. 3. Distributions of velocity of air and gases mixture (a, b) and dangerous velocity (c, d) because of hazard of endogenous fires in goaf with caving in a distance of 0.5 m from slough of exploited seam for remaining rocks in goafs with weighted means values of destratification resistance amounted to 3 MPa (a, c) and 6 MPa (b, d).

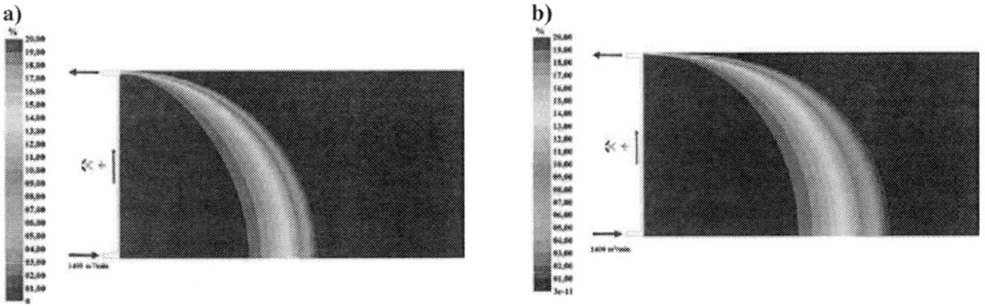

Fig. 4. Oxygen concentration distribution in goaf with caving in a distance of 0.5 m from slough of exploited seam for remaining rocks in goafs with weighted means values of destratification resistance amounted to 3 MPa (a) and 6 MPa (b).

In Figure 5 and 6 there are presented the same distribution as in Figures 3 and 4 for goaf with caving at a distance 2.0 m from slough of exploited seam.

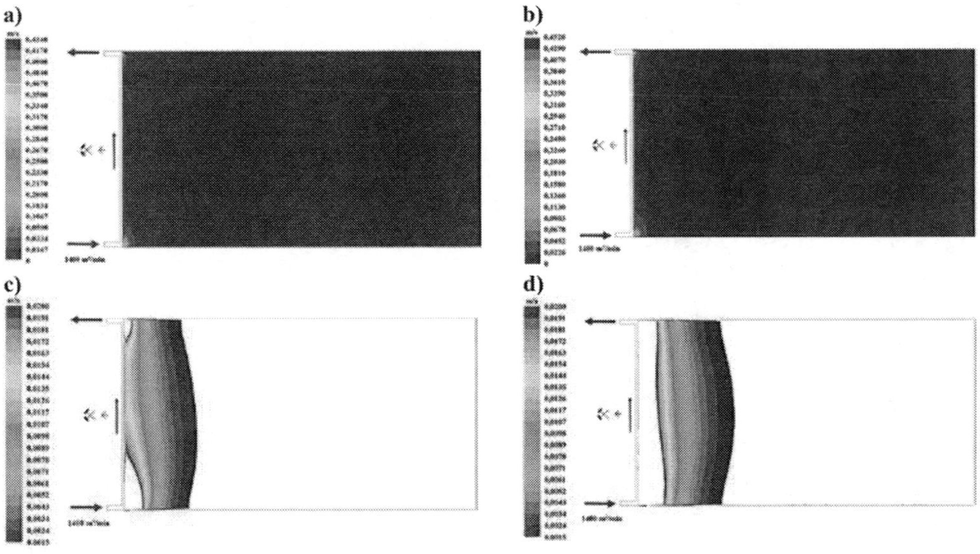

Fig. 5. Distributions of velocity of air and gases mixture (a, b) and dangerous velocity (c, d) because of hazard of endogenous fires in goaf with caving in a distance of 2.0 m from slough of exploited seam for remaining rocks in goafs with weighted means values of destratification resistance amounted to 3 MPa (a, c) and 6 MPa (b, d).

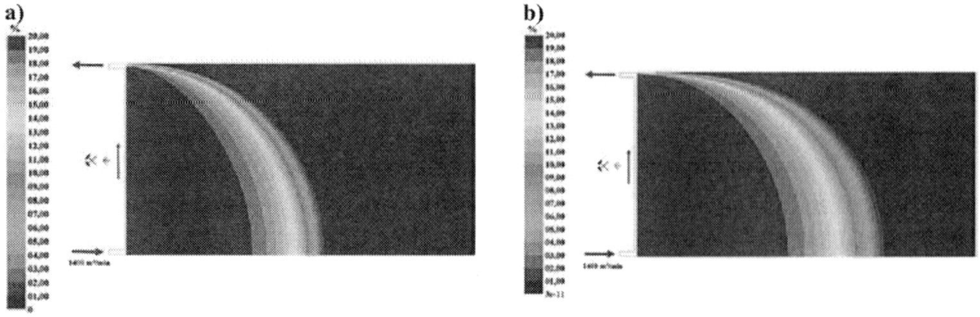

Fig. 6. Oxygen concentration distribution in goaf with caving in a distance of 2.0 m from slough of exploited seam for remaining rocks in goafs with weighted means values of destratification resistance amounted to 3 MPa (a) and 6 MPa (b).

Based on the obtained results courses of changes in velocity of air flowing through goaf with caving and oxygen concentration in this air depending on the distance from the longwall front for different weighted average values of destratification resistance of roof rocks forming this heap were determined.

In Figures 7 and 8 there are presented comparative characteristics velocity of air flowing though goaf with caving (Fig. 7) and oxygen concentration in this air (Fig. 8) for goafs formed by different types of rocks (characterized by weighted average value of destratification resistance). In these figures, dashed lines indicate dangerous values of air velocity and minimum oxygen concentration due to possibility of occurrence of endogenous fire. These data enable to determine the location and range of zones in which endogenous fire can occur.

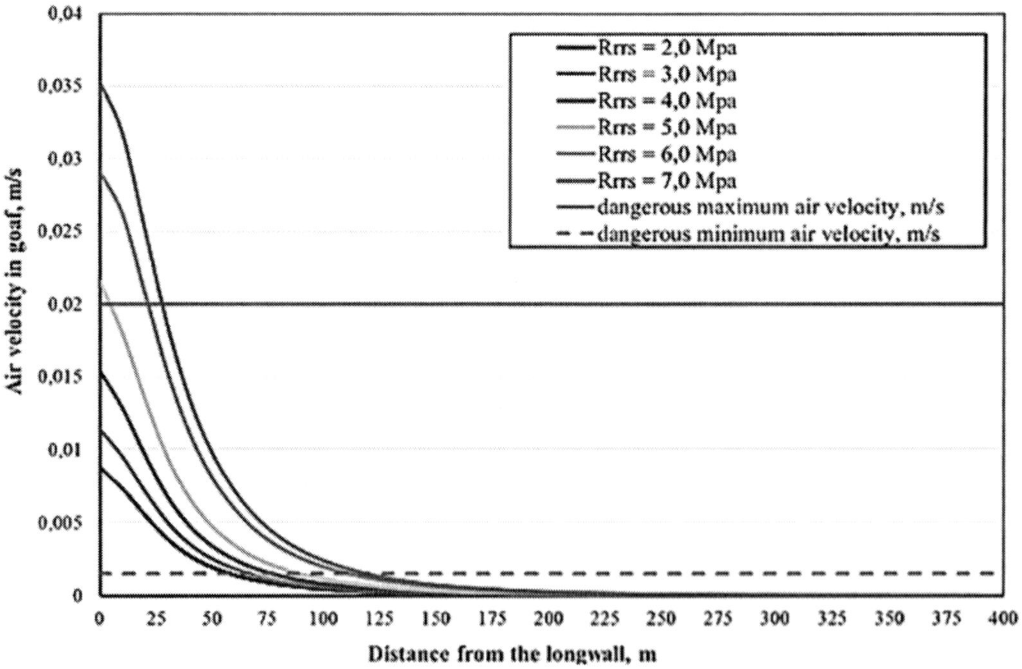

Fig. 7. Comparative characteristics velocity of air flowing though goaf with caving.

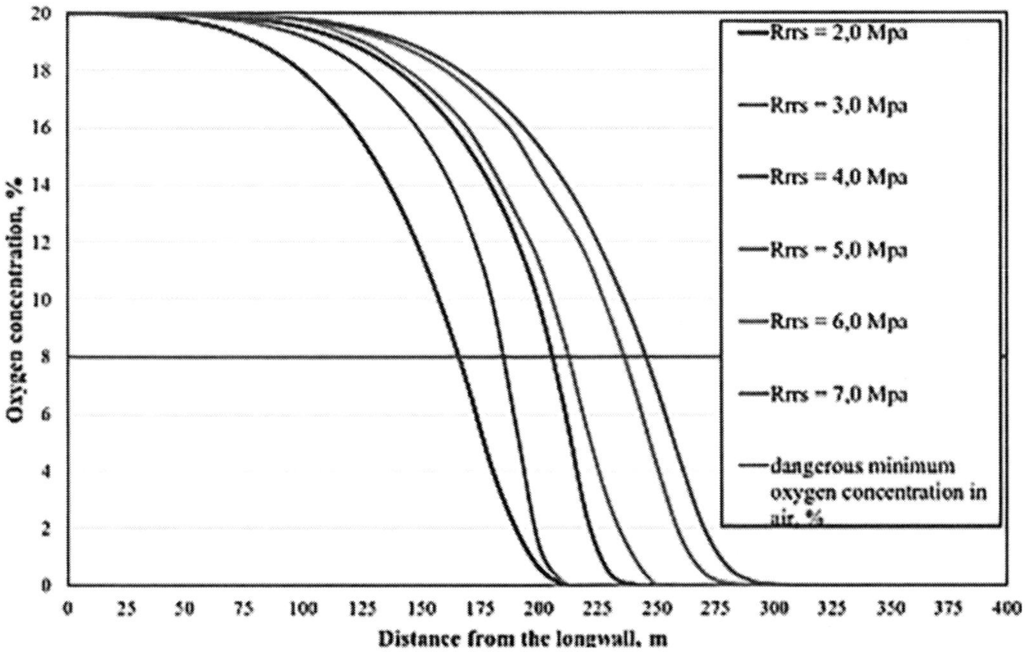

Fig. 8. Comparative characteristics oxygen concentration in air flowing though goaf with caving.

Based on the obtained results location and range of endogenous fires particular hazard zone in goaf with caving depending on weighted mean values of destratification resistance of rocks forming these goafs were determined. Obtained results are shown in Table 2.

Table 2. Location and range of endogenous fires particular hazard zone in goaf with caving depending on weighted mean values of destratification resistance of rocks forming goafs.

The resistance of roof rock stratification, MPa	Critical air velocity zone, m	Critical oxygen concentration zone, m	Particular endogeneous fires hazard zone, m
2	0 – 54.0 m	0 – 154.0 m	0 – 54.0 m
3	0 – 69.0 m	0 – 175.0 m	0 - 69.0 m
4	0 – 84.0 m	0 – 200.0 m	0 – 84.0 m
5	11.0 – 101.0 m	0 – 210.0 m	11.0 – 101.0 m
6	21.0 – 115.0 m	0 – 230.0 m	21.0 – 115.0 m
7	25.0 m – 125.0 m	0 – 242.0 m	25.0 m – 125.0 m

7 Conclusions

Model developed and used for flow analysis of mixture of air stream and mining gases through the goaf with caving enabled to determine the distribution of physical and chemical parameters of this mixture for different destratification resistances of rocks forming goaf with caving.

Obtained results enabled to determine of zones particularly endangered by endogenous fires in goafs, in which velocity of airflow equals from 0.0015 m/s to 0.02 m/s, oxygen concentration – min. 8%. Based on that, one can conclude, that location and range of this zone in significantly way depends on a type of roof rocks forming goaf with caving.

As measure characterizing particular types of rocks forming goafs was assumed destratification resistance of these rocks.

In goaf with caving created from rocks with low value of destratification resistance, the particular hazardous zone of fires forms just behind the front of the longwall, however in goafs created from rocks with high value of this resistance – this zone is distant from the longwall front. This rule regards both to velocity distribution in goafs, and oxygen concentration.

Knowledge of range of this zone has very important practical meaning. It allows to take preventive actions, in order to prevent occurrence of endogenous fire.

Obtained results indicate, that numerical methods can be successfully used for variant analyses of processes connected with ventilation of underground mine headings, including analyses connected with airflow through the goaf, and also in the analyses of emergency states, like initialization of oxidation process in favorable conditions leading to self-heating and self-ignition. Therefore, it can be assumed that simulation methods based on computational fluid dynamics (*CFD*) could be significant tool in improvement of the occupational safety in the mining.

This article is the result of the research project **No. 06/030/BK_17/0024** realized in 2017, financed by MNiSW.

References

1. A. Lisowski, *GIG Works,* **2001** (1959)
2. V. Palchik, Env. Geol. **44**, 1, DOI: 10.1007/s00254-002-0542-y (2003)
3. J. Szlązak, Arch. Mining Sci. **46** (2001)
4. J. Brodny J. Arch. Mining Sci., **55**, 4 (2010)
5. J. Brodny J. Arch. Mining Sci., **56**, 1 (2011), DOI: 10.2478/v10267-012-0015-4
6. J. Brodny J. Arch. Mining Sci., **57**, 4 (2012)
7. J. Brodny, M. Tutak, *16th International Multidisciplinary Scientific GeoConference.* (2016), DOI: 10.5593/SGEM2016/B12/S03.115
8. D. Szurgacz, J. Brodny: *17th International Multidisciplinary Scientific GeoConference.* (2017), DOI: doi.org/10.5593/sgem2017/13
9. Y. J. Song, *Beijing: The Press of Coal Industry*, (2002)
10. K. Cui, J. Shan. Univ. Sci. Techn., **21** (2002)
11. K. Cui, Univ. Sci. Techn., **21** (2002)
12. G. Dai, S. Zhang, M. Tang, AGH J. Min. Geo. **36**, 3, (2012)
13. J. Gao, H. Wang, China Saf. Sci. J. **20**, 03, (2010)
14. J. Gao, J. Liu and X. Zhang, China Saf. Sci. J., **20**, 09, (2010)
15. X. Zhou, Liaoning Technical University (2016).
16. K. K. Veersteg, W. Malalasekera, *An Introduction to Computational Fluid Dynamics: The Finite Volume Method* (Pearson Education 2007)

Analysis of rock mass dynamic impact influence on the operation of a powered roof support control system

Dawid Szurgacz[1,*], and *Jarosław* Brodny[2]

[1]Center of Hydraulics DOH Ltd, 147 Konstytucji St., 41-906 Bytom, Poland
[2]Silesian University of Technology, 26-28 Roosevelta St., 41-800 Zabrze, Poland

Abstract. A powered roof support is a machine responsible for protection of an underground excavation against deformation generated by rock mass. In the case of dynamic impact of rock mass, the proper level of protection is hard to achieve. Therefore, the units of the roof support and its components are subject to detailed tests aimed at acquiring greater reliability, efficiency and efficacy. In the course of such test, however, it is not always possible to foresee values of load that may occur in actual conditions. The article presents a case of a dynamic load impacting the powered roof support during a high-energy tremor in an underground hard coal mine. The authors discuss the method for selecting powered roof support units proper for specific forecasted load conditions. The method takes into account the construction of the support and mining and geological conditions of an excavation. Moreover, the paper includes tests carried out on hydraulic legs and yield valves which were responsible for additional yielding of the support. Real loads impacting the support unit during tremors are analysed. The results indicated that the real registered values of the load were significantly greater than the forecasted values. The analysis results of roof support operation during dynamic impact generated by the rock mass (real life conditions) prompted the authors to develop a set of recommendations for manufacturers and users of powered roof supports. These include, inter alia, the need for innovative solutions for testing hydraulic section systems.

1 Introduction

The number of mining tremors occurring in Polish mines has significantly increased in recent years. It is caused by mining exploitation carried out at grater depths. In 2016, at output of 70.4 million tonnes of hard coal, as much as 36.9 million was acquired from longwalls exploited in hazardous conditions of rock mass tremors. With regard to total production, it was over 50% [1]. A comparison of the amount of coal extracted with the number of high energy tremors and accidents in the Polish hard coal mines in 2006-2016 is presented in Table 1.

* Corresponding author: dawidszurgacz@doh.com.pl

Table 1. Extraction, number of high-energy tremors, rock bursts and accidents in Polish hard coal mines [1].

Year	Extraction [mln Mg]	Output form seams with tremor hazard, mln Mg		No. of tremors $\geq 1 \times 10^5$ J (acc. to GIG)	No. of rock burst	Accidents caused by rock bursts	
		[mln Mg]	[%]			fatal	total
2006	95.2	42.1	44.2	1,177	4	4	20
2007	87.5	40.5	46.3	939	3	0	10
2008	83.6	41.9	50.1	917	5	0	26
2009	77.4	34.3	43.8	772	1	0	6
2010	76.1	35.8	47.1	1,206	2	2	15
2011	75.5	34.2	45.3	1,034	4	1	17
2012	79.2	37.6	47.47	1,022	1	1	2
2013	76.47	36.9	48.25	1,427	1	0	5
2014	72.5	36.0	49.58	1,765	1	0	0
2015	72.2	37.1	51.39	1,322	2	2	2
2016	70.4	36.9	52.41	1,640	1	1	3

The analysis of data presented in Table 1 allow to conclude that despite the decrease in level of coal mining extraction, the number of recorded tremors increases. As a consequence, the level of hazard of coal mining is greater. This in turn forces the need to apply appropriate protection of mining excavations against these dynamic impacts. The equipment used to secure mining excavations is a mining roof support. In the case of longwall excavations, these are powered roof supports. Their task, apart from ensuring the safety of the crew and the protection of the equipment in the mining area, is to facilitate the process of mining the rock mass.

One of the important factors influencing the reduction of effects caused by the dynamic impact of the rock on the support is the proper co-operation of the roof support with the rock mass [2-4]. During the dynamic impact of the rock mass, the powered roof support should ensure the support and prevent loss of contact with the roof and excessive convergence of the excavation [5, 6, 7, 8].

Hydraulic legs with the safety system are the most important factors impacting on safety and proper operation of the powered roof support. The safety system is the basic support element of the roof support, which determines its performance [5, 7, 8].

The requirement that must be met in order for a powered roof support to be accepted for mining exploitation is the provisions of the Regulation of the Minister of Energy on Occupational Safety and Health dated November 23, 2016 § 523 sec. 2. The provisions state that a powered roof support designed to operate at longwalls exploited in the areas of mining tremors occurrence shall be adapted by yielding to absorb dynamic loads. The hydraulic system of the leg and the control system should conform with the requirements included in the Machinery Directive and harmonised PN-EN 1804 standards [9].

The system was analysed taking into consideration the role, importance and functions that must be fulfilled in the excavation by the powered roof support and the importance of the hydraulic system. The analysis included tests on hydraulic legs and a control system dynamically impacted by free falling impact mass. The time course of the pressure in the leg was determined for different values of force of impact mass (different heights of impact mass drop). The tests were aimed at determining the impact of dynamic loads on the characteristics of leg's operation. It was assumed that the energy and load waveforms during field tests are similar to those in real conditions.

One of the aims of the study is also to approximate the problem associated with the proper selection of the system protecting the leg mounted in a roof support against the negative effects of the dynamic impact of the rock mass.

2. Analysis of dynamic rock mass impact

Rock mass impacts underground mining excavations as a result of mining operation. This interaction can be static or dynamic, and is directly transferred to mining roof support.

An example of the static impact of a rock on a powered roof support is the phenomenon of loading it with the weight of overlying rocks. It occurs in the case of roofs with very low durability or with completely or partially destroyed structure. The reaction of the roof support in such conditions is largely dependent on the weight of the rock mass impacting on it. When static value of impact is lower than working support (bearing) capacity of the roof support, no dynamic phenomena occur. In the case of exceeding the value of the working capacity, however, the roof support tightens. This is a result of activation of a safety system, which in this case allow to change the dimensions of the roof support. Such system in a roadway support is based on a sliding connectors [10-12], while in a longwall powered roof support, it is based on fluid flowing out of spaces under the leg's piston through a working or bypass valve, and unit convergence (decreasing its height). The tightening process of the longwall support due to the weight of loosely deposited rock masses in the roof is difficult to stop because the hysteresis of the operation of the working or relief valves. An additional obstacle is the need to slow down the movement of rock masses. To interrupt the tightening process for this type of load is only possible if the weight of the rock masses decreases as a result of the spontaneous formation of the natural roof. In this case, the rock mass moves. If the load is above the yield capacity, the roof support is fully tightened which leads to all overload consequences [5, 6, 13].

The consequences of such an impact recorded in the mining excavations indicate repeated violation of the permissible load values that can be transferred by the applied support system.

In the case of coarse solid and strong rock deposits in the layers above the operating areas, apart from the static pressure, manifested by the settling of the roof and the heave of the floor in the excavation, other phenomena include rock mass tremors. They are mainly caused by layers cracking. These are sudden and dynamic phenomena that may lead to dangerous consequences. In a short time, there is a significant increase in the rock mass load impacting on the excavation, and consequently also on the roof support.

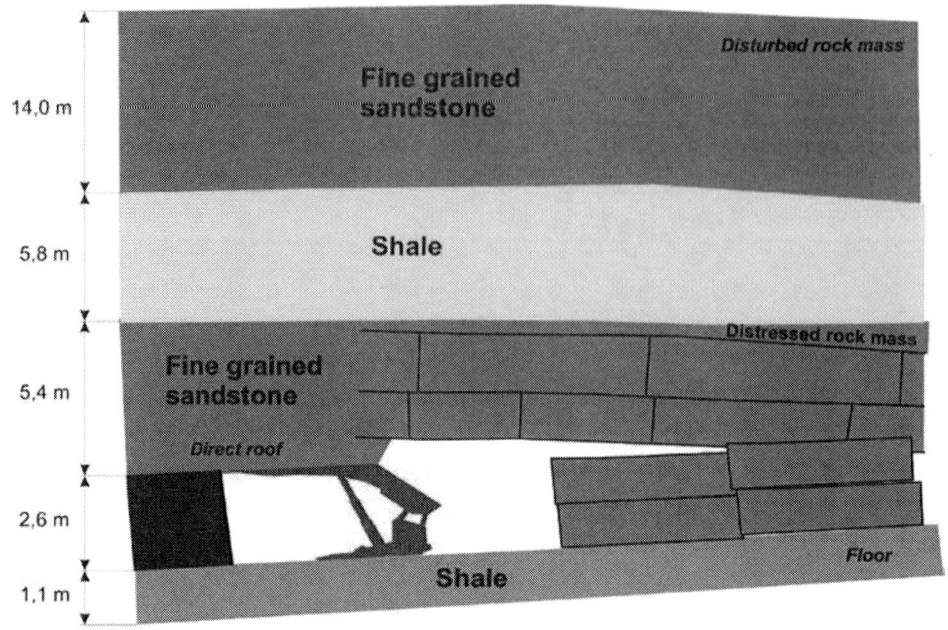

Fig. 1. Distressed rock mass.

Practical tests show that the presence of strong mining tremors during mining operation or after its completion, are likely to cause rock bursts due to the presence of thick, uniform rock layers of high strength deposited above the rock mass. The thickness of those layers (not less than 20 m) and their uniformity, which are expressed primarily by the lack of distinct bottoms surfaces, are essential [5, 13].

The analysis presented in the paper includes the influence of the dynamic impact of the rock mass in one of the hard coal mines in Poland.

The dynamic impact of the rock mass was recorded in the longwall 7, seam 409, in the Wujek Śląsk Mine, located at a depth of about 985 to 1,025 m.

The seam in that area is characterized by thickness of 2.4 ÷ 2.6 m and is inclined towards the south and south west at approx. 3 ÷ 8° angle. There is a fine-grained sandstone with a thickness of 5.4 m deposited in the roof of the subject seam with liner sandy shale in the floor. Above it is a layer of clay slate with a thickness of 1.1 m, below which alternately occur layers of sandstone and sandy shale with a thickness of 11.0 m. The layers are presented in Figure 1. A high energy tremor, $4x10^9$ J, occurred in that area and was located in the distance about 120 m before the front areas of the longwall area 7, and about 30 m south of south heading. The analysis was conducted for these real rock dynamic conditions and involved leg's control system of a roof support operating in the subject area. The results are presented herein.

3 Leg‘s control system test

Flow studies in hydraulic systems protecting the legs against overloads were first applied with the introduction of power hydraulics, mainly in powered roof supports. The research was aimed at determining the dependency of the volume flow of the working fluid depending on the pressure. The tests were carried out for individual components as well as for the whole hydraulic system. Particularly important in this system are the systems protecting the legs against dynamic overload caused by rock mass. The analysis and flow tests concerned systems operating at extremely high pressures up to 100 MPa. These flows are classified as violent, as in such cases the Reynolds number exceeds 10^5, and in occasional cases even 10^6. For flows with such a high Reynolds number in hydromechanics, the current calculation methods based on the Bernoulli equation cannot be used. In the case of dynamic load impacting on the support unit, the fluid flow rates in its hydraulic systems are estimated to be several dozen or even hundreds of ms^{-1}. In addition, there are difficulties in assessing the phase of such flow. This is related to the operation of pumps that pump the fluid along with the sucked air, and therefore the fluid should be treated as a two-phase system.

Consequently, the process of studying such phenomena is very complex. From a practical point of view, the most credible is bench testing. It guarantees reliable reproduction of the actual operating conditions of the roof support or hydraulic leg. It also seems appropriate to supplement the field research with model studies. Numerical computational methods using ANSYS software [14, 15,] provide great possibilities for simulation studies of flow phenomena. They can be applied to model a number of phenomena occurring during the dynamic loading impacting on the roof support, such as the effect of the boundary layer, cavitation, biphasic liquid, and to a large extent change the physical and chemical properties of the liquid. Simultaneous use of computational methods based on specialized software and bench testing provides optimal cognitive effects and enables the utilization of utilitarian applications.

An example of a bench test that reliably reproduces the dynamic impact of a rock mass on a powered support section is the load-bearing test on a hydraulic leg.

By impacting the leg with impact mass it is possible to obtain certain flow values in the safety system. By selecting the proper diameter of the leg and the impact mass, the flow characteristics in the entire leg's control system can be determined.

The tests were based on a control system consisting of a DOH DN12 valve block equipped with a pressure control valve to which the SP10.C valve was connected via a DN19 connection type ZRP and a DOH valve connected via a M45x2 connection, as presented in Figure 2.

The purpose of the study was to evaluate the hydraulic system of the leg with the control system to avoid damage caused by high-energy rock mass tremors [2]. These tests were aimed at checking the correctness of the control system of the support's leg with the hydraulic system to meet the requirements of yielding resulting from the Regulation of the Minister of Energy on OSH §523 sec. 2. The subject leg and the hydraulic system was spread between a traverse and the foundation with the pre-support and then loaded with a free-fall impact mass. The setting and yielding was conducted by supplying the leg with hydraulic fluid from the pump station. Measurement and registration of selected physical parameters necessary to assess the dynamics of the leg were performed using specialized apparatus [16].

The method of mounting the leg and the layout of its control system during the dynamic test of the free fall impact mass is shown in Figure 2.

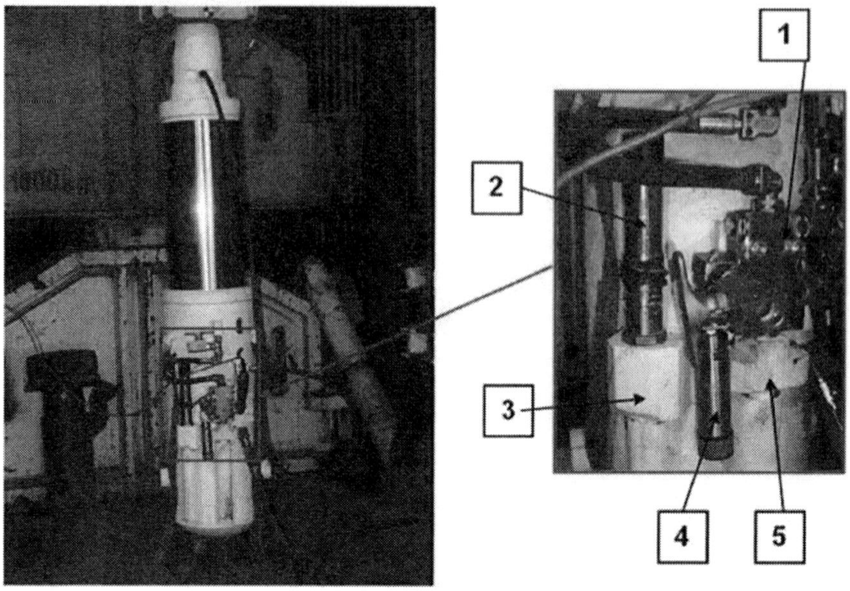

Fig. 2. The Ø320 two-telescopic hydraulic leg and the control system during the dynamic impact mass test.

The tests involved a hydraulic leg type 320/305/250/230x1351, applied to a KHW-12/28 POz roof support with a support system (DOH DN12 valve block type 2010 0312 0502 with a SP10.C valve, and a DOH valve type 2020 0345 1000) [16]. The impact mass test was carried out in a Central Mining Institute's testing rig located in Łaziska, Poland. The leg was hit with a free falling impact mass of 20 t, from six different heights (h=0.20 m; h=0.230 m; h=0.35 m, h=0.40 m; h=0.45 m). The study of the hydraulic support system excitation of the subject leg was based on the spreading of the rack in a rigid frame to its yielded load-bearing capacity (34 MPa). Then, the leg was yielded by applying to the connected DOH DN12 valve block, according to the hydraulic pattern, the set pressure set at 20, 25 and 30 MPa. Pressure values were measured and recorded in the space under the piston and the space over the piston in the DOH DN 12 valve block [16].

The effectiveness of the support system of the test leg is shown in Figure 3. The system comprised the DOH DN 12 valve block type 2010 0312 0501 (with a check valve controlled with a pressure; item 1, Fig. 2) to which the SP10.C valve was connected (item 4, Fig. 2) with the DN19 connection type ZRP (item 5, Fig. 2) and the DOH valve type 2020 0345 1000 (item 2, Fig. 2) connected with the M45x2 terminal (item 3, Fig. 2).

The time waveforms of the pressure changes shown in Figure 3 confirm the effectiveness of the DOH DN12 valve type 2010 0312 0502 with the valve SP10.C (valve setting p_r= 34 MPa) and the DOH valve 2020 0345 1000 (valve setting p_r = 37 MPa).

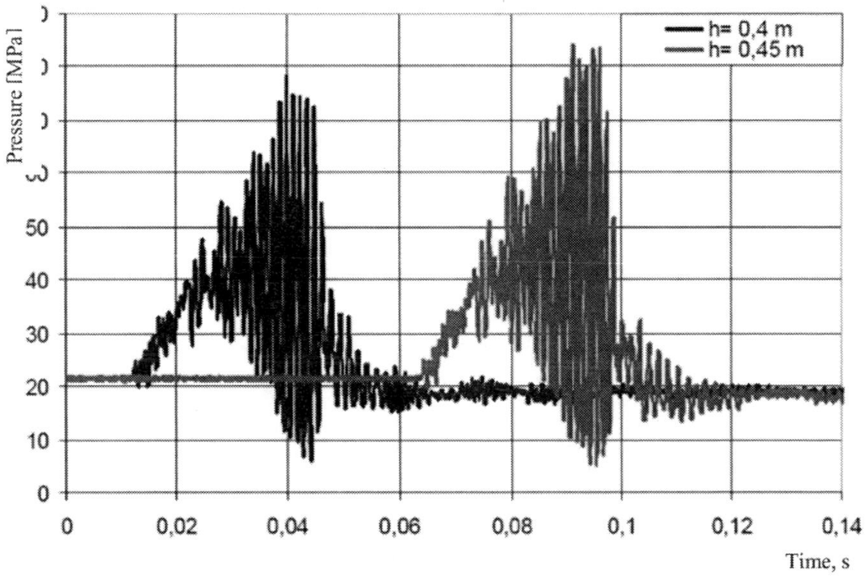

Fig. 3. Time change waveforms of pressure during dynamic load impacting on the hydraulic leg with a control system for two different heights of impact mass drop [10].

The obtained results indicate that the tested support system of the leg along with the increase of the impact energy, stabilizes the pressure in the leg at level about 22 MPa. At the same time, the DOH valve type 2020 0345 1000 reducing the pressure in the leg releases a much greater amount of fluid at the impact mass drop from h = 0.45 m than when released from 0.4 m. All hydraulic components of the leg as well as the entire control system retained its functionality and tightness.

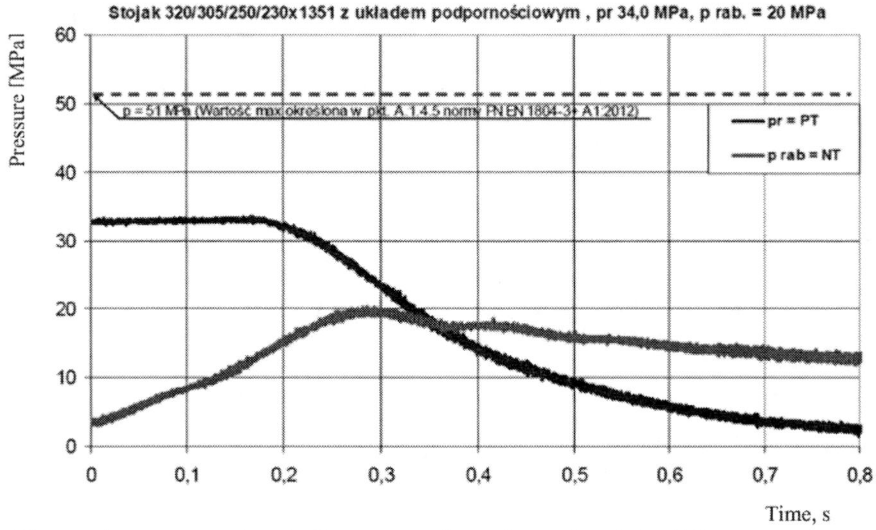

Fig. 4. Changes in pressure values in the space under and over the piston [10].

Pressure values were also recorded over and under the space of the leg's piston. Time waveforms of these values are presented in Figure 4. The analysis of the waveforms suggest that the recorded values fall within the acceptable range. Pressure of 34 MPa is generated in the valve block during dynamic tests, while at drawing off the leg, the pressure in the space over the piston is 30 MPa. Table 2 lists the maximum pressures and delay times during the overload test.

Table 2. Maximum pressure values and delay times during the overload test.

Drop height [m]	Energy of impact mass, Eu [kJ]	Maximum recorded pressure (MPa)	Overload test time [s]
0.20	4	35	0.02
0.30	5	45	0.08
0.35	6	43	0.12
0.40	7	78	0.04
0.45	8	84	0.1

4 Conclusion

When operating in underground excavations, the powered roof support is exposed to external static and dynamic loads. In either case, its response can be dynamic. In the case of static load of the roof support resulting mainly from the gravitational pressure of the rocks deposited over the subject area, the dynamic response can occur only when the critical mass exceeds the working load. On the other hand, the occurrence of dynamic phenomena in the rock mass (bursts, tremors) directly causes a dynamic response of the roof support.

The most important element of the powered roof support that carries these loads is the hydraulic leg. As the basic support element of the powered support unit, it also determines its safety and performance. It is a double acting actuator, which mainly works as a compression element. Selection of technical parameters of the leg for specific mining and geological conditions is extremely significant due to requirements for the acquisition of static and dynamic load, and permissible values of overload. This particularly refers to dynamic loads.

The research allowed to determine the characteristics of its work during dynamic load and to determine whether the tested leg is characterized by an overload greater than the maximum weight with which the rock mass will impact on the roof support. The maximum overload capacity of the leg was analysed for the case of clamping of the excavation at the maximum speed resulting in the liquid flowing through the hydraulic valve from the space under the piston.

The results indicated that the valve block used with the SP10.C valve was more effective than expected. The effectiveness of the safety valve with the preset pressure p_r = 37 MPa increases with the increase of the impact energy.

It can be assumed that the results obtained are very good and the methodology used enables an objective assessment of applied legs. However, it should noted that in real conditions may occur loads with different characteristics (e.g. longer operating time) than load values simulated during the tests.

The results obtained prove that the research is valuable, justified, and that it is necessary to continue the research.

The work was carried out within the project 'Innovative electrohydraulic control system for powered roof support' No. POIR.01.01.01-00-1129/15. Intelligent Development Operational Programme 2014–2020 carried out by the National Centre for Research and Development.

References

1. R. Patyńska, *Zagrożenia tąpaniami (2008 – 2017). Raport roczny o stanie zagrożeń naturalnych i technicznych w górnictwie węgla kamiennego*, The Central Mining Institute, Katowice (2017)
2. S. Prusek, S. Rajwa, A. Walentek, W. Masny, 3rd International Symposium on Mine Safety Science and Engineering, Paper No. 148, Montreal (2016)
3. S. Prusek, S. Rajwa, A. Wrana, A. Krzemień, International Journal of Mining, Reclamation and Environment **31/8** (2017)
4. S. Prusek, M. Płonka, A. Walentek, Arabian Journal of Geosciences **3/9** (2016)
5. A. Biliński, Prace Naukowe – Monografie CMG KOMAG, Gliwice (2005)
6. S. Prusek, *Stability of longwall workings during the operation of coal deposits based on cave mining* (in Polish), The Central Mining Institute, Katowice (2016)
7. D. Szurgacz, Polské hornictví v souèastné dobì, VIII. Mezinárodní konferencje, *Strojní zařízení používaná při hornické činnosti a činnosti prováděné hornickým způsobem, jejich bezpečnost a nové trendy* č. 101, 27-28.4.2017 v Ostravici (2017)
8. D. Szurgacz, Wiadomości Górnicze **10** (2011)
9. K. Stoińsk, J. Gil, XXII International Conference TEMAG 2014, Gliwice – Ustroń 249-255 (2014)
10. J. Brodny, Archives of Mining Sciences **56/2**, 303-318 (2011)
11. J. Brodny, Archives of Mining Sciences **57/1**, 209-227 (2012)
12. M. Tutak, A. Dymark, T. Dzitkowski, MECHANIKA 2017, Proceedings of the 22nd International Scientific Conference, Kaunas 2017, 383-386 (2017)
13. K. Stoiński, Napęd i Sterowanie **7/6** (2006)
14. L. Doległo, J. Gil, K. Stoiński, Maszyny Górnicze **4** (2009)
15. D. Szurgacz, Mining Science **22/2**, 171-179 (2015)
16. GIG Test Report no. 57205506-182. Technical test of a hydraulic leg type 320/305/250/230x1351, applied to a KHW-12/28 POz roof support with a support system (DOH DN12 valve block type 2010 0312 0502 with a SP10.C valve, and a DOH valve type 2020 2020 0345 1000) (2015)

Analysis characteristics determination of electrohydraulic control system operation to reduce the operation time of a powered roof support

Dawid Szurgacz[1,*]

[1]Center of Hydraulics DOH Ltd, 147 Konstytucji St., 41-906 Bytom, Poland

Abstract. The article discusses basic functions of a powered roof support in a longwall unit. The support function is to provide safety by protecting mine workings against uncontrolled falling of rocks. The subject of the research includes the measures to shorten the time of roof support shifting. The roof support is adapted to transfer, in hazard conditions of rock mass tremors, dynamic loads caused by mining exploitation. The article presents preliminary research results on the time reduction of the unit advance to increase the extraction process and thus reduce operating costs. Conducted stand tests showed the ability to increase the flow for 3/2-way valve cartridges. The level of fluid flowing through the cartridges is adequate to control individual actuators.

1 Introduction

Modern powered roof supports are extremely complex and require the most technically advanced solutions and tests concerning parameters of friction joints [1-4]. The research concerning the selection of the roof support which parameters would be suitable for given mining and geological conditions does not include the control system based on electrohydraulics [5-7]. At the same time it is necessary to pay particular attention to the reliability in terms of power supply, so that the individual elements of the longwall complex function properly [8, 9]. The electrohydraulic control system for powered roof supports has a number of advantages and positive characteristics [10]. The research showed that a longwall equipped with the electrohydraulic control system has a greater level of safety than the longwall without such system [11, 12]. Even if the capabilities of the automatization of the system are not fully used, the system is a valuable tool supporting top productive results. Such control system, however, has to be set and configured in a way that allows a steady shift from manual control to fully automated control.

The tests carried out in the past of the section advance time showed that the manual shift time was about 22 seconds, while the best result ever achieved with electrohydraulic control has been about 12 seconds [11-14]. The use of the electrohydraulic control for the implementation of the powered roof support involves also the exploitation of the longwall

* Corresponding author: dawidszurgacz@doh.com.pl

without the participation of people. This solution mainly used in intact rock conditions [15, 16].

The article discusses basic functions of a powered roof support in a longwall unit. The subject of the research includes the measures to shorten the time of roof support shifting. Due to the specific nature of the process of coal mining, the powered roof support performs the function of the wall advance as safety function, protecting mine workings against uncontrolled falling of rocks. In order to introduce an innovative electrohydraulic control system for the powered roof support it is required to conduct an analysis and tests aimed at optimization of working parameters of the unit. The article presents preliminary research results on the time reduction of the unit advance to increase the extraction process and thus reduce operating costs. Conducted stand tests showed the ability to increase the flow for 3/2-way valve cartridges. The level of fluid flowing through the cartridges is adequate to control individual actuators.

2 State of the art

A powered roof support as an element of a longwall complex is set by means of legs placed between a canopy and base. The lemniscate mechanism is used to mount the gob shield to the base. The structure of the longwall roof support during its operation takes over loads in the form of roof rocks and moves it to the floor. The stringer transfers the load from the roof to the floor [3].

The powered roof support unit consists of a canopy (1), legs (2), a lemiscate mechanism (3), a shifter with a beam connected to the stringers (4), a front wall shield, a cross-bar support, a gob shield (5). The stringer is the base of the support unit on which the hydraulic legs are set. The canopy is tightened against the roof with legs. The canopy supports the roof, while the gob shield protects it from caving. The force supporting the roof is generated by the pressure of the liquid in the space under the piston of the legs. The space under or over the piston of the hydraulic cylinder must be protected against overloads by a yield valve. Support unit's shifter, connected cylinder with a shifter and piston rod with a conveyor [17].

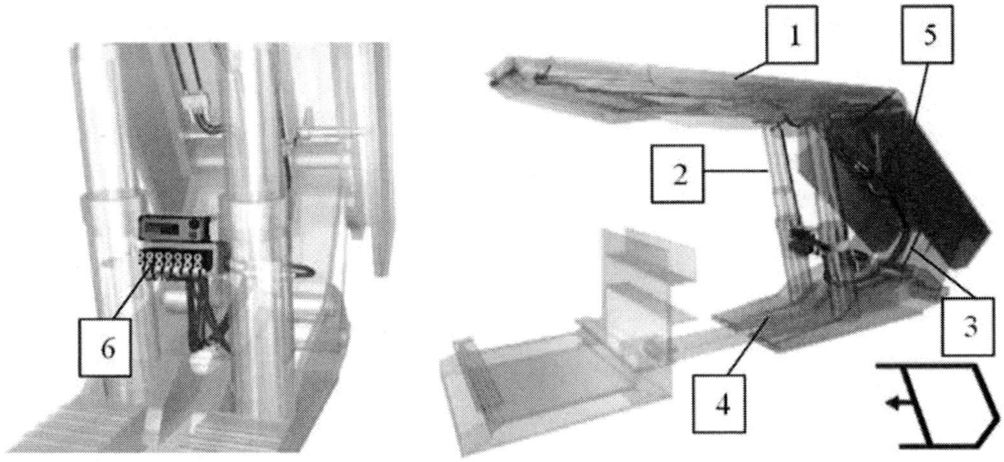

Fig. 1. Powered roof support unit.

The entire control cycle of the powered roof support unit begins with caving in. When the roof support moves, the piston rod is compressed and the sliding beam is stretched. Moving action of the conveyor causes the pressure to flow into the space under the piston, and as a result, the conveyor moves, while the fluid's pressure in the space over the piston causes tightening of the support unit against the conveyor. The moving unit must be caved in, the fluid pressure in the space under the piston of its legs must be reduced to the minimum. All functions are controlled by the control system (6). Each hydraulic cylinder is controlled with a control system consisting of a hydraulic block connected to it by hydraulic hoses. Roof support's hydraulic cylinders with the control block and the power source form the whole unit [17].

3 Research analysis of the electrohydraulic system

The research carried out for many years [10] shows the average recorded advance time of the powered roof support unit of about 12 seconds in the electro-control mode. For comparison, the manual unit advance times recorded were about 22 seconds. Recorded pressure measurement results for stressed legs generated by the roof and floor in the actual operation of the support. The analysis of the charts from the conducted tests of the pressure distribution in a leg allows to diagnose the technical condition and damage of the support hydraulics of the longwall support. Monitoring the pressure in the space under the piston of the legs and the operation of the pressure limiting valves allowed the study focused on the reduction of time of the unit movement to 9 seconds. The tests began with 3/2-way valve cartridge measurement to examine its operability, resulting in a reduction in unit advance time.

The tests were carried using a testing bench that allows the development, commission and modification of the software, and to create virtual drivers. Moreover, it is possible to carry out measurements of the energy required to control selected types of solenoid valves that initiate chosen machine functions. Software of the test stand was created based on LabView application and allows conducting tests of control elements and parameters visualization and their operation. It is also possible to modify the functionality of the test stand, depending on ongoing research. The initial testing included the electrohydraulic part of the powered support control system, consisting of solenoid valves each comprising 4/3-way control valves. A built-in set of 3/2-way valve cartridges DN12 or DN19 is responsible for the execution of tasks. The 4/3-way control valves receive a mechanical impulse generated by the dual electromagnet, and as a result they clip the 3/2-way valve cartridges, through which a suitable amount of fluid flows to control the individual actuators. Figure 2 shows the electrohydraulic part of the control system.

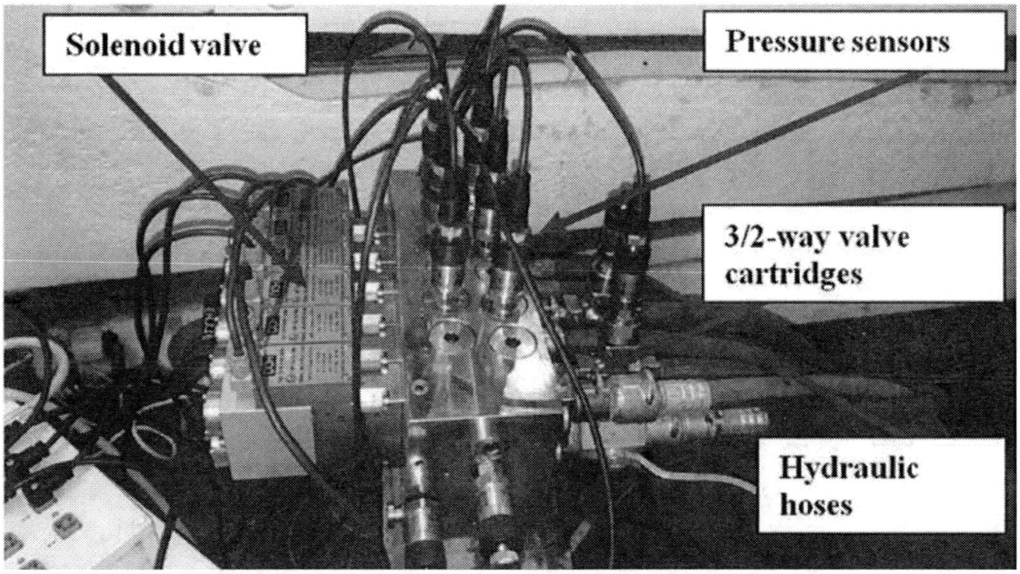

Fig. 2. Subject model of the roof support electrohydraulic control system.

The study included measurements of 3/2-way valve cartridges through which flows the fluid to control the valves of individual actuators. A set of pressure sensors and pressure gauges enables continuous monitoring of the pressure in the elements responsible for execution of tasks.

Preliminary studies on the reduction of unit advance time during the mining were carried out to examine the potential of way valve cartridges. Measurement of pressure values included 3/2-way valve cartridges for DN 10, DN 12 and DN 19, which performs an actuator function. The level of fluid flowing through the cartridges is adequate to control individual actuators. The tests included 18 measurements (as shown in the figures below) for different pressure values and DN values, as shown in Table 1.

The diagram shown in Figure 3 shows the performance characteristics for time waveforms the impacting on 3/2-way valve cartridges. The time waveforms obtained for different pressure values are shown in Table 1. Determined values for different pressures allowed for optimal timing characteristics.

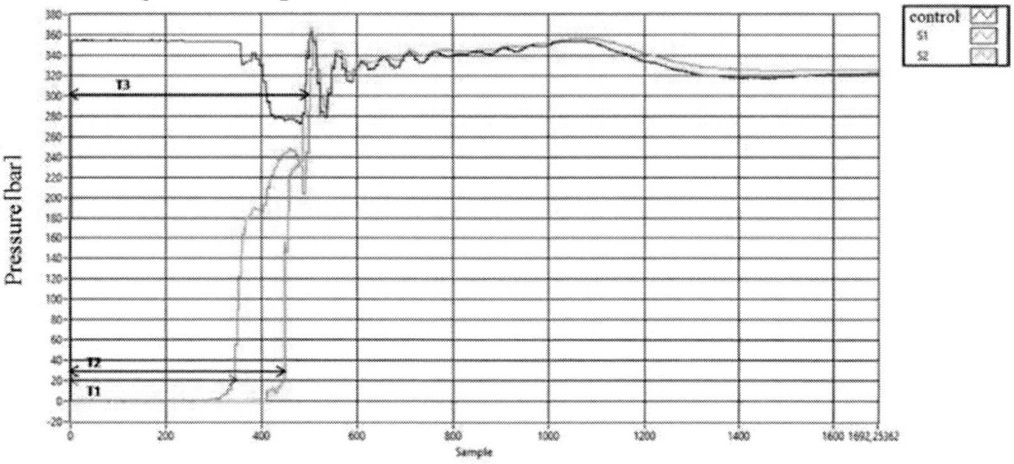

Fig. 3. Setting the response time for determining the characteristics of time waveforms.

Table 1. Results of response time measurements for determining time waveform characteristics.

Measur.	Pressure for T1	Pressure for T2	Pressure for T3	Pressure for T1-T2	T1 ms	T2 ms	T3 ms	T2-T1
1	335	450	507	115	33.5	45	50.7	11.5
2	336	492	528	127	33.5	49.2	52.8	12.7
3	438	487	700	49	43.8	48.7	70	4.9
4	412	460	696	48	41.2	46	69.6	4.8
5	516	617	6,000	101	51.6	61.7	600	10.1
6	536	620	6,000	84	53.6	62	600	8.4
7	314	422	523	108	31.4	42.2	52.3	10.8
8	350	492	541	142	35	49.2	54.1	14.2
9	451	526	1,188	75	45.1	52.6	118.8	7.5
10	406	467	1,286	61	40.6	46.7	128.6	6.1
11	425	526	1,065	101	42.5	52.6	106.5	10.2
12	470	577	1,035	107	47	57.7	103.5	10.7
13	411	507	X	96	41.1	50.7	X	9.6
14	458	470	6,956	112	35.8	47	695.6	11.2
15	466	474	X	108	36.6	47.4	X	10.8
16	359	453	X	94	35.9	45.3	X	9.4
17	281	418	593	137	28.1	41.8	59.3	13.7
18	320	450	642	130	32	45	64.2	13

The characteristics shown below depend on the different pressure values in the 3/2-way valve cartridges for DN 10, DN 12 and DN 20, which perform an actuation function. The level of fluid flowing through the cartridges is adequate to control individual actuators. The tests were conducted for three DN sizes. They showed that with the increase in pressure flowing, its initial impulse impacts the value of the force transmitted by the control system of the roof support. The obtained characteristics were compared with the pressures of different DN. These tests presented in the graphs below were used to indicate the significant impact of a given DN for the same pressure values. Characteristics charts refer to the measurements in Table 1.

The analysis of pressure characteristics shown in the figures 4,5,6 below for different DNs indicates that the increase in the pressure in individual DN diameters can have a significant effect on the speed of the advance of the unit in the longwall system. Determining the right pressure and selecting the right DN diameter can reduce the time required for the proper operation of the powered roof support.

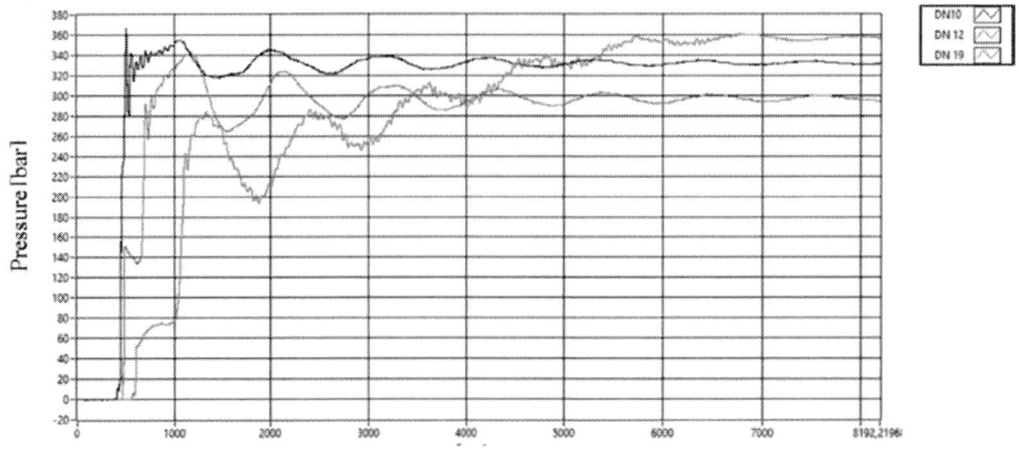

Fig. 4. 350 bar pressure characteristics.

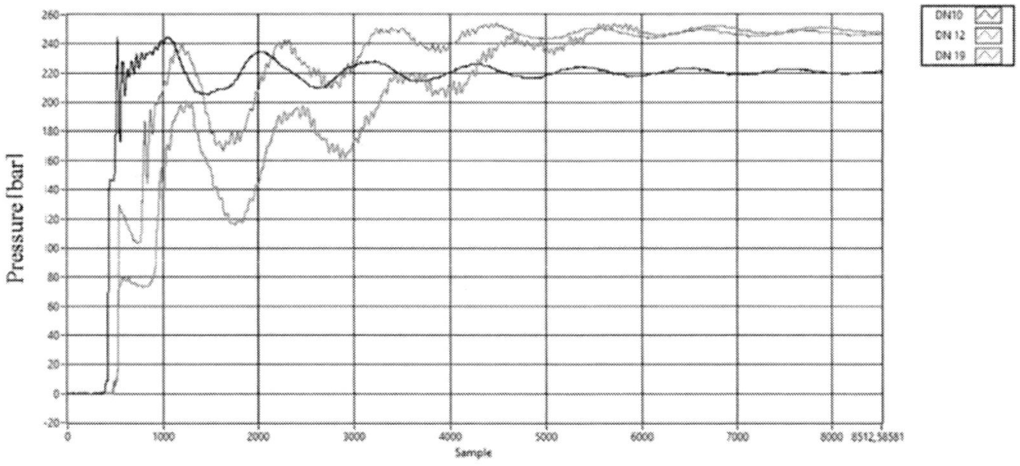

Fig. 5. 250 bar pressure characteristics.

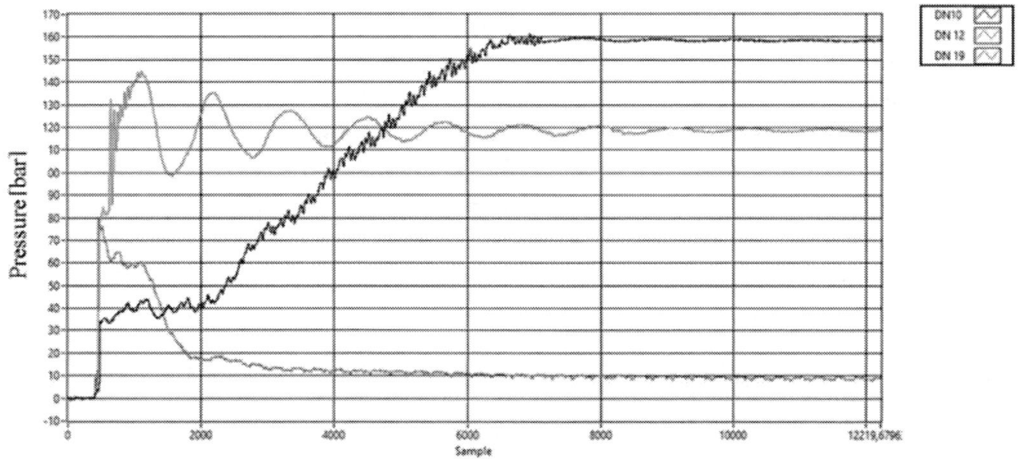

Fig. 6. 150 bar pressure characteristics.

4 Conclusion

Constantly worsening mining and geological conditions cause that the actions focused on costs optimization may affect work safety. The control system of the powered roof support, in spite of significant structural improvements and compliance with the principles of proper operation, does not protect the excavation sufficiently nor does it reduce the time required for the advance. The powered roof support has significant intrinsic value compared to other machines and equipment in which the mine is equipped. The longwall roof support is the basic protection of mining operation.

The comparison of the characteristics presented in the paper allow to conclude that the application of the DN19 diameter in the roof support's control block significantly increase the pressure value transmitted by the fluid flow. It can therefore be stated that the application of DN19 for a pressure of 350 bar for 3/2-way valve cartridges clearly indicates that a shorter adjustment time can be achieved. The research analysis made it possible to conclude that for higher pressures DN value must be increased. In the case of low pressure value, the 3/2-way valve cartridge is unable to transfer the energy required for the advancement of the section. Determining the maximum pressure and selecting the appropriate DN diameter will allow to obtain shorter times of unit advance during the mining operation.

The results obtained including the pressure increase for the optimal DN indicate that at its initial part the force needed to advance the section increases. The preliminary results of the study should be extended by developing an optimum electrohydraulic control of a powered roof support that will not only reduce the section advance time but will also improve safety.

The work was carried out within the project 'Innovative electrohydraulic control system for powered roof support' No. POIR.01.01.01-00-1129/15. Intelligent Development Operational Programme 2014–2020 carried out by the National Centre for Research and Development.

References

1. J. Brodny, Archives of Mining Sciences **56/2**, 303-318 (2011)
2. J. Brodny, Archives of Mining Sciences **57/1**, 209-227 (2012)
3. J. Brodny, Arch. Min. Sci., Vol. **55**, No 4, 733-746 (2010)
4. M. Tutak, A. Dymark, T. Dzitkowski, MECHANIKA 2017, Proceedings of the 22nd International Scientific Conference, Kaunas 2017, 383-386 (2017)
5. S. Prusek, S. Rajwa, A. Walentek, W. Masny, 3rd International Symposium on Mine Safety Science and Engineering, Paper No. 148, Montreal (2016)
6. S. Prusek, S. Rajwa, A. Wrana, A. Krzemień, International Journal of Mining, Reclamation and Environment **31/8** (2017)
7. S. Prusek, M. Płonka, A. Walentek, Arabian Journal of Geosciences **3/9** (2016)
8. A. Kozieł, D. Jasiulek, K. Stankiewicz, S. Bartoszek, Napędy i Sterowanie **2** (2012)
9. S. Mikuła, M. Wojtas, M. Skrabaka, M. Janik, P. Świecza, J. Kuska, K. Fitowski, Napędy i Sterowanie **1** (2013)
10. D. Szurgacz, Journal of Sustainable Mining **14/ 4**, 157-163 (2015)
11. M. Janik, J. Kuska, P. Świeczak, M. Wojtas, K. Fitowski, Napędy i Sterowanie **7-8** (2011)
12. M. Janik, J. Kuska, P. Świeczak, M. Wojtas, K. Fitowski, Napędy i Sterowanie **9** (2012)
13. A. Kasprusz, S. Mikuła, M. Wojtas, Wiadomości Górnicze **5** (2013)
14. M. Schaeffer, *Longwall Automation: State of the Art.*, Joy Corp., Mine Expo International, Las Vegas (2008)
15. J. Wen, Z. Lian, Web Information Systems and Mining Lecture Notes in Computer Science, **6987**, 73-78 (2011)
16. M.X. Lu, M.F. Lv, J.G. Wang, M. Li, J. Zhao, Y.Y. Liu, Y. Zhao, J. Wang, Advanced Manufacturing Systems, ICMSE (2011)
17. J.B. Gwiazda. *Tremor resistant mining roof support* (in Polish), Wydawnictwo Śląsk, Katowice (1997)

Monitoring of the stability of underground workings in Polish copper mines conditions

Krzysztof Fuławka[1,*], *Piotr* Mertuszka[1,] and *Witold* Pytel[2]

[1]KGHM CUPRUM Ltd. Research and Development Centre, 2-8 Gen. Sikorskiego St., 53-659 Wrocław, Poland
[2]Wroclaw University of Science and Technology, Faculty of Geoengineering, Mining and Geology, 27 Wyb. Wyspianskiego St., 50-370 Wroclaw, Poland

Abstract. One of the problems associated with the excavation of deposit in underground mines is the local disturbance in a state of unstable equilibrium results in the sudden release of energy, mainly in the form of roof falls. The scale and intensity of this type of events depends on a number of factors. To minimize the risk of instability occurrence, continuous observations of the roof strata condition are recommended. Different roof strata observation methods used in the Polish copper mines have been analysed within the framework of presented paper. In addition, selected prospective methods, which could significantly increase efficiency of rock fall prevention are presented.

1 Introduction

Natural hazards associated with the underground exploitation of copper ore deposit are ranked among the most serious threats in the mines belonging to KGHM Polska Miedź S.A. The largest risk for the mining operation are created by seismic activity with their potential effects such as rockbursts or roof falls. In addition, the increasing surface of mined out areas causes a local disturbance in a state of unstable equilibrium, which may lead to rock migration towards the working. The level and intensity of this phenomena depends on a number of factors, the most important of which are [1]:

- depth of mining,
- geometry of the mining drifts,
- distance to the disturbed zones,
- proximity to mined out areas,
- physical and mechanical properties of the surrounding rocks,
- applied support,
- presence of tectonic disturbances, etc.

The economic aspect should be also considered, since the instability occurrence may result in machinery and underground infrastructure damage and consequently lead to stoppages of mining operations. The greatest difficulties with the roof control are observed in the vicinity of the exploitation front, where the highest deformations occur [2]. Monitoring of the condition of workings in Polish underground mines is conducted in a very wide range [3,4]. Different types of observations and measurements covering all mining districts are carried out by specialized mining services in accordance with the internal guidelines.

* Corresponding author: kfulawka@cuprum.wroc.pl

This study reviews the current observation methods of underground workings stability and presents selected prospective methods that can significantly improve the safety of work in Polish copper mines.

2 Current state of safety in the context of instability risk

Most of currently applied observations of the immediate roof strata cannot trigger any alert prior to the event occurrence. This means, that the crew is not always informed of the emergency. This situation significantly affects the accident rates in underground mining. Figure 1 summarizes the number of accidents in KGHM's mines that occurred between 2010 and 2016, including fatal and serious accidents.

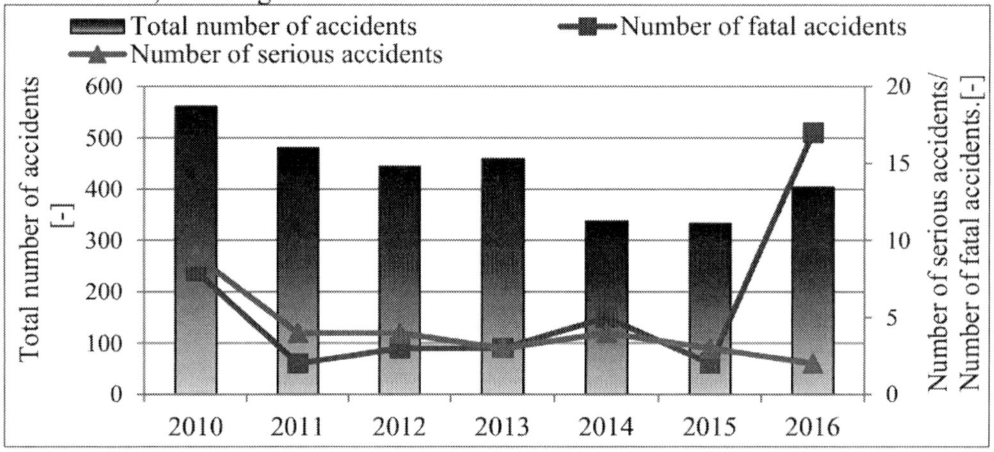

Fig. 1. Number of accidents in Polish copper mines between 2010 and 2016.

In 2016, the number of fatal accidents dramatically increased compared to the previous year. Fourteen of the seventeen fatalities were caused by roof falls and rockbursts. Those events were also the cause of two serious and 49 light injuries (see Fig. 2). It is worth highlighting that these statistics refer only to the incidents, the consequences of which were fatal, serious or collective accidents. In fact, the number of light accidents caused by falling rocks can be much higher, since they are reported as "caused by other factors", so their exact number is difficult to determine.

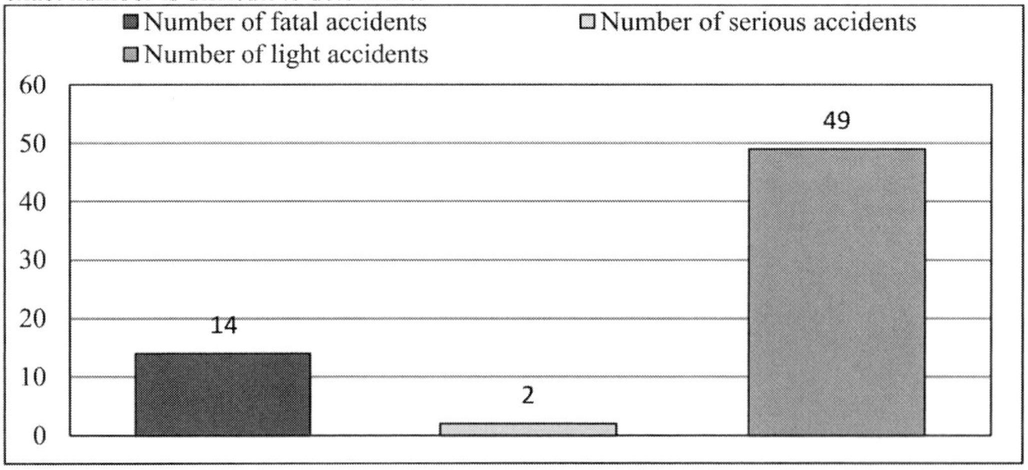

Fig. 2. Number of people injured in accidents caused by roof falls and rockbursts in KGHM's mines in 2016 (based on [6])

3 A review of observation methods of mine workings used in KGHM's mines

Apart from technical solutions of the roof falls prevention (see Fig. 3), observations of the current condition of the excavations are required.

Roof fall hazard and its consequences severity are mitigated by the following safety measures:
- bed separation gauges installed at all crossing of production excavations,
- endoscopic examination within boreholes in immediate roof strata
- identifying of fissures, bedding planes, and other important characteristics of rock mass,
- verification of the tightening torque, bearing capacity measurements,
- analysis of drill holes deformation using different sensors,
- measurements of convergence of mine workings.

In addition, the following visual observation are carried out:
- monitoring of vertical cracks within the immediate roof strata,
- bolt plates' deformation inspection,
- breaking of timber posts,
- survey of artificial support individual elements performance.

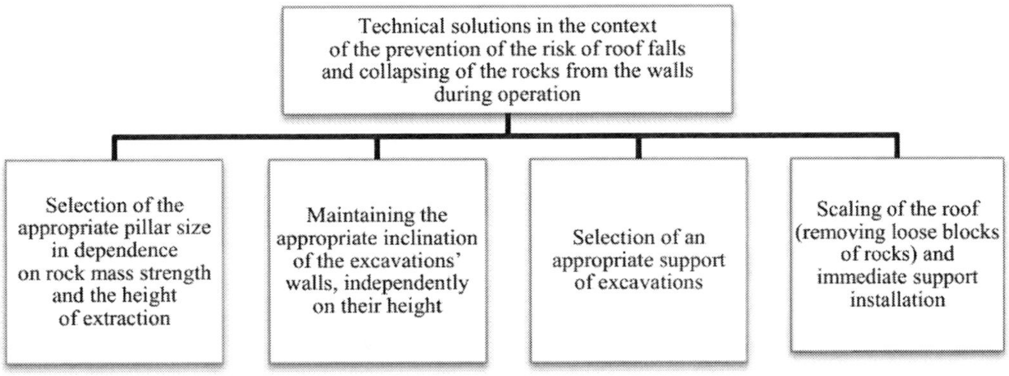

Fig. 3. Technical solutions of roof falls prevention applied in Polish copper mines.

3.1 Observations of bed separation within the immediate roof strata

Roof bed separation gauges called SRS are installed at all crossings of production excavations. It is primary tool for regular observation of the roof stability. Roof bed separation gauges are available in both mechanical and electronic versions [5]. Due to the ease of installation, mechanical gauges are the most commonly used in Polish copper ore mines. They are usually installed using a single resin cartridge. Their length depends on the height of the opening and the length of the rock bolts used. Generally, the length of the SRS sensor's rod ranges from 3 to over 7 meters. It is assumed that the depth of installation of this type of a sensor should be equal to double length of the regular support. This allows to install SRS sensors above the weak layers of immediate roof. These sensors are manufactured in several variants, which differ in the number of signalling plates, their thickness and shape. The most common, however, is a five-plate sensor with a single plate's thickness of 5 mm, what is presented on Fig. 4.

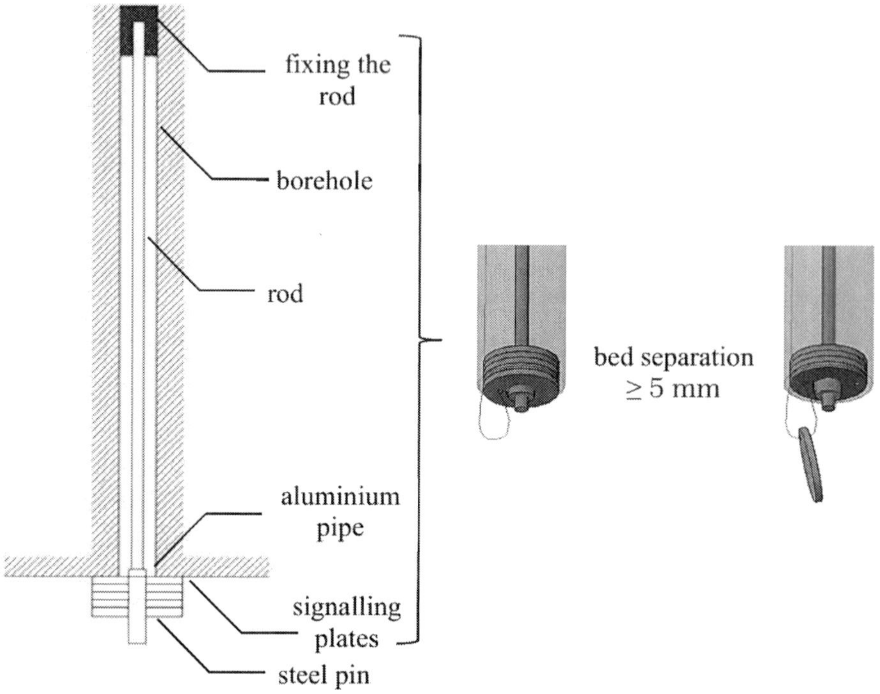

Fig. 4. Principle of operation of the bed separation gauges (based on [7]).

The many advantages of this method include the ease of results interpretation. As the stratification of the rocks in the roof progresses (above 5 mm), the first plate is being squeezed through the steel pin and hangs on the safety rope (see Fig. 4). The falling of the subsequent plates indicates the progressing of stratification process: 10÷15 mm (2 plates); 15÷20 mm (3 plates); 20÷25 mm (4 plates) and above 25 mm (5 plates).

SRS gauges, despite their many advantages, do not provides information about the scale of the hazard. This means that identification of the exact location of bed separation due to excessive shearing is impossible. Another disadvantage of this type of signalling device is the high probability of damage due to mining operations, e.g. as a result of a loop breaking, on which the signalling plates should hang.

More detailed information on the condition of immediate roof strata can be obtained by application of electrical extensometers. Currently, several variants of this type of sensors are used, including:
- multi point cable extensometers,
- CRN-type extensometers,
- tell-tale type extensometers.

The CRN extensometer (Fig. 5) determines any unstable trends between two anchored points in the roof strata. This system accuracy is 0.03 mm. The idea of measurement using this type of sensor is to determine the displacement of the sliding transducer as a result of the movement of the top and lower anchors. The advantage of this type of sensors is their high durability, which determines their suitability in excavations designed as a long-term use, e.g. functional chambers, workshops etc.

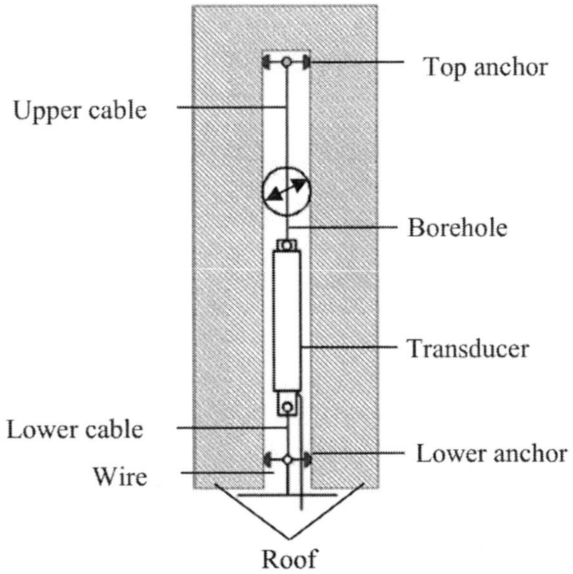

Fig. 5. Scheme of CRN-60 extensometer (based on [2]).

Another type of ground movement sensor used in Polish copper mines is flexible sonic probe extensometer (Fig. 6), which are used to measure displacements and horizons of strata. It allows for location of bed separation that occurs along the borehole, typically around 7.0 m.

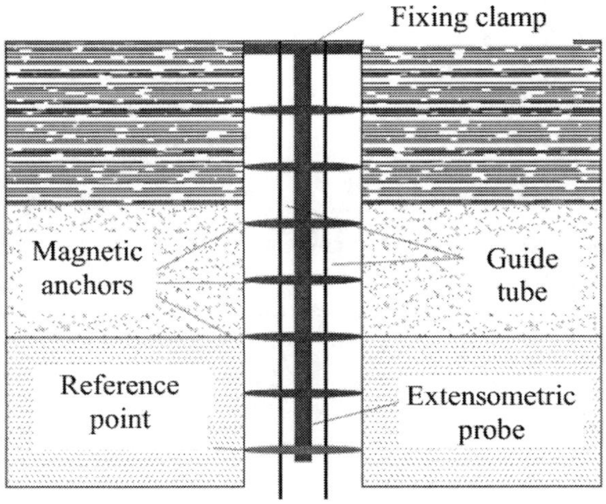

Fig. 6. Principle of operation of sonic probe extensometer [8]

3.2 Measurements of convergence in underground excavations

Convergence measurement is one of the simplest and most common methods of monitoring the underground workings. The concept of convergence refers to the reduction of workings cross-section, which may be vertical (the height of the working reduces) or horizontal (the width of the working reduces). In general, horizontal convergence is many times smaller than convergence in the vertical direction, therefore in practice, vertical convergence measurements are mostly used [9-11]. The main objective of the convergence monitoring is to locate the areas of increased stress generated by mining operations or geology [12]. The results of such measurements are therefore crucial in the assessment of geomechanical hazards in underground mines and are often the basis for determining the working life of the given mining drift [13, 14]. A number of different types of convergence meters are currently available, but the principle of operation in all types is similar. In general, convergence sensors are permanently fixed between the roof and the floor. They consist of two coaxial sleeves (Fig. 7) of different diameters. One of sleeve moves inside the other. In addition, the inner sleeve is fitted with a millimetre scale that allows the distance between the roof and the floor of the working to be read.

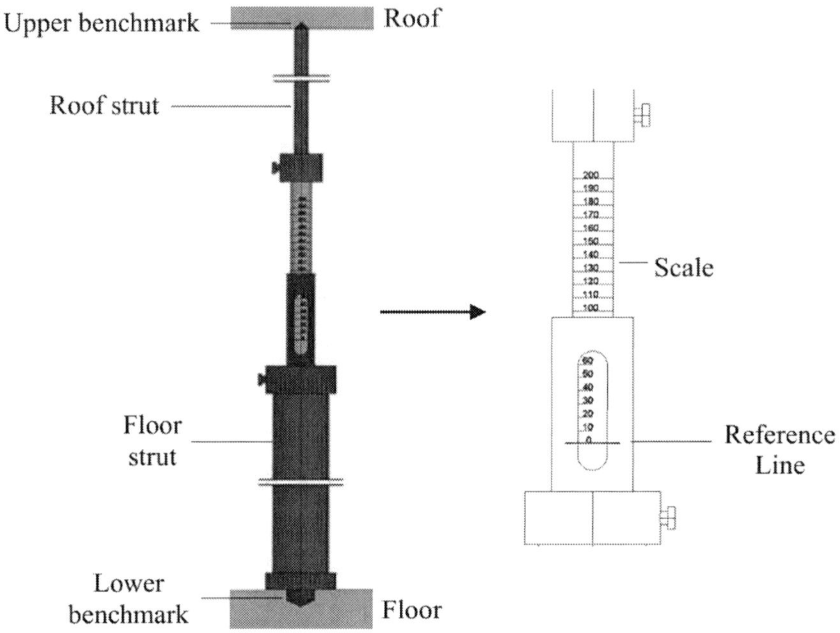

Fig. 7. Scheme of a telescopic convergence indicator [7].

The basic disadvantage of this type of measurement is the low reading frequency and accuracy of the results obtained. It is assumed that the readings are carried out once a day, most often on the first shift. Finally, it is difficult to determine an alarm condition caused, for example, by a sudden increase in convergence on the three preceding shifts.

3.3 Visual observations of boreholes

Irrespective of the applied mining method, excavation of the rock mass generates the cracks with a range, which is usually larger than the size of the working [15]. Observations of the cracked zone within the roof strata in KGHM mines are conducted mainly in areas of increased exploitation pressures by the use of borehole endoscope or infrared camera [16]. The optical endoscope is composed of several tubular elements with optical circuits. The principles of roof control using the borehole endoscope is shown in Figure 8.

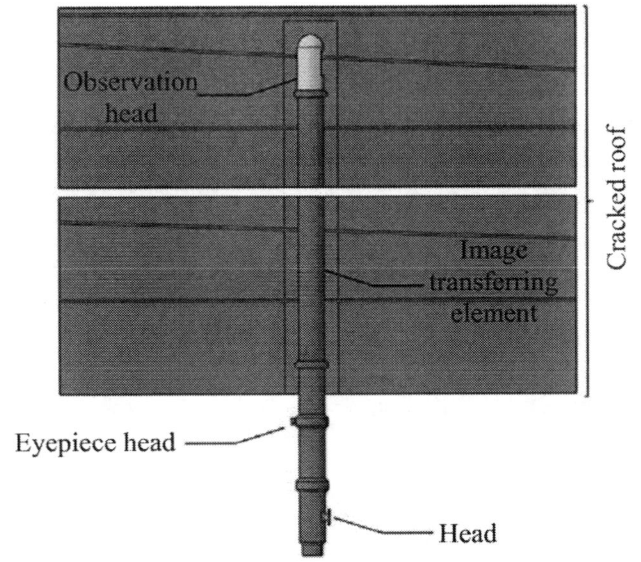

Fig. 8. Diagram of a borehole endoscope [7].

Endoscopic studies provide information on the range of the cracked zones and their distribution within the observation hole [17]. This information may be used at the roof support selection stage [18]. It should be noted, however, that despite the valuable data about the condition of the roof, such measurements are not often carried out. This method is both cumbersome and time-consuming. In general, roof observation using the borehole endoscope is carried out at all intersections of mining drifts on which the bed separation gauges have already signalled the roof fall hazard.

3.4 Measurements of borehole deformation

The analysis of borehole deformation can be carried out using different types of sensors. Currently in KGHM's mine, two types of sensors are used, i.e. DDN (Diameter – Diameter) and DLN (Diameter – Length) sensors, allowing to track the changes of hole diameter and changes of the distance between two points anchored in axis of the drill hole [19]. Sudden changes in diameter or length of inspected hole will follow the alert to take preventive actions. These sensors can be used to monitor the condition of the roof and to observe the transition of the pillars into the post-critical phase (Fig. 9) [20]. It is generally recommended that DD sensors should be applied in vertical holes. This is caused by the lack of sensitivity to delamination in the roof layers associated with the separation of thin rock layers. DL-type sensors, in turn, should be used in horizontal holes, because deformations in the walls and pillars are mainly vertical, which is directly related to the phenomenon of pillar's compression.

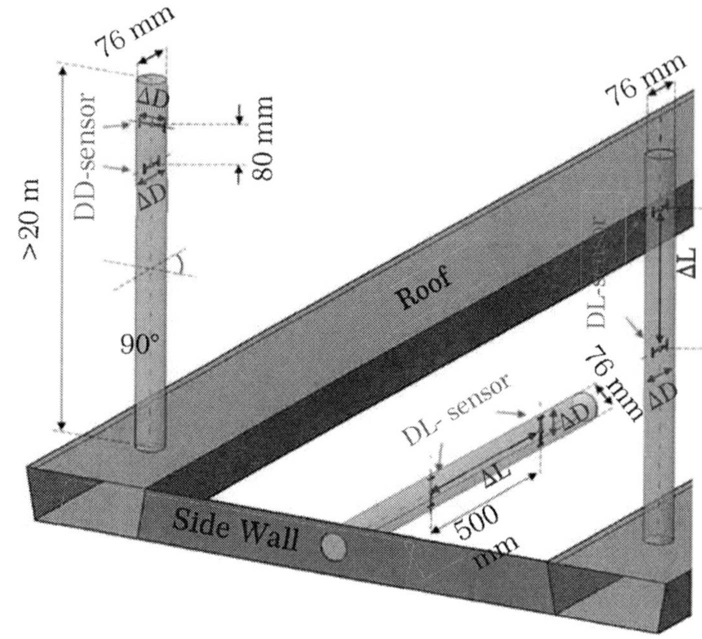

Fig. 9. Application of borehole deformation sensors.

3.5. Measurement of anchor tensioning

Among the many methods used to assess the condition of the rock mass, the tests of the load-bearing capacity of the rock bolts (Fig. 10) are performed. Checking the torque tensioning of the rock bolt indicates the degree of fastening of the roof layers. Mechanical expansive shell bolts should have a tensioning torque of at least 250 Nm, while grouted rock bolts should have 150 Nm [1].

The rock bolt pull-out test procedure is following: after installation, the rock bolt is loaded with the static tensile strength generated by the hydraulic cylinder, gradually increasing until it is ripped out from the hole. The maximum amount of tensile force measured at failure is considered as a parameter of the rock-bolt load bearing capacity. Unlike laboratory tests, underground tests do not raise any significant objections to the required stiffness of the measuring system. This type of tests represents rather static and slow physical processes in the rock mass associated with the progress of mining works or excavation of deposits in adjacent mining parcels and processes of squeezing when we are dealing with inelastic-plastic type of rocks walls.

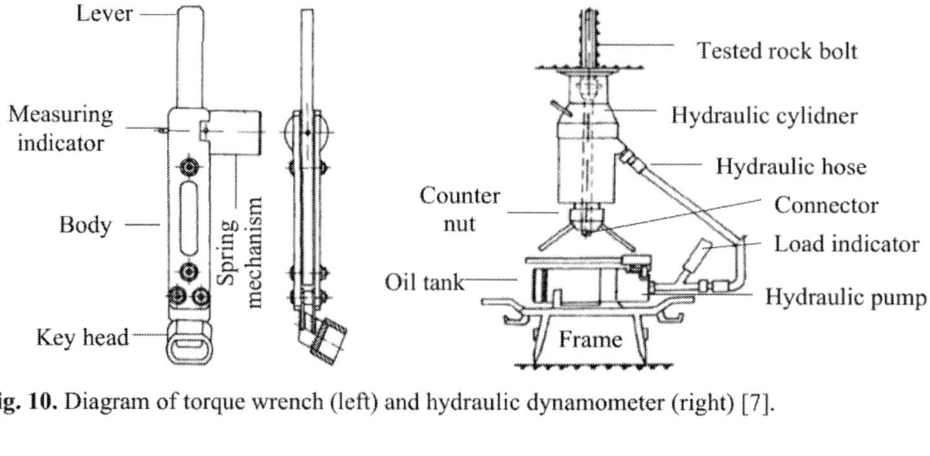

Fig. 10. Diagram of torque wrench (left) and hydraulic dynamometer (right) [7].

4 Prospective methods of monitoring of mine workings conditions

All of the above-mentioned monitoring methods, beside SRS bed separation gauges, have one common disadvantage - they do not automatically alert the crew of the possible hazard. Moreover, presented methods do not allow for continuous analysis of the data collected due to low frequency of measurements. The amount and quality of collected information and data is essential in contemporary monitoring [20]. Properly selected frequency and resolution of measurements are the basis of appropriate assessment of the rock mass condition, what can be directly translated into the crew's safety (alerting of the possible instability) and measurable financial benefits. Therefore, it seems important to undertake additional actions aimed at improvement of the existing monitoring systems, e.g. by implementing new and more advanced measuring tools, which can lead to the minimization or elimination of the most common error, i.e. human error. The following part of the study presented selected innovative systems of monitoring of underground workings.

4.1 Measurements of mine workings inclination

Inclinometric measurements allow to determine the angle of inclination of all elements of the geomechanical system, i.e. floor, roof and pillars [21]. Measurements are carried out using a special devices that consists of inclination sensor(s) and a communications terminal (Fig. 11). This type of measurements is characterized by high resolution and quasi-continuous character (frequency $\geq$ 1 sample per 1 min). Data are collected from two perpendicular directions X and Y, what allows to determine the resultant direction of deformation [22].

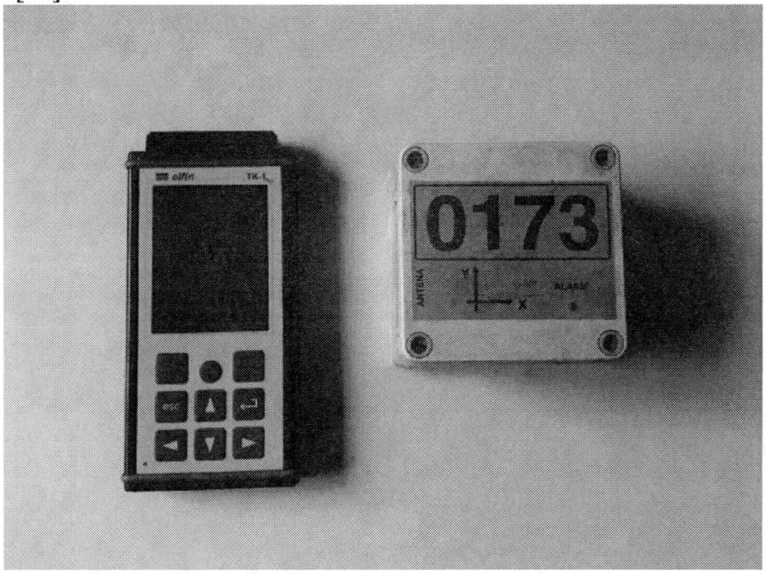

Fig. 11. Components of inclinometric measurements system: communication terminal (left) and a CNS sensor (right).

The measurements of the changes of inclination angle of roof, floor or walls may be useful for the local assessment of mining drifts stability and for monitoring of the rock mass deformation. The CNS sensor is equipped with a signalling diode, which inform, that the threshold values of the inclination angle have been exceeded. Communication with the CNS sensor is done wirelessly. Previous experiences of the application of inclinometric measurements indicate that it is a very useful tool for the stability assessment of the roof stratum and for tracking of large scale deformation. Example of data gathered with inclinometric measurements system is shown on Fig. 12.

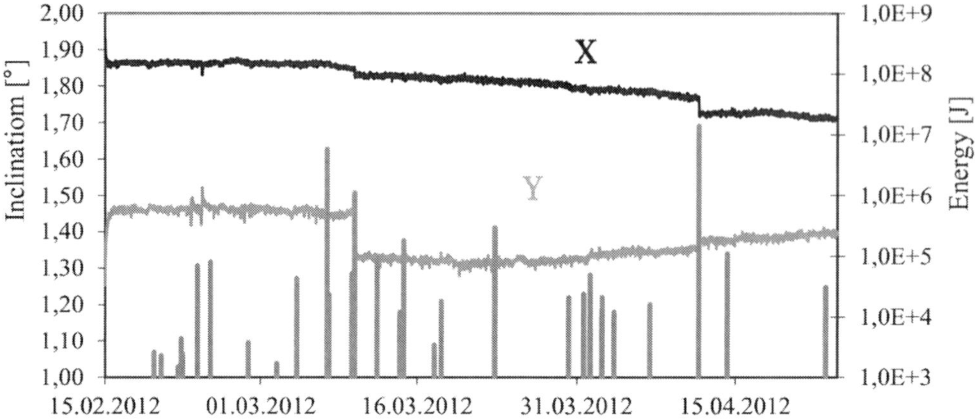

Fig. 12. Changes of the roof inclination in time.

4.2 Spatial laser scanning

The convergence monitoring described in Section 3.2 are point-based measurements, which directly translates into the assessment of the hazard level in the given mining parcels. Detailed information on the convergence of mine workings can be obtained by the use of laser scanners. Laser scanning technology has already been used successfully in the open-pit mining industry. Application of this method in the monitoring of underground workings conditions is still being developed.

In 2016, laser scanning technology of underground excavations was tested in one of the mine belonging to KGHM. As a result, two-dimensional models describing the convergence process and a spatial model of the whole mining division were obtained (Fig. 13).

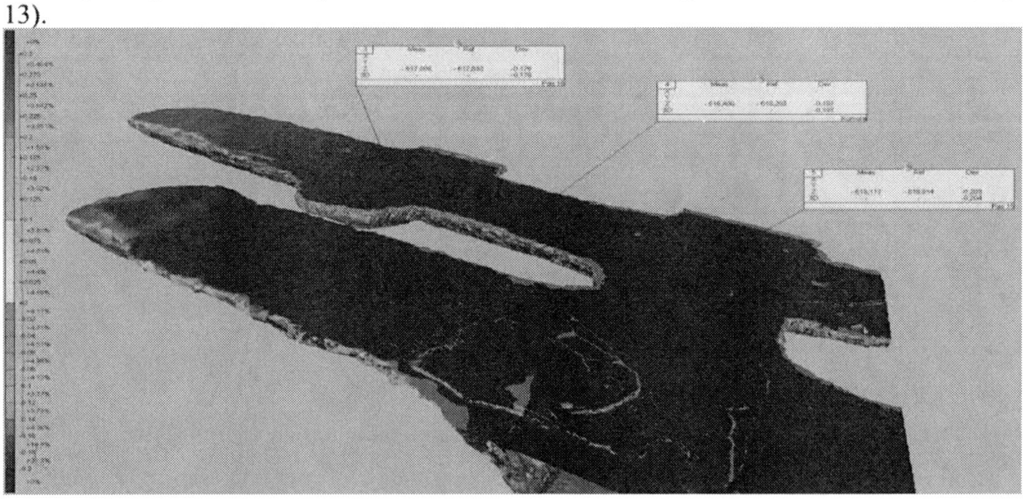

Fig. 13. Spatial model of mine workings generated using laser scanning [23].

The major disadvantage of this type of measurement is the disturbance of the mining process within mining parcel being measured. Obviously, during the scanning of workings, any movement of machinery is highly unwanted. However, a regular laser scanner allows for precise determination of the deformations and displacements of the mine workings. This information may be useful while generating the model of rock mass behaviour subjected to changes due to mining operations.

4.3 Instrumented rock bolts

Observations of engineering structures using instrumented rock bolts have been carried out for many years, but so far, they have not been widely used in the Polish copper ore mining industry. The most popular are the rock bolts equipped with strain gauges [24, 25], but they are also available with fibre optic sensors [26]. Unfortunately, the solutions used so far in the Polish copper ore mining industry did not meet the requirements of measurement accuracy and sampling frequency. In addition, it was impossible to perform the long-term measurements without external power supply. In 2016, KGHM CUPRUM collected and tested the equipment dedicated for the observations of immediate roof strata. The pilot stress measurements were performed using a four-groove bolt with a highly-advanced data collecting system. The rock bolt used allowed for the stress measurements on five levels of the bolt along four orthogonal directions. The four-way system allowed for a full tracking of the stresses acting vertically and in horizontal directions [27] (Fig. 14). Moreover, the continuous recording may be conducted at a frequency adjusted to the requirements of the user. During the pilot underground tests, stress changes were recorded at a frequency of 1 Hz. The operating time of one battery lasted 2 months. Reducement of the sampling rate allows for extension of the operating time). Another important feature is also the autonomous nature of the data collection system.

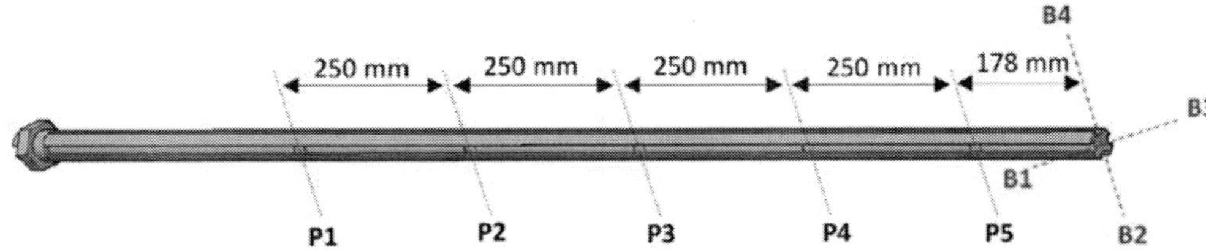

Fig. 14. General view of the instrumented rock bolts (based on [27]).

Measurements using electrical resistance strain gauges are based on the increase or decrease of conductor resistance as a result of a change in cross-sectional area due to tension or compression. Installation of the instrumented rock bolt itself does not differ from the standard bolt installation. The recorder with battery should be placed in cylindrical borehole next to the rock bolt (Fig. 15).

Fig. 15. Installation of instrumented rock bolt.

The results of the measurements confirmed the validity of the use of this type of instruments for the monitoring of underground workings. The location of the strain gauges in four vertical grooves allows for the axial and shear stresses determination. Rapid increase in one of these values indicates that the instability may occur. In addition, determination of the scale of hazard and, therefore, indication of the height at which instability has occurred or will occur may be reached by the application of few measuring levels. Based on measured values, the following parameters may be determined:

- displacements,
- bending moments,
- axial stresses,
- shear stresses,
- von Misses stresses

on each measuring level (Fig. 16).

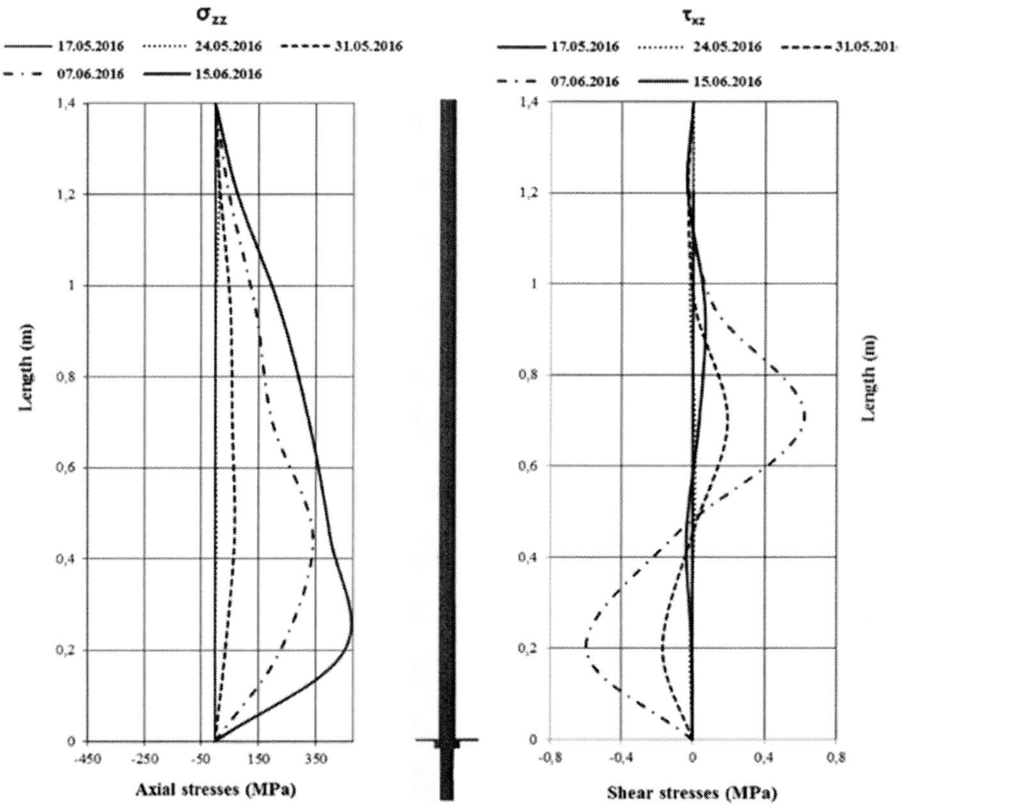

Fig. 16. Distribution of axial stresses (left) and shear stresses (right) on different measuring levels.

5 Summary

The stability of underground workings in Polish copper ore mines is monitored using a number of measuring instruments. These tools have relatively simple construction, however allow for the quick assessment of the current state of the rock mass in the area of existing mining drifts. Despite the obvious advantages, currently used instruments do not allow for a comprehensive assessment of the scale of the hazard, what is limited due to obsolete technology, which requires involving additional personnel. Bearing in mind, that the exploitation of the flat copper ore deposits in KGHM's mines is conducted in relatively difficult mining and geological conditions, new technological solutions for the observation of rock mass behaviour subjected to both static and dynamic loading should be considered. The current knowledge on applicable mining methods and access to advanced measuring devices should result in the introduction of newer and more effective observation technologies, that will be able to inform on the level of local hazards as soon as safety thresholds are exceeded. An appropriate interpretation of the collected data, in turn, will allow a detailed analysis of the rock mass state in the vicinity of active mining parcels prior to the roof fall or rockburst occurrence.

The perspective observation methods described within the framework of this paper will provide precise information on the rock mass behaviour during mining operations within given mining parcel, which in the longer term may contribute to the implementation of more effective preventive measures, what was confirmed by the results of performed measurements. Inclinometric measurements conducted in the vicinity of exploitation front enable evaluation of the behaviour of all elements of the geomechanical system. i.e. roof, floor and walls. On the other hand, laser scanning at regular time intervals shall produce the deformation profile throughout entire mine workings, which may allow to identify zones that are prone to instability. The basic purpose of monitoring of the rock mass behaviour using the instrumented rock bolts is to determine the scale of movement in the horizontal and vertical direction of the roof and floor stratum around the yielding pillar. When considering the roof fall hazard, instrumented rock bolt seems to be most suited to mining and geological conditions in KGHM's mines. Alerting on the increased risk of the occurrence of instability followed by a critical condition of the rock mass gives an opportunity to take appropriate preventive measures.

This work was financed by the Grant no. 0401/0128/17.

References

1. W. Pytel, *Geomechanical problems of the selection of mining excavation casing (*Wyd. KGHM CUPRUM, Wrocław 2012)
2. J. Butra, R. Dębkowski., D. Pawelus, M. Szpak, Cuprum, no. **58,** 43-71 (2011)
3. T. Majcherczyk, P. Małkowski, Górnictwo i Geoinżynieria, no. **29** z. 3/2, 61-76 (2005)
4. G. Mutke, W. Masny., S. Prusek, Acta Geodynamica et Geomaterialia, Vol. **13**, No. 4 (184), 367-378 (2016)
5. C. Matusz, K. Szczerbiński, Cuprum no. **66,** 33-44 (2013)
6. website www.wug.gov.pl
7. *Monograph of KGHM Polska Miedź S.A.*, Lubin 2007
8. S. Fabich S., W. Pytel., L. Cuprum no. **17,** 73-94 (2000)
9. T. Majcherczyk, P. Małkowski, Z. Niedbalski, Górnictwo i Geoinżynieria, no. **3/1** (2005)
10. S. Piechota, Przegląd Górniczy, no. **12** (2001)
11. P. Małkowski, Z. Niedbalski, T. Majcherczyk, Górnictwo i Geoinżynieria, no. **32** (2008)
12. R. Hejmanowski, A. Malinowska, K. d'Obyrn, Przegląd Górniczy, vol. **70**, no. 7, 51-55 (2014)
13. E. Hrubešova, Z. Kaláb, R. Kořinek, P. Žůrek, Kwartalnik UWND AGH, Górnictwo i Geoinżynieria, no. **3** (2007)

14. T. Majcherczyk, T. Małkowski, Z. Niedbalski, Materiały Zimowej Szkoły Górotworu i Geoinżynierii, Krynica, 509-517 (2006)
15. P. Małkowski, Wiadomości Górnicze, no. **5**, 259-269 (2014)
16. J. Parchanowicz, S. Laskownicki, K. Szczerbiński, Cuprum, no. **19**, 27-32 (2001)
17. P. Małkowski, Kwartalnik AGH Górnictwo i Geoinżynieria, no. **3-4**, Uczelniane Wydawnictwa Naukowo-Dydaktyczne AGH, 419-425 (2003)
18. T. Majcherczyk, P. Małkowski, WUG: Bezpieczeństwo Pracy i Ochrona Środowiska w Górnictwie, no. **2** (2002)
19. S. Orzepowski, J. Butra, Rudy i metale nieżelazne, no. **2** (2011)
20. S. Orzepowski, WUG: Bezpieczeństwo Pracy i Ochrona Środowiska w Górnictwie, no. **9** (2002)
21. W. Grzebyk, L. Stolecki, WUG: Bezpieczeństwo Pracy i Ochrona Środowiska w Górnictwie, no. **8**, 17-22 (2013)
22. W. Grzebyk, L. Stolecki, WARSZTATY 2014 z cyklu: Górnictwo - człowiek - środowisko: zrównoważony rozwój, Mat. Symp., 92-100 (2014)
23. *Detailed project of exploitation in field XIII/4 of G-2 mining district with information about the exploitation progress*, KGHM Polska Miedź S.A., Oddział Zakłady Górnicze „LUBIN" (2016)
24. W Pytel, K. Szeptun, XXVII Zimowa Szkoła Mechaniki Górotworu, Kraków (2004)
25. T. Majcherczyk, Z. Niedbalski, Ł. Bednarek, Przegląd Górniczy, no. **3**, 41-47 (2014)
26. S. Spearing, *The Precise Monitoring Of Rockbolt Performance Underground Final Technical Report*, Short-Term Innovative And Exploratory Research Project Alpha Foundation For The Improvement Of Mine Safety And Health (2015)
27. W. Pytel, P. Mertuszka, E. Fabiańczyk, K. Fuławka, Wiadomości Górnicze, no. **12**, 654-664 (2016)

Method of operator safety assessment for underground mobile mining equipment

Paulina Działak[1,*], *Jacek* Karliński[1], and *Eugeniusz* Rusiński[1]

[1]Wrocław University of Science and Technology, Mechanical Faculty, Department of Machine Design and Research, 27 Wyb. Wyspianskiego St., 50-370 Wroclaw, Poland

Abstract. The paper presents a method of assessing the safety of operators of mobile mining equipment (MME), which is adapted to current and future geological and mining conditions. The authors focused on underground mines, with special consideration of copper mines (KGHM). As extraction reaches into deeper layers of the deposit it can activate natural hazards, which, thus far, have been considered unusual and whose range and intensity are different depending on the field of operation. One of the main hazards that affect work safety and can become the main barrier in the exploitation of deposits at greater depths is climate threat. The authors have analysed the phenomena which may impact the safety of MME operators, with consideration of accidents that have not yet been studied and are not covered by the current safety standards for this group of miners. An attempt was made to develop a method for assessing the safety of MME operators, which takes into account the mentioned natural hazards and which is adapted to current and future environmental conditions in underground mines.

1 Introduction

All structures (cabins) designed to protect operators of underground mining machinery that are to be put into operation in mines must comply with the requirements of the Machinery Directive 2006/42/EC. This document includes, among other things, two standards that describe tests performed on cabins and the criteria that must be met for the machine to be authorized for use. The purpose of these tests is to ensure the protection of operators of mobile mining equipment. These standards are: ISO 3449:2005 (FOPS – Falling-Object Protective Structures) and ISO 3471:2009 (ROPS – Roll-Over Protective Structures) [1]. In addition, Polish legislation also applies the PN-92/G-59001 standard (RSPS – Rock Slide Protective Structures; Fig. 1). It is designed to protect machine operators from the effects of roof falls. The first one (FOPS) is associated with the protection of operators against falling objects. The test consists in the drop of a cylindrical object onto the roof of the protective structure from a height sufficient to develop the required impact energy of 11 600 J. The testing procedure for RSPS is very similar to the FOPS test, with the difference that RSPS is performed with greater impact energy of at least 60 kJ [2]. The ROPS standard, however,

* Corresponding author: paulina.dzialak@pwr.edu.pl

is required for all self-propelled machines with a mass greater than 700 kg (described in ISO 6165), designed for an on-board seated operator. It is a static test in which the protective structure is subjected to standardized load that represents roll over situations [3].

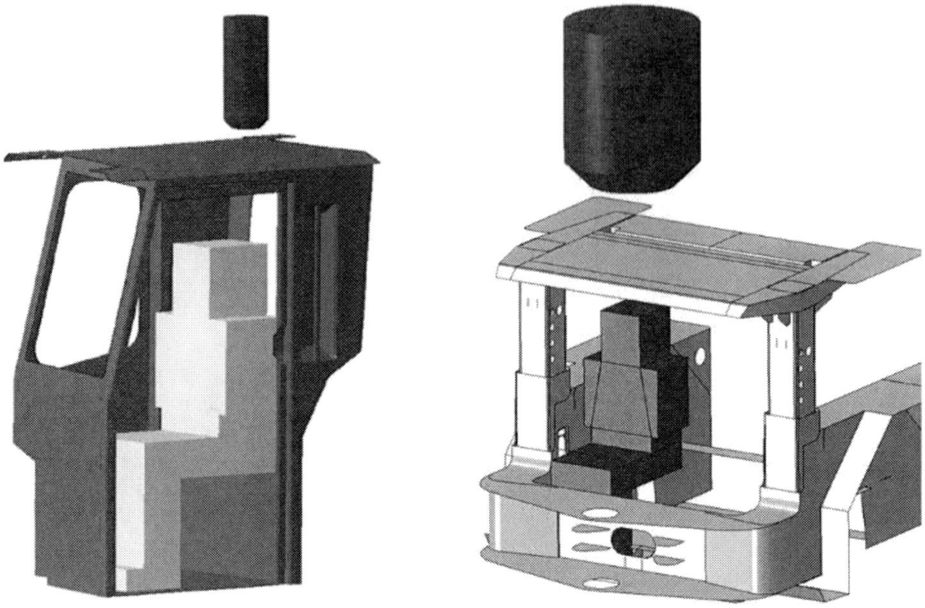

Fig. 1. FOPS test (left) and RSPS test (right); DLV placed inside the protective structure.

In standard tests, the DLV (Deflection Limiting Volume) is used to represent the operator (Fig. 1). It is an orthogonal approximation of a large seated operator that defines the deformation limits of the cab [4]. Tests are considered positive if none of the elements of the protective structure that has been punctured or deformed in the elastic and plastic range enter the protective space defined by the DLV [5]. Due to the required distance between the model and the structure, the sufficient condition is that no part exceeds the dynamic deflection value of 50 mm. However, no tests are performed to determine the loads acting on the body of the operator or the injuries that an operator may sustain during the accident.

On the other hand, an analysis of accidents in underground mining with respect to machine operators indicates that there are accidental situations in which the DLV is not penetrated but the operator still sustains severe or even fatal injuries. Once the deposit is depleted on the current level, mining proceeds to deeper levels where geological and mining conditions are different. These conditions activate natural hazards which differ in intensity and range, resulting in events that have thus far been considered extraordinary [6].

The authors have decided to propose a new method of testing the safety of operators of machines intended for underground mining. The method takes into account both the injuries that could be sustained by the operator located in the protective structure as well as the effects of dynamic phenomena, which have not been analysed yet, but which are becoming a major threat to work safety in underground mines as the depth of mining increases.

2 Methodology of research

The proposed method is based on numerical analysis of events that are dangerous to operators of underground mining machinery. It involves tests on a complete numerical model of the protective structure with a full-size operator dummy as well as a seat with a lap belt and a simplified model of its mount. This specially devised model is then subjected to loads that represent those generated by dangerous phenomena acting on the cab. First it is necessary to identify the phenomena that may be hazardous or fatal to operators and this should be followed by the formulation of appropriate boundary conditions that best reflect the actual event [7].

2.1 Identification of hazards

Based on the results of a statistical analysis of underground mining accidents, with special consideration of copper mines, it was possible to identify the most dangerous phenomena that threaten the safety of machine operators. Apart from the dangers associated with roof collapses and rocks falling from the ceiling and sidewalls, which have already been identified by current safety standards, new phenomena are emerging that have previously been unheard of. One such new danger that has appeared in recent years is the upthrow of machinery caused by sudden floor uplift or by the rapid filling of the loader bucket with the extracted material. Due to the low height of the mine gallery the machine that is ejected upwards often hits the ceiling. The same happens to the operator protective structure on the machine. However, neither the cabin load-carrying structure nor its attachment are adapted to such loads. This is also true for the operator's seat and the safety systems inside the protective structures. Such incidents can be harmful or fatal to operators, which has been confirmed by the accidents that have taken place thus far [8] (Fig. 2).

Fig. 2. Damage resulting from the upthrow of machinery caused by a rock burst.

2.2 Positioning

The developed numerical model combines the Finite Element Method (FEM) [9] with MultiBody (MB) models and uses the advantages of both numerical codes [10]. By combining these two systems it is possible to perform a complete strength analysis [11] of the protective structure and of the injuries to the operator, while significantly reducing the required calculation time. The numerical model prepared by the authors is based on three basic components: the MADYMO dummy, the operator protective structure and the operator's seat with the mount and a two-point seat belt.

2.2.1 Dummy

The authors have decided to abandon the standard approach to operator safety tests and include a full-scale human model in the testing procedure. This was achieved by selecting the appropriate numerical model of the operator, whose design allowed for the collection of all the required data resulting from specific dynamic loads generated by rock bursts in the mine. A comparison was made between numerical equivalents of three dummies that are used in vehicle and road transport safety tests – Hybrid II and Hybrid III, as well as aircraft tests – Hybrid III FAA (Federal Aviation Administration). The simulations were based on ellipsoid MADYMO dummies (Fig. 3). This system offers access to a complete database of validated numerical dummies that are ready for testing [12].

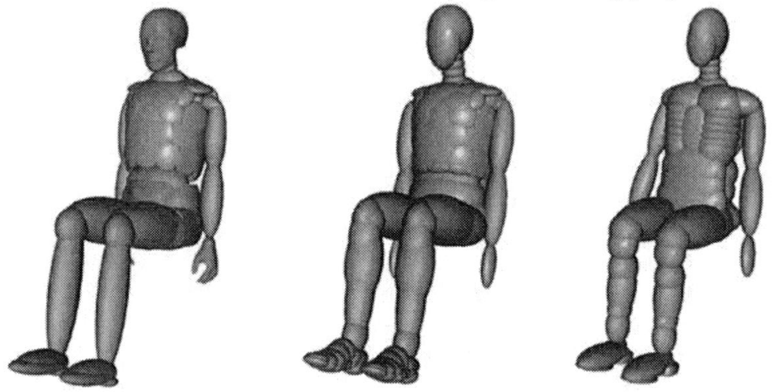

Fig. 3. MADYMO dummies used in the comparative analysis; from left to right: Hybrid II, Hybrid III and Hybrid III FAA.

Given the typical size and gender of operators of self-propelled machinery designed for underground mining, the studies compared 50th percentile male dummies (with approximate height and weight of 175 cm and 75 kg respectively). For the purpose of this comparative analysis, a simplified model of the seat with cabin elements has been developed, which the operator's body can come into contact with during the incident. This model was based on perfectly rigid surfaces so that results can be compared between different human body models without the necessity to analyze additional variables (Fig. 4). The dummies were also fitted with a lap belt, which should be used by every operator at all times.

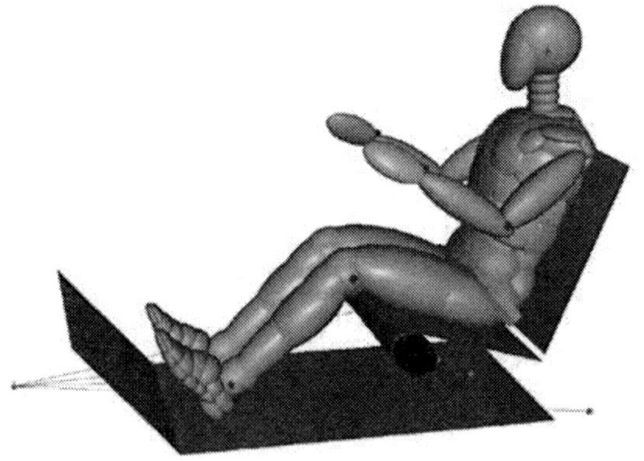

Fig. 4. Model developed for comparative analysis of dummies.

Eventually, the Hybrid III FAA dummy was selected for further analysis as it is best suited to respond to and process vertical loading, particularly in sections of the spine and upper legs [13], and it is those vertical forces that will be dominant in cases analyzed by the authors.

2.2.2 Seat

The geometric model of the operator's seat was not available and was therefore obtained by capturing the shape of the actual seat. This was achieved by means of TRITOP and ATOS optical systems. TRITOP is a portable device that measures the coordinates of selected three-dimensional objects. This system was used to represent the geometry of the foam seat and backrest. ATOS, on the other hand, is a 3D scanner which delivers accurate and complete measurement data, even in the case of complex and glossy elements [14]. In this case it was used to obtain the geometry of the frame of the seat (Fig. 5). By combining the abovementioned systems a full and detailed geometry of the operator's seat was obtained, which, after discretization, was included in the simulation.

a) b) c)

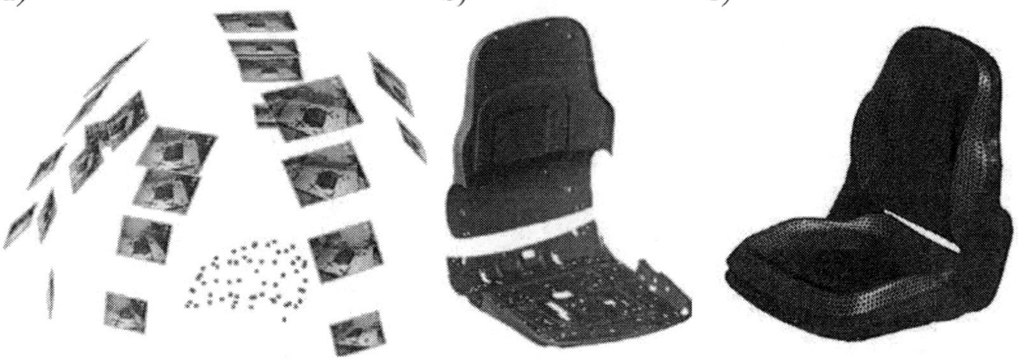

Fig. 5. Development stages of the operator seat model: a – measuring with the TRIROP system, b – scan made with the ATOS system, c – final numerical model of the seat.

2.2.3 Protective structure

The protective structures that were analyzed are cabins which are already in use and thus meet the current safety standards for machine operators. Based on the geometry, a discrete model was developed, which, in this instance, was tested in accordance with the new method proposed by the authors. Discretization was performed using triangular and rectangular low-order shell elements of the *explicit* type [15], with edge length of about 15 mm. The final computational model of the cabin with the dummy and operator's seat is shown in Figure 6.

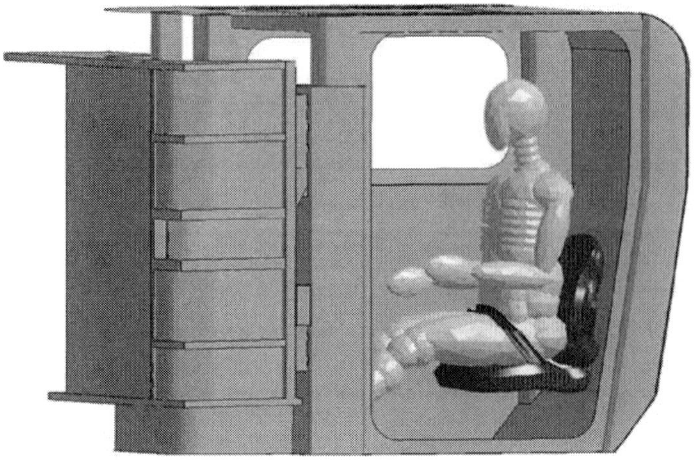

Fig. 6. Final model of the studied cabin with the dummy and operator seat.

3 Analysis of results

The proposed method includes an analysis of injuries sustained by biomechanical parts of the body, whose damage can lead to death or disability (physical impairment of the organism leading to limitation of bodily functions without a reasonable chance for recovery) [16]. Given the loads acting on machine operators during different accidents, the parts of the body most vulnerable to injury are the spine, head, and neck. The authors have decided to take into account the following criteria of injury:
- vertical load measured in the lower section of the lumbar spine;
- HIC (Head Injury Criterion);
- Nij (Neck Injury Criterion).

The US Federal Aviation Administration (FAA) established the safety testing conditions for aircraft and seats in the event of an emergency landing. One of the criteria for spinal injuries that is included is the vertical force acting on the spinal column, measured in the lumbar section of a 50th percentile male dummy. This criterion states that such force cannot exceed 6672N [17]. On the other hand, the US military regulations regarding loads acting on the spine in combat helicopters define the permissible forces on the basis of the maximum acceleration value that is safe for the spinal column (14.5g). For a standard male dummy the permissible force level is as high as 9185N [18]. In the case of mining machinery operators this criterion should be verified by accurately adapting it to this group adapted to this particular group. However, assuming that operators are mostly men at a relatively young age, the authors, in their studies, have adopted the criterion of the American army as the limit value.

HIC is a criterion which describes the likelihood of a head injury during an accident. The time-history of head accelerations that is necessary to calculate the HIC value is most often read from the accelerometer located at the center of gravity of the head of the dummy, which is subjected to appropriate loads. This criterion is described by the following formula (1)

$$HIC = \left(\left[\frac{1}{t_2 - t_1} \int_{t_1}^{t_2} a(t)dt \right]^{2.5} (t_2 - t_1) \right)_{max} \tag{1}$$

where t_1 and t_2 are respectively the initial and final times (expressed in seconds) of the interval during which the load acting on the head attains maximum value whereas a is the acceleration (expressed in g). The maximum time duration (t_2-t_1) is generally limited to 3-36ms, and is typically 15 ms. This means that HIC takes into account both the value and duration of the loading. The permissible HIC level at which no severe injury or disability occurs is 1000 [19].

The Nij criterion is the likelihood of injury to the cervical spine. It combines injuries associated with neck axial loads (F_z) and bending moment to the neck (M_{OCy}) compared to the critical values (F_{zc}, M_{yc}) (2) defined by the standard [20] and can be crucial when analyzing severe injuries to operators resulting from strong jolts. According to the literature [21], for Nij equal to 0.5 the probability of sustaining AIS≥3[†] injury is minimized to 0.1, which is why the authors have adopted this value as appropriate for the health and safety of operators.

$$N_{ij} = \frac{F_z}{F_{zc}} + \frac{M_{OCy}}{M_{yc}} \quad (2)$$

All of the mentioned criteria can be obtained directly in the MADYMO software, which acquires them from appropriate dummy sensors. Table 1 presents a comparison between the results obtained from the analysis of a sample protective structure during an upward thrust of the machine (in the worst case when machine ejection velocity is 10 m/s) and the limits of the injury criteria.

Table 1. Comparison between limit values and values obtained during a sample simulation.

Criterion	Limit value	Obtained value
Vertical load on the spine [kN]	9,2	30.3
HIC	1000	1440
Nij	0.5	1.4

The cabin analyzed in this case complies with all the standard requirements and is implemented for operation in an underground mine. However, the results presented in Table 1, which were obtained using the original method proposed by the authors, suggest that this structure is not safe for the operator.

4 Summary and conclusions

The analysis of accident incidence rate in underground mining over the last dozen years, mainly in terms of machine operators, indicates an urgent necessity to improve their safety. As extraction reaches into deeper layers of the deposit and conditions become increasingly difficult, the number of uncontrolled phenomena occurring inside the rock mass continuously increases and poses a serious threat to the health and safety of miners. Previous studies of protective structures focusing on machine operator safety have been carried out using the DLV, which is a volume that simulates the safety zone of a seated

[†] AIS (Abbreviated Injury Scale) – a six point scale that determines the severity of injury in individual regions of the human body; 1 – minor, 2 – moderate, 3 – serious, 4 – severe, 5 – critical, 6 – maximal (untreatable) [22, 23].

operator that must remain intact during the test. The DLV is not equipped with any sensors, nor does it reproduce the shape of the human body. Hence it is impossible to analyze the loads that act on the operator's body and the injuries that are sustained in accidents caused by various phenomena occurring in underground mines. However, the situations that have occurred thus far indicate the need to monitor injuries sustained by operators, which may be very severe even when protective structures comply with standard safety requirements. Studies should also include other dynamic phenomena resulting from rock bursts, which are not included in the current analyses, but which pose a significant threat to the health and safety of operators.

The authors have proposed a new safety testing method which, apart from evaluating the entire protective structure, also includes a full model of the operator's body as well as the operator seat with its geometrical and material properties and a simplified model of its attachment to the cabin. The prescribed motion has been defined based on a review of literature. The results of tests performed with the application of this approach will provide a better understanding of the biomechanics of injuries that operators may sustain and will enable the continuation of research aimed at improving their safety, e.g. by introducing new protective systems.

This study described it this article involved a simulation of a dangerous accident where a machine is ejected upwards. Thus far, this phenomenon has not been taken into account in operator safety studies, but it remains a serious threat, as evidenced by accidents that have already occurred in mines. Tests were performed using a cabin that meets the requirements of current regulations. The results indicate that injuries suffered in the event of such phenomena significantly exceed the permissible values of criteria and further research is needed to improve the safety of operators of underground mining machines.

Research is co-financed by the National Science Centre Poland within the framework of the project „The assessment of the seismic phenomena effects in the aspect of the operator safety" no. 2015/17/N/ST8/01212.

References

1. J. Karliński, E. Rusiński, T. Smolnicki, Autom. Constr. **17**, 232 (2008)
2. M. Kalita, Min. Mach. **1**, 16 (2013)
3. J. Karliński, M. Ptak, P. Działak, Arch. Civ. Mech. Eng. **13**, 57 (2013)
4. S. Kaneda, T. Tamagawa, Komatsu Tech. Rep. **49**, 2 (2003)
5. N. J. Perera, D. P. Thambiratnam, B. Clark, Australas. J. Min. Saf., **3**, 68 (2011)
6. J. Butra, *Eksploatacja złoża rud miedzi w warunkach zagrożenia tąpaniami i zawałami*, KGHM Cuprum, Wrocław (2010)
7. A. Goszcz, *Elementy mechaniki skał oraz tąpania w polskich kopalniach węgla i* miedzi, SEP, Kraków (1999)
8. WUG, *Raport o stanie bezpieczeństwa pracy w polskich kopalniach* (2017)
9. E. Rusiński, P. Moczko, D. Pietrusiak, J. Mech. Eng. **59**, 556 (2013)
10. M. Ptak, E. Rusiński, J. Karliński, S. Dragan, Arch. Civ. Mech. Eng. **12**, 68 (2012)
11. T. Smolnicki, M. Stańco, D. Pietrusiak, Tech. Gaz. **20**, 831 (2013)
12. https://www.tassinternational.com/madymo-dummy-models, (2017)
13. FAA, *Hybrid II and Federal Aviation Administration Hybrid III Anthropomorphic Test Dummy Dynamic Evaluation Test Series*, NTIS, Springfield, Virginia (2013)
14. http://www.gom.com/metrology-systems.html, (2017)

15. T.J.R. Hughes, *The Finite Element Method: Linear Static and Dynamic Finite Element Analysis*, Dover, New York (2000)
16. https://www.mpips.gov.pl/ubezpieczenia-spoleczne/ubezpieczenia-wypadkowe/definicje-zawarte-w-ustawie-o-ubezpieczeniu-spolecznym-z-tytulu-wypadkow-przy-pracy-i-chorob-zawodowych/uszczerbek-na-zdrowiu (2017)
17. FAA Regulation 29 – *Airworthiness standards: Transport category rotorcraft, Emergency landing conditions* (2017)
18. S.P. Desjardins, AHS J. **51**, 150 (2006)
19. K. Jamroziak, Modell. Eng. **42**, 179 (2011)
20. NHTSA, *Development of improved injury criteria for the assessment of advanced automotive restraint systems*, Washington, DC (2011)
21. B.D. Stemper, N. Yoganandan, F.A. Pintar, B.S. Shender, G.R. Paskoff, Aviat. Space Environ. Med. **80**, 489 (2009)
22. T.A. Gennarelli, E. Wodzin, Inj. **37**, 1083 (2006)
23. P.M. Cartera, C.A.C. Flannagana, M.P. Reeda, R.M. Cunninghama, J.D. Rupp, Accid. Anal. Prev. **72**, 146 (2014)

Nonnegative Matrix Factorization of time frequency representation of vibration signal for local damage detection – comparison of algorithms

Jacek Wodecki[1,*]

[1]Wroclaw University of Science and Technology, Faculty of Geoengineering, Mining and Geology, 27 Wyb. Wyspianskiego St., 50-370 Wroclaw, Poland

Abstract. Local damage detection in rotating machine elements is very important problem widely researched in the literature. One of the most common approaches is the vibration signal analysis. Since time domain processing is often insufficient, other representations are frequently favored. One of the most common one is time-frequency representation hence authors propose to separate internal processes occurring in the vibration signal by spectrogram matrix factorization. In order to achieve this, it is proposed to use the approach of Nonnegative Matrix Factorization (NMF). In this paper three NMF algorithms are tested using real and simulated data describing single-channel vibration signal acquired on damaged rolling bearing operating in drive pulley in belt conveyor driving station. Results are compared with filtration using Spectral Kurtosis, which is currently recognized as classical method for impulsive information extraction, to verify the validity of presented methodology.

1 Introduction

In mining industry all over the world, belt conveyors are one of the most essential assets of horizontal transportation [1, 2]. In underground mining they span over the entire mine forming complex network transporting excavated material towards the shaft, while in surface mining they can also transport the material over very long distances, up to many kilometers. Since conveyors are typically set up in series, failure of one of them can effectively stop large portion, or even entire transportation network [3]. Most of the failure events are connected to belt itself, which is the most expensive part of the conveyor, or to elements of driving station, which includes engines, clutches, gearboxes and numerous pulleys that interact with the belt. Most of machines included in driving station are made of rotating elements installed on shafts, and such architecture requires using bearings.

Rolling element bearings are prone to three main types of failure: damage of rolling element, inner race or outer race. In most cases local damage of such types results in the presence of cyclic impulses in vibration signal, very often invisible in time domain. Literature worldwide presents many different techniques for analysis of such signals, e.g.

* Corresponding author: jacek.wodecki@pwr.edu.pl

classification with artificial neural networks [4], supervised and unsupervised learning techniques [5], filtration with informative frequency band selectors [6, 7], or other data exploration techniques [8].

Previous research showed that such impulsive component can be successfully extracted using Nonnegative Matrix Factorization (NMF) approach [9-11]. There are however several types of NMF algorithm that are characterized with different assumptions and constraints, therefore they can produce different results. In practical case it is very useful to be aware of such differences and be able to choose the most appropriate method for the analysis of such type of data.

In this paper authors analyze real and simulated vibration data measured on ball bearing installed in drive pulley. Data is processed with proposed algorithm using three different NMF methods. Results are compared based on visual inspection and quantitative statistics proposed for the comparison as quality measures. Additionally, performance of algorithms is compared with the classical method for vibration signal processing, which is the filtration with spectral kurtosis [6].

2 Methodology

Fig. 1 presents the flowchart of the proposed methodology for all investigated NMF algorithms. After selecting real or simulated signal as input, method begins with transformation of the provided signal into a time-frequency representation. Authors use the spectrogram, which is a square of absolute value of the short-time Fourier transform (STFT). The STFT for discrete signal $x_0, \ldots, x_{N-1}$ is defined as follows [12]:

$$STFT(t,f) = \sum_{k=0}^{N-1} x_k w(t-k) e^{-j2\pi fk/N}. \tag{1}$$

where $w(t-\tau)$ is a shifted window. Next, we apply the selected NMF algorithm to classify vectors of spectra for all timestamps from the spectrogram, which is performed based on so called encoding matrix generated by NMF. Such clustering provides K spectrograms where K is number of clusters. Finally, method applies the inverse short-time Fourier transform (ISTFT) [13] to all selected time-frequency map to obtain the informative signal. To prove that the proposed methodology is superior to the classical approach, authors propose to compare obtained signals and their envelope spectra with signals and envelope spectra after Spectral Kurtosis filtration.

NMF algorithms authors selected to be compared in this paper are: Convex NMF [14], Semi-Binary NMF [15] and classic NMF [16].

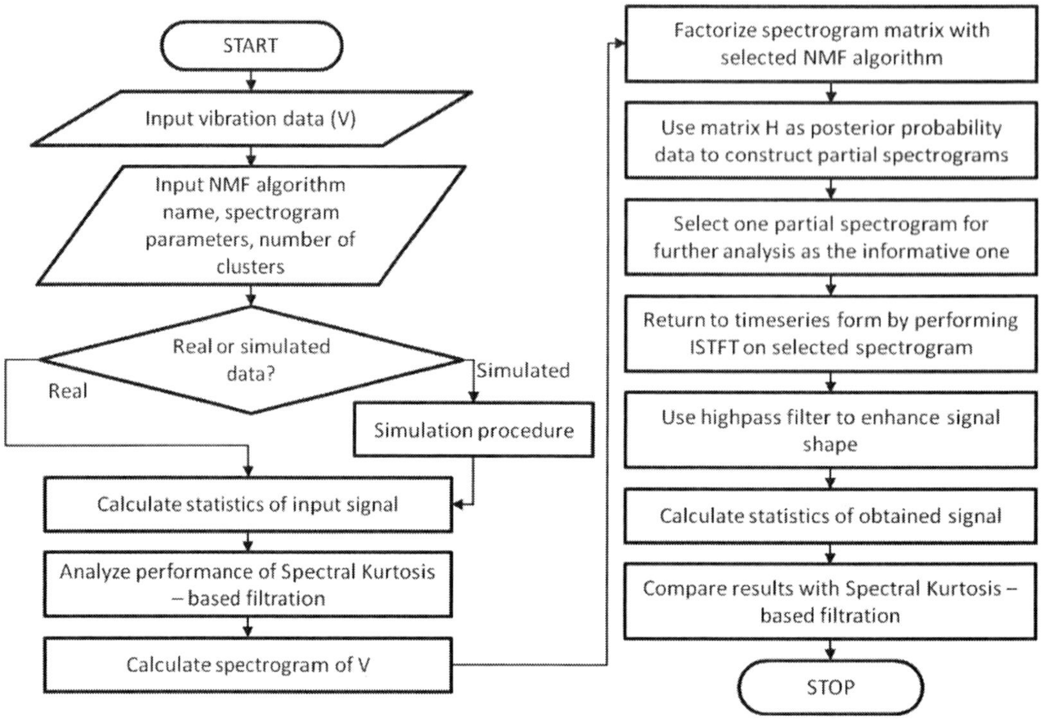

Fig. 1. Procedure flowchart.

2.1 General model of NMF

Nonnegative Matrix Factorization (NMF) is a family of algorithms in linear algebra and multivariate analysis where a nonnegative matrix **V** is factorized into two lower-rank nonnegative matrices **W** and **H** (see Fig. 2), hence general model is defined as $\boldsymbol{V} \approx \boldsymbol{WH} = \widehat{\boldsymbol{V}}$.

Since this problem is not exactly solvable, it is approximated numerically. Such matrix multiplication can be understood as computing the column vectors of estimated matrix $\widehat{\mathbf{V}}$ as linear combinations of the column vectors in matrix **W** using coefficients held by column vectors of matrix **H**. If orthogonality constraint is imposed on matrix **H**, e.g. $\boldsymbol{HH}^T = \boldsymbol{I}$, its columns behave as vectors of posterior probability in the sense of clustering problem. Even if the orthogonality constraint is not explicitly imposed, orthogonality holds to a large extent, as well as clustering effect. Hence, it is recognized that NMF algorithms reveal internal clustering properties, and output matrices are called *base matrix* (**W**) that contains clusters' centroids, and *encoding matrix* (**H**) containing posterior probabilities of data points (columns of **V** or $\widehat{\mathbf{V}}$) belonging to appropriate clusters. Additionally, under certain assumptions some NMF algorithms become mathematically or algorithmically equivalent to known clustering methods as k-means or probabilistic latent semantic analysis (PLS) [17].

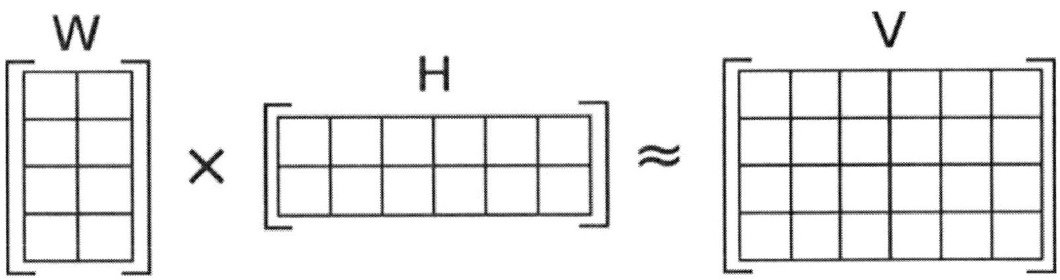

Fig. 2. Illustration of approximate Nonnegative Matrix Factorization model.

Several types of NMF algorithms have been developed over the years, that differ in constrains and implementation methods. Multiplicative update rule by Lee and Seung [16] gained popularity due to simplicity of implementation. Many successful implementations are based on alternating nonnegative least squares method, but specific algorithmic approaches use projected gradient descent methods [18], active set method [19], optimal gradient [20] or block principal pivoting [21].

First of used algorithms is called Convex NMF. It imposes the constraint that centroids contained by the column vectors of base matrix base matrix:

$$\boldsymbol{W} = (w_1, \dots, w_k) \tag{2}$$

are limited to conic combinations of column vectors of the input matrix:

$$w_l = v_1 a_{1l} + \dots + v_n a_{nl} = V a_l, \qquad a_{ij} > 0 \ \forall i, j, \qquad \sum_{i=1}^{n} a_{il} = 1 \tag{3}$$

Geometrically it can be interpreted as centroid of a cluster being placed within the convex hull of a cluster in p-dimensional Euclidean space.

Second algorithm is called Semi-Binary NMF and is very useful for hard clustering of nonnegative data assuming non-overlapping clusters. It assumes that $\boldsymbol{H} \in \mathbb{R}_+^{J \times T}$ is characterized with the property that:

$$\forall t, \exists j: h_{jt} = 1 \ and \ h_{st} = 0 \ \forall s \neq j \tag{4}$$

Hence, assuming that $J<T$, **H** is a row-orthogonal matrix. In the notion of clustering it can be interpreted as assigning data points to appropriate clusters with 100% certainty.

Third and final approach is a classic, unconstrained implementation of NMF with Euclidean distance as the objective function, using multiplicative update rules.

3 Simulated data experiment

Data simulation was aiming to capture general structure of the signal, while being able to control the parameters of impulsive component (see Fig. 3). Firstly, Autoregressive Moving Average model (ARMA) was used to model the behaviour of the healthy machine. As a next step, convolution was used to create damage signal combining healthy component with impulsive excitation, which is a 10 Hz comb of impulses contaminated with small amount of Gaussian noise to prevent the simulated signal from taking zero values between impulses. Amplitudes of impulses were also randomized to introduce more natural diversity.

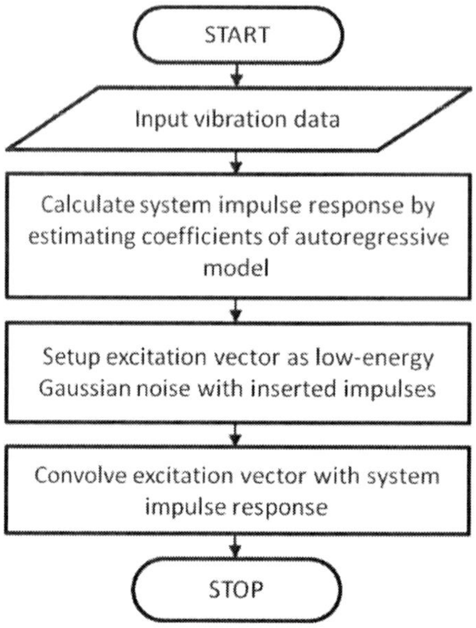

Fig. 3. Simulation procedure flowchart.

Signal was prepared in a way that it wouldn't be easier to process than the real one. Average energy of the impulses was adjusted so that they are visible but not standing out very much in the time series (see Fig. 4). We were also evaluating the spectrogram of the simulated signal, so that it wouldn't end up being easier to work with than the one of the real signal (see Fig. 5).

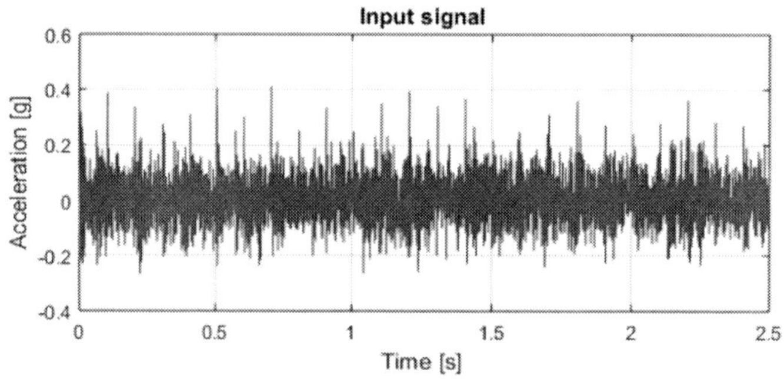

Fig. 4. Time series of the simulated signal.

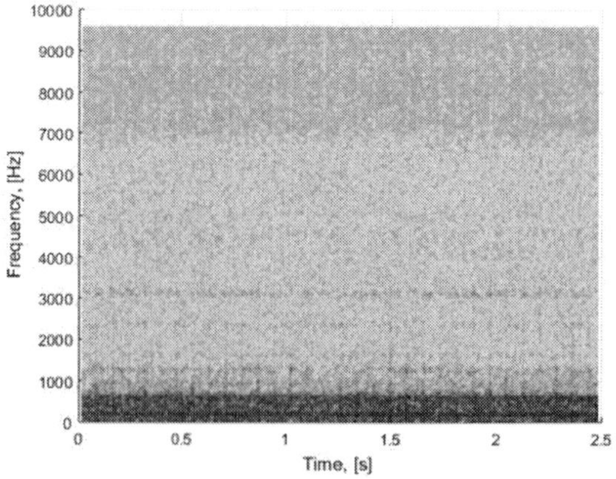

Fig. 5. Spectrogram of the simulated signal.

Signal prepared that way was fed into the procedure. Selected partial spectrograms as well as recovered time series of informative signal for all algorithms are presented in Fig. 6. Filtered signals are presented in Fig. 7. It can be seen that along with very satisfying results, algorithms perform with similar precision.

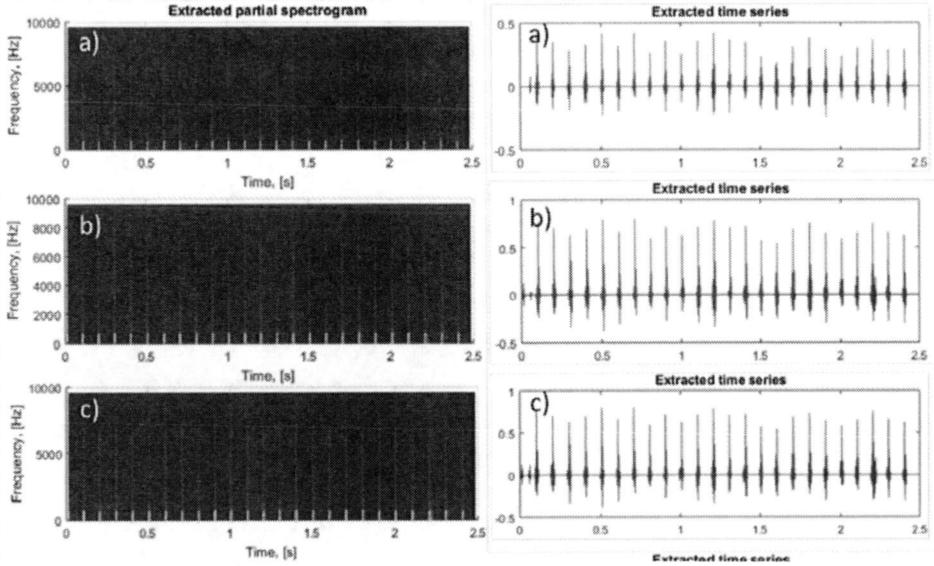

Fig. 6. Selected partial spectrograms and extracted time series for all algorithms. a) Convex NMF, b) Semi-Binary NMF, c) Classic NMF.

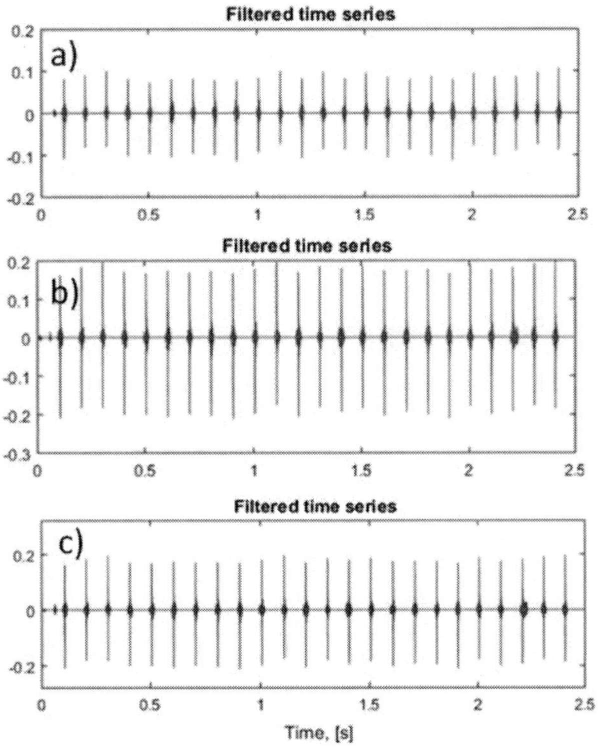

Fig. 7. Filtered time series for all algorithms. a) Convex NMF, b) Semi-Binary NMF, c) Classic NMF.

Another observation regards the amplitudes of output signals. It is clear that for classic Convex NMF impulses are roughly 50% weaker in comparison to the other two, which are quite consistent in terms of the amplitude. It comes from the fact that in all cases single impulse is not constructed from one spectra vector of the spectrogram, but always from a few of them (typically 3-8 slices per impulse). Convex NMF selected less slices per impulse than the other algorithms, hence ISTFT procedure produced signal with lower amplitude of impulses. It makes this algorithm more precise.

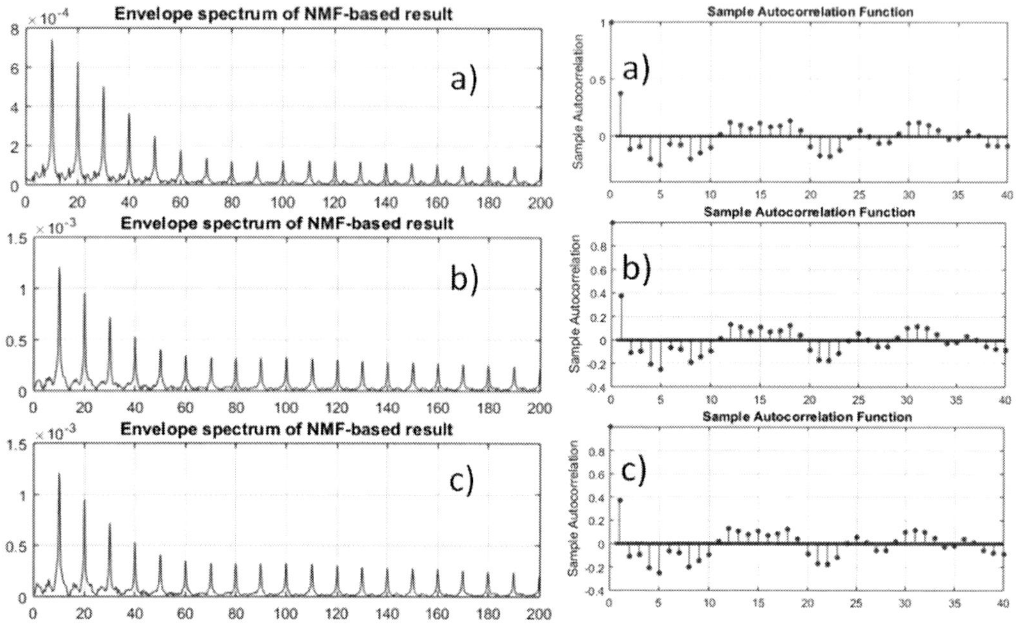

Fig. 8 Envelope spectra (left panels) and autocorrelation functions (right panels) for all algorithms. a) Convex NMF, b) Semi-Binary NMF, c) Classic NMF.

Fig. 8 provides functions useful for cyclicity analysis of the obtained signals. Envelope spectra reveal a lot of very clear peaks at harmonic frequencies (multiples of 10 Hz), which is typical for cyclic wideband-impulsive signals. Low energy of components between the peaks informs about signal clarity and high SNR. Similarly, autocorrelation functions are very similar. They reveal cyclicity which can be noticed based on first ~20 lags.

3.1 Comparison to Spectral Kurtosis

Presented algorithms have been compared with well established method of signal filtration with the Spectral Kurtosis (SK) that is used as the filter amplitude characteristic. The general principle of operation is to calculate kurtosis value for each frequency bin of the spectrogram. As a result, one can obtain indicator of impulsiveness across the signal spectrum, which allows to determine which frequency bands carry the most visible information about the damage-related impulsive behavior. Vector of spectral kurtosis can be then used as a filter to extract those impulsive components from the signal. Calculated SK is shown in Fig. 9, and filtered signal is presented in Fig. 10. Signal is very similar to the input in terms of shape, but its amplitude is significantly lower. It proves that SK-filtration is not an appropriate method for impulses recovery.

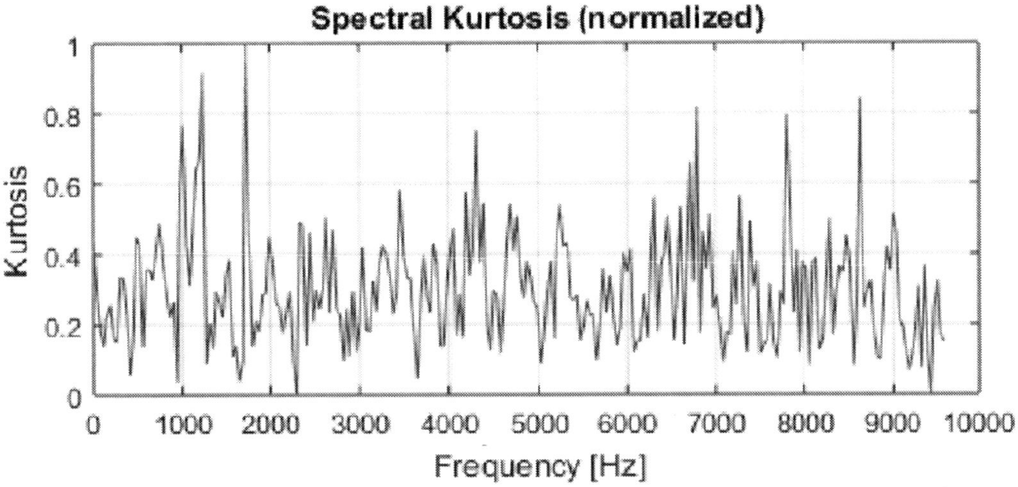

Fig. 9. Normalized Spectral Kurtosis for simulated signal.

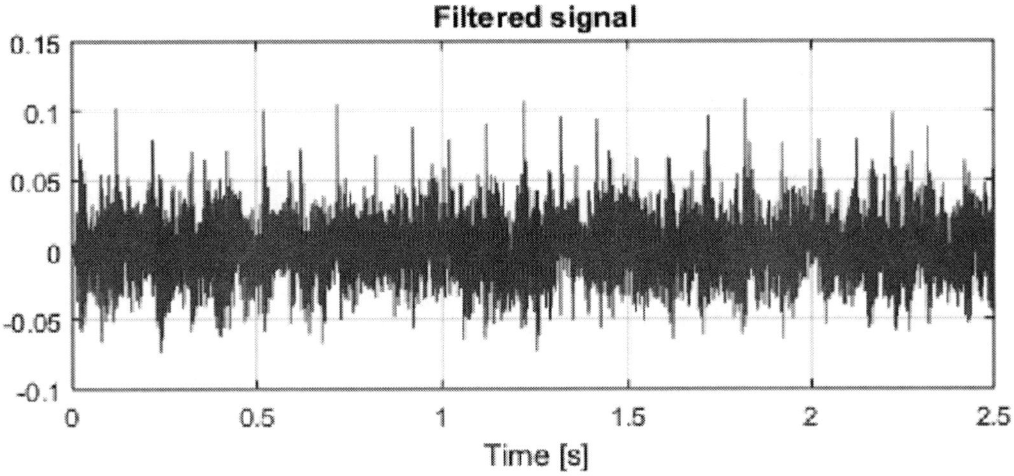

Fig. 10. Simulated signal filtered with Spectral Kurtosis.

Fig. 11 presents functions for cyclicity analysis. Autocorrelation function doesn't reveal any particular behaviour and shows that signal is highly correlated. Although some of the peaks in envelope spectrum are visible, their amplitudes are not so consistent and monotonic as for any of NMF-based results.

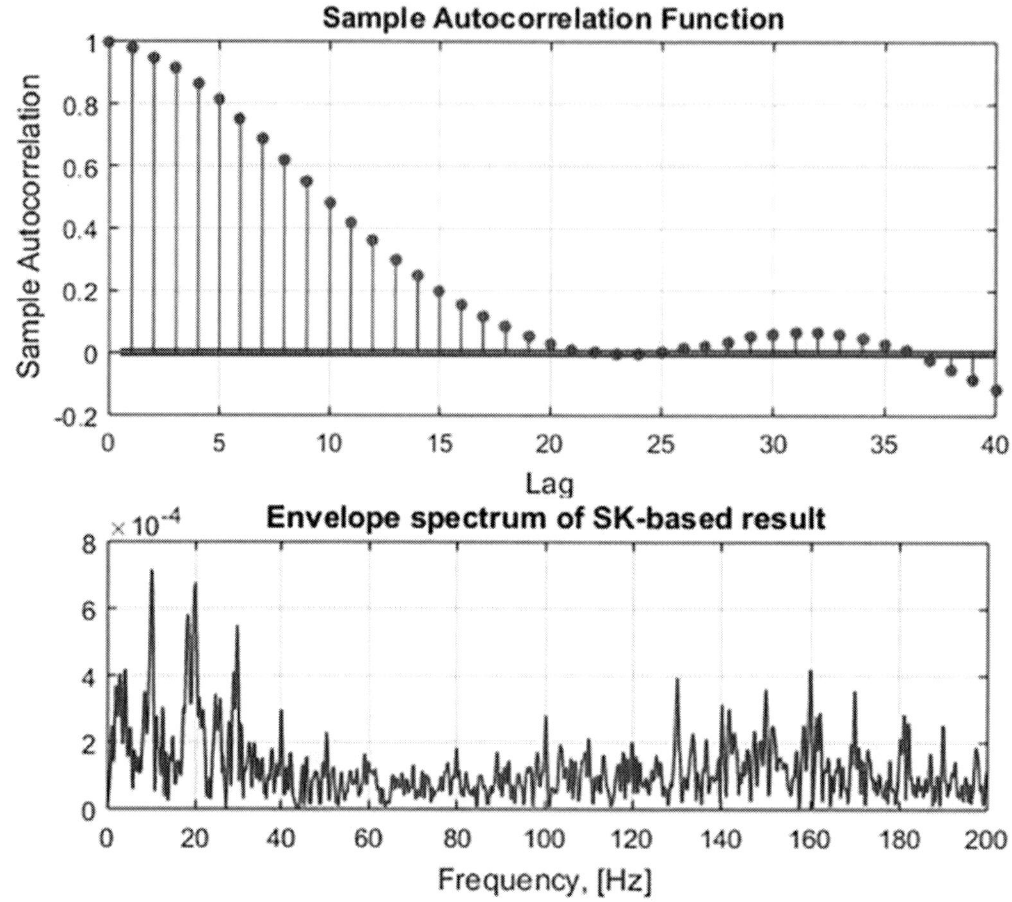

Fig. 11. Autocorrelation function and envelope spectrum for SK-filtered result.

Frequency components between peaks are also stronger and disrupt the quality and clarity of the spectrum. It is also noticeable that after disappearing at about 60 Hz, peaks then return at about 130 Hz, just to decrease the amplitude again after a few harmonic components.

4 Application to real data

Presented method has been tested using vibration data from damaged rolling bearing of the drive pulley operating in belt conveyor driving station in mining industry. Fig. 12 presents time series of recorded signal, and its spectrogram is shown on Fig. 13. Sampling frequency is equal to 19.2 kHz for 2.5 seconds of the measurement. Several wide-band excitations are visible on the spectrogram at carrier frequency band 1-6 kHz that occur with the modulation frequency of 12.7 Hz.

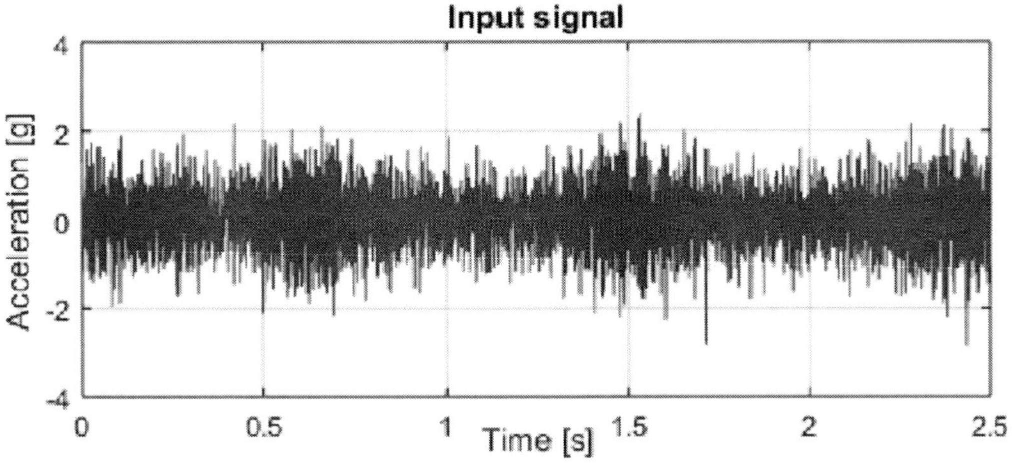

Fig. 12. Raw input signal.

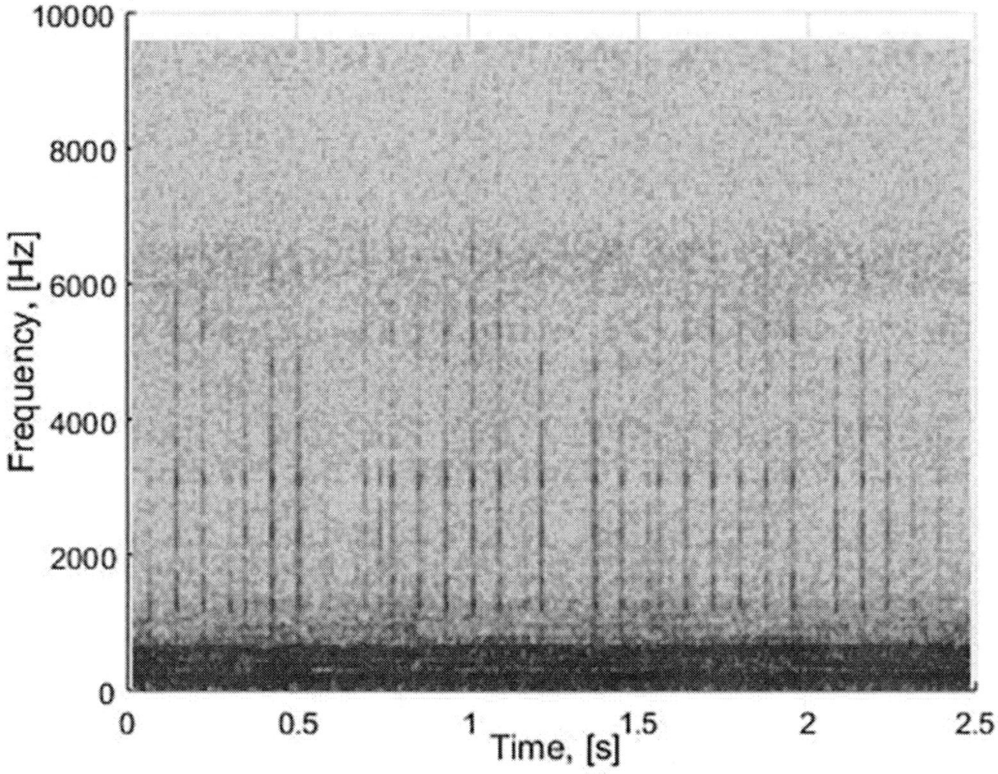

Fig. 13. Spectrogram of the input signal.

Similarly to the simulation case, signal was analyzed using presented method with all of the NMF algorithms. Results are presented in Fig. 14 in the form of constructed selected partial spectrograms (left panels) and extracted time series (right panels).

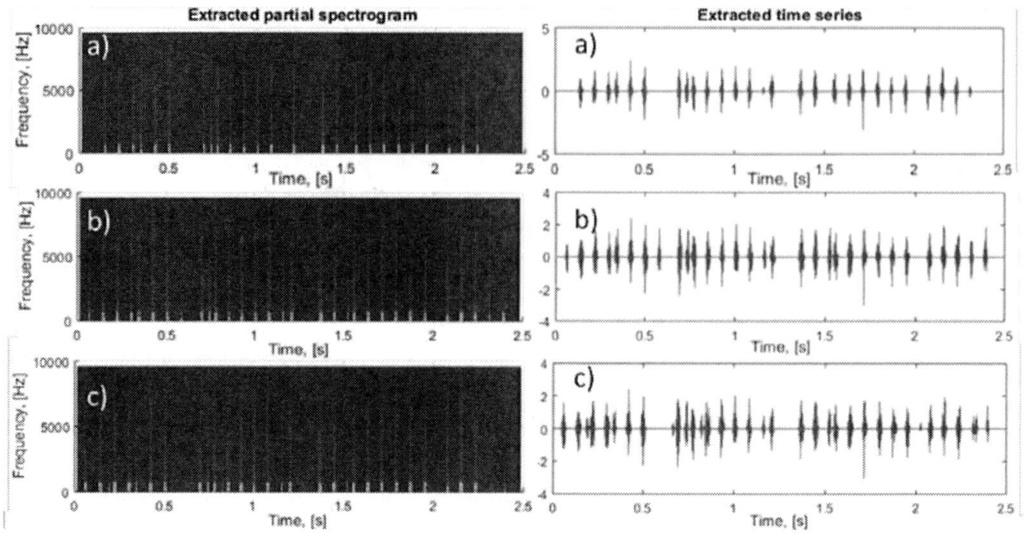

Fig. 14. Selected partial spectrograms and extracted time series for all algorithms. a) Convex NMF, b) Semi-Binary NMF, c) Classic NMF.

Fig. 15 presents output signals filtered the same way as in simulated case. This time results are not the same. Convex NMF appears to be the least sensitive in comparison to other two algorithms. It classifies the smallest amount of impulses and some of them are weaker than for other algorithms. On the other hand, Classic NMF is the most sensitive. In case of real data, filtered signals remain on similar level of impulses' amplitude, while in simulated case, Convex NMF produced significantly weaker signal.

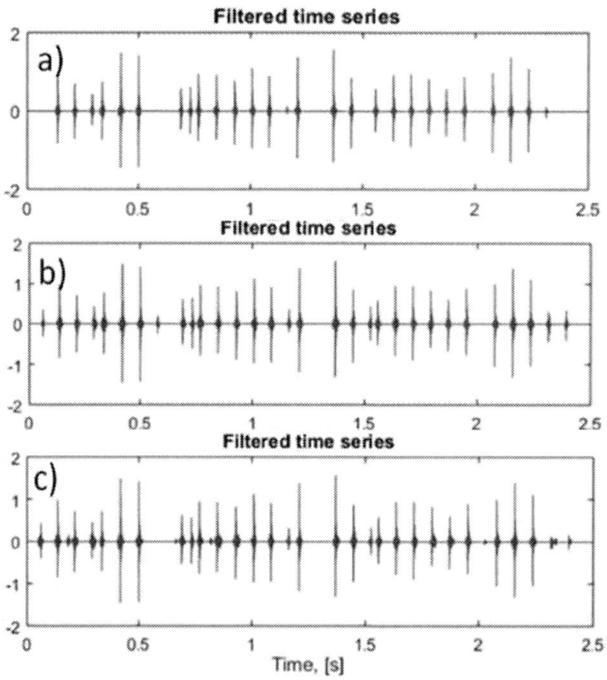

Fig. 15. Filtered time series for all algorithms. a) Convex NMF, b) Semi-Binary NMF, c) Classic NMF.

Fig. 16 provides the functions to determine cyclicity of the output signals. Period is frequency detected based on spectra is equal to 12,7 Hz. Autocorrelation functions reveal common cyclicity as well.

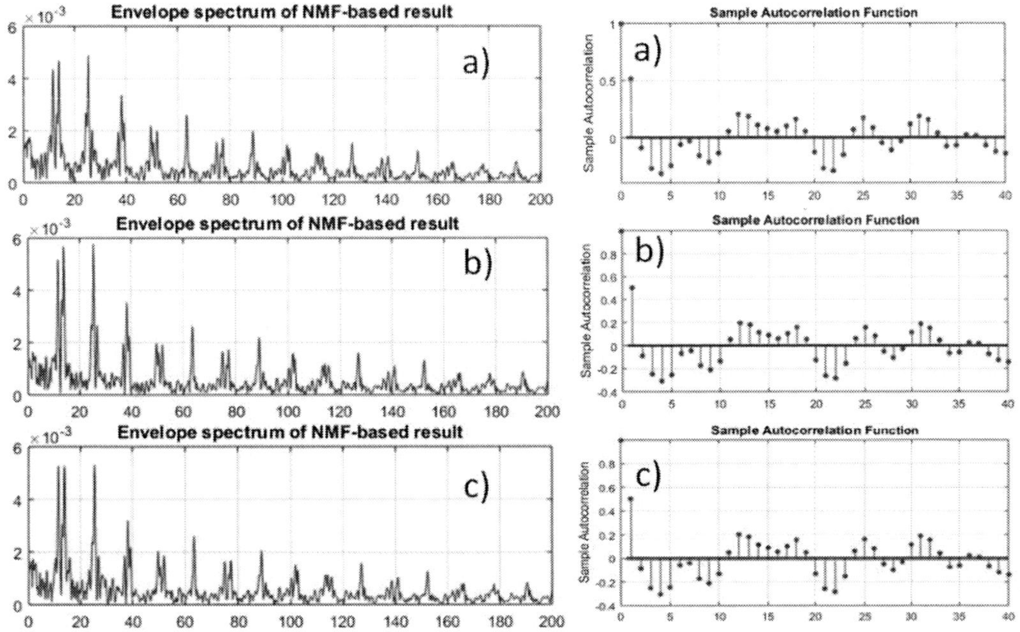

Fig. 16. Envelope spectra (left panels) and autocorrelation functions (right panels) for all algorithms. a) Convex NMF, b) Semi-Binary NMF, c) Classic NMF.

4.1 Comparison to Spectral Kurtosis

As in the case of simulated signal, we compare the results to the Spectral Kurtosis filtration. Vector of SK is presented in Fig. 17 and filtered signal is shown in Fig. 18.

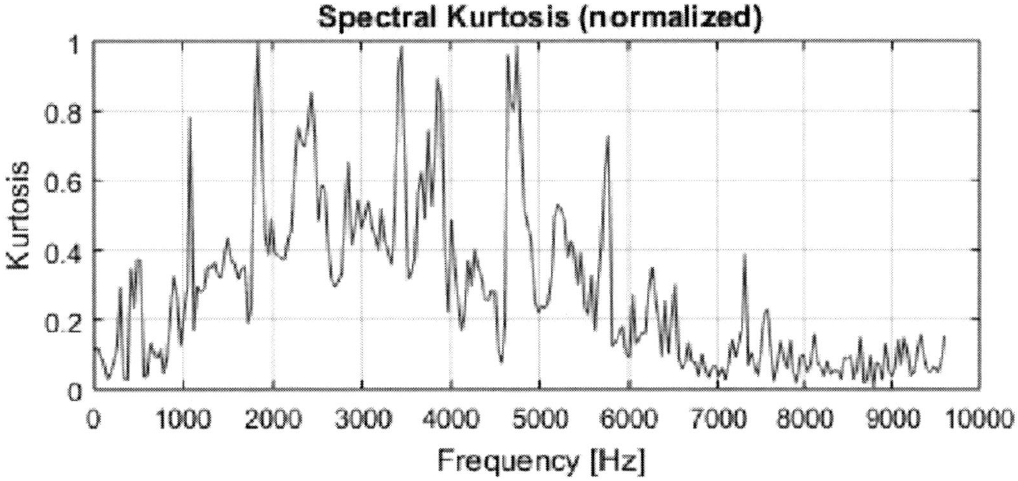

Fig. 17. Normalized Spectral Kurtosis for real signal.

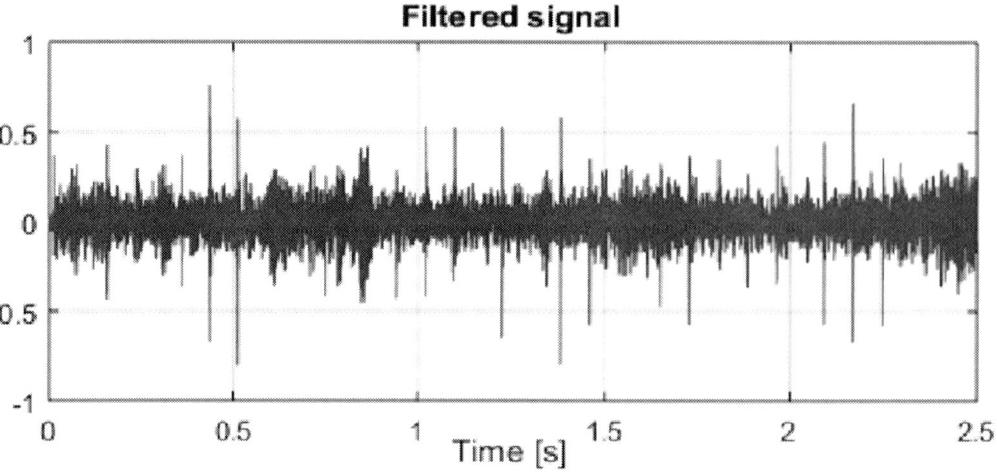

Fig. 18 Real signal filtered with Spectral Kurtosis.

In this case, SK filtration made a significant difference to the signal. Most of the strongest impulses became visible but energy of the noise is clearly too high, hence a lot of impulses are lost in the noise. However, cyclicity analysis does not look that promising (see Fig. 19). Behavior of autocorrelation function is not consistent with obtained results and its cycle denotes low-frequency modulation visible in the original signal, that is probably related to shaft misalignment. Envelope spectrum does not provide satisfying result either. Peak of fundamental frequency component has very low amplitude and is difficult to notice and interpret, especially for an automatic procedure.

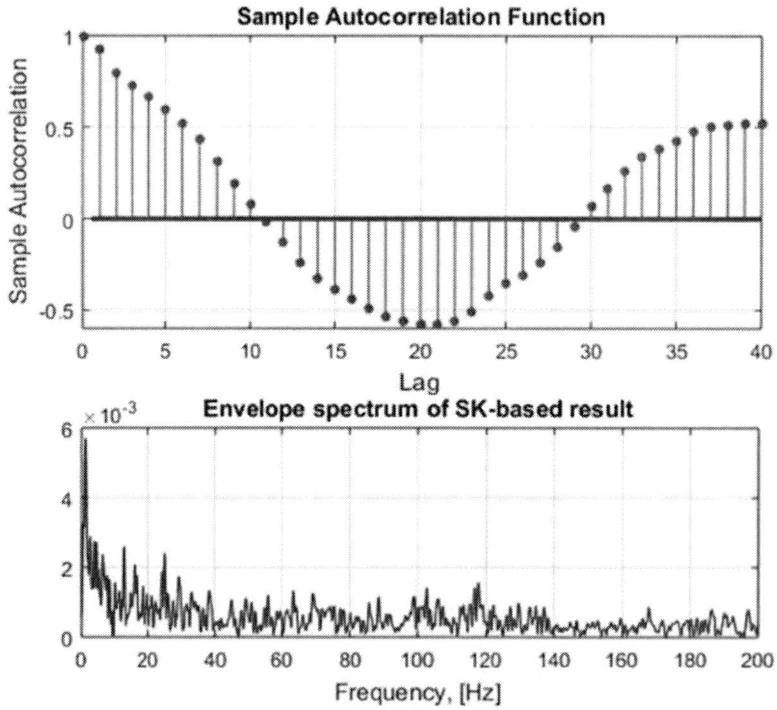

Fig. 19. Autocorrelation function and envelope spectrum for SK-filtered result.

5 Discussion

To be able to come to objective conclusions, input signals and obtained results have been described with quantitative measures presented in Tab. 1., that is:

- Signal-to-Noise Ratio (SNR),
- Signal-to-noise-and-distortion ratio (SINAD),
- Total harmonic distortion (THD),
- Kurtosis,
- Factorization error defined as Frobenius norm of difference between input spectrogram matrix and full spectrogram matrix reconstructed after factorization, divided by Frobenius norm of input spectrogram matrix. Of course, for input signals and results obtained by SK filtration this error is not defined and denoted as 0 value.

6 CONCLUSIONS

In this paper authors present a comparison of selected NMF algorithms' performance for the application of local damage detection based on vibration data. Algorithms have been tested using simulated as well as real-life data from damaged rolling bearing operating in belt conveyor driving station in mining industry. Authors use previously developed processing method that can employed three different NMF algorithm in the same way. Results show that algorithm to be selected for such task can be neither too sensitive because of false-positive impulses occurrence, nor too resistant, since the information can be lost. In considered scenario Semi-Binary NMF algorithm turned out to be optimal for discussed analytical strategy.

Table 1. Quantitative measures of obtained results' quality.

	SNR [dBc]	SINAD [dBc]	THD [dBc]	Kurtosis	Error
Raw signal	-1,7589	-2,4163	-6,1024	3,0978	0
Spectral Kurtosis	-11,3322	-11,335	-19,7888	15,0115	0
Convex NMF	-20,0208	-20,0211	-16,5702	147,1919	0,9288
Semi-binary	-21,4136	-21,4136	-20,6512	123,0783	0,9108
Classic NMF	-21,3803	-21,0060	-20,4106	128,015	1,1102

	SNR [dBc]	SINAD [dBc]	THD [dBc]	Kurtosis	Error
Simulated signal	-9,1961	-9,2	-18,1437	3,7729	0
Spectral Kurtosis	-11,4008	-11,4016	-21,8665	10,0268	0
Convex NMF	-20,3613	-20,3659	-8,8795	130,2016	0,9632
Semi-binary	-20,4019	-20,4065	-8,8422	204,9253	0,8361
Classic NMF	-20,4245	-20,4291	-8,8076	203,5364	0,8933

This work is supported by the Framework Programme for Research and Innovation Horizon 2020 under grant agreement n. 636834 (DISIRE - Integrated Process Control based on Distributed In-Situ Sensors into Raw Material and Energy Feedstock) and Grant no. 0401/0128/17.

REFERENCES

1. M. Kacprzak, P. Kulinowski, D. Wędrychowicz, Ekspl. i Niezaw. – Maint. and Reliab., **2**, 81-93 (2011)
2. M. Hardygóra, W. Bartelmus, R. Zimroz, R. Król, R. Błażej, Transp. & Logist. (Belgrade), **6**, 351-354 (2009)
3. P.K. Stefaniak, R. Zimroz, W. Bartelmus, M. Hardygóra, *Appl. Mech. and Mat.*, **68**, 108-113 (2014)
4. T. Barszcz, A. Bielecki, M. Wójcik, M. Bielecka, Diag., **14 (4)**, 679-688 (2013)
5. M. Strączkiewicz, P. Czop, T. Barszcz, Diag., **17 (2)** (2016)
6. R.B. Randall, J. Antoni, Mech. Sys. and Sig. Proc., Vol. **25**, Issue 2, 485-520 (2011)
7. G. Żak, J. Obuchowski, A. Wyłomańska, R. Zimroz, Diag. **15 (3)** (2014)
8. J. Wodecki, P. Stefaniak, J. Obuchowski, A. Wyłomańska, R. Zimroz, Jour. of Vibroeng., **18 (4)** (2016)
9. A. Cichocki, R. Zdunek, A.H. Phan, S. Amari, *Nonnegative Matrix and Tensor Factorizations: Applications to Exploratory Multi-way Data Analysis and Blind Source Separation* (John Wiley and Sons, 2009)
10. Y.-X. Wang, Y.-J. Zhang, IEEE Trans. Know. and Data Eng., **25(6)**, 1336-1353 (2013)
11. J. Wodecki, R. Zdunek, A. Wyłomańska, R. Zimroz, Diag., Vol. **18**, No. 1, 3-9 (2017)
12. A.V. Oppenheim, R.W. Schafer, J.R. Buck, *Discrete-Time Signal Processing,* 2nd Ed, (Upper Saddle River, NJ: Prentice Hall, 1999)
13. J. Benesty, M. Sondhi, Y. Huang, *Springer Handbook of Speech Processing*, Berlin, Springer (2008)
14. C. Ding, T. Li, M.I. Jordan, IEEE Trans. Patt. Anal. and Mach. Intel., **32(1)**, 45-55 (2010)
15. R. Zdunek, Proc. ICAISC, **5097**, Springer, 705-716 (2008)
16. D. Lee, S. Seung, Adv. in Neur. Inf. Proc. Sys., 556-562 (2001)
17. C. Ding, T. Li, W. Peng, Comp. Stat. & Data Anal. **52**, 3913-3927 (2008)
18. C.-J. Lin, Neur. Comp., **19 (10)**, 2756-2779 (2007)
19. H. Kim, H. Park, SIAM Jour. on Mat. Anal. and Appl., **30 (2)**, 713-730 (2008)
20. G. Naiyang, T. Dacheng, L. Zhigang, Y. Bo, IEEE Trans. on Sig. Proc., **60 (6)**, 2882-2898 (2012)
21. J. Kim, H. Park, SIAM Jour. on Sci. Comp., **33 (6)**, 3261-3281 (2011)

Review of smoothing methods for enhancement of noisy data from heavy-duty LHD mining machines

Jacek Wodecki[1]*, *Anna* Michalak[2], and *Paweł* Stefaniak[2]

[1] *Machinery Systems Division, Wroclaw University of Science and Technology, Wroclaw, Poland*
[2] *KGHM Cuprum R&D Ltd., Wroclaw, Poland*

Abstract. Appropriate analysis of data measured on heavy-duty mining machines is essential for processes monitoring, management and optimization. Some particular classes of machines, for example LHD (load-haul-dump) machines, hauling trucks, drilling/bolting machines etc. are characterized with cyclicity of operations. In those cases, identification of cycles and their segments or in other words – simply data segmentation is a key to evaluate their performance, which may be very useful from the management point of view, for example leading to introducing optimization to the process. However, in many cases such raw signals are contaminated with various artifacts, and in general are expected to be very noisy, which makes the segmentation task very difficult or even impossible. To deal with that problem, there is a need for efficient smoothing methods that will allow to retain informative trends in the signals while disregarding noises and other undesired non-deterministic components. In this paper authors present a review of various approaches to diagnostic data smoothing. Described methods can be used in a fast and efficient way, effectively cleaning the signals while preserving informative deterministic behaviour, that is a crucial to precise segmentation and other approaches to industrial data analysis.

1 Introduction

Load-Haul-Dump machines (LHD) are common assets used for ore transportation in production area of a mine with room-and-pillar system. Their character of the work seems very simple – machine loads blasted ore in mining face using bucket, next transports it to dumping point where dumps material onto conveyor belt, and finally returns to the mining face with empty bucket. Each of these sub-processes often differs from cycle to cycle and, consequently, takes up more or less time according to operating conditions in production area [1]. Depending on length of haulage route LHD machine may work alone as well as may operate in the configuration with haul truck.

Assessment of LHD machine is crucial from view point of production planning and optimization of haulage of ore from production areas [2, 3]. In the literature, this subject is well-known – in global mining, monitoring solutions are used on a large scale for tracking machine operating parameters, both for efficiency and damage detection [4–7]. Algorithms for identification of haulage sub-processes are also developed for both LHD and haul truck. In practice, it is still needed to develop a comprehensive

*e-mail: jacek.wodecki@pwr.edu.pl

algorithm for operating cycles identification [8, 9]. Arguably, the easiest way is the analysis of the most informative signals acquired from the machine.

One of the most interesting signals in terms of its cyclic structure is the pressure signal measured at a bucket's hydraulic cylinder. From view point of recognition of LHD's operations pressure signal is much more informativeness than signal of bucket weight mesured when bucket is raised to unload material [10]. High and low levels of pressure indicate driving with full or empty bucket. It is even possible to notice the underlying tasks of loading and unloading. Such signal is unfortunately highly contaminated with noise as well as fluctuations coming from the fact that the bucket, either full or empty, is prone to wobbling when the machine is moving, very often while driving on uneven surfaces. Considering described facts, it can be said that the first task required for any algorithm for analysis of the operational cycles is smoothing out the raw signal while preserving its informative structure. In this paper authors compare the performance of six basic smoothing methods for real-life signal denoising and presents the segmentation based of them.

2 Methodology

Signal smoothing can be realized using many different techniques, e.g.: filtering using lowpass or custom filters [11], additive smoothing [12], kernel smoothing [13] or many others. In this paper six methods are compared in terms of quality of denoised signal defined as a readiness for the segmentation, which is the next step in processing of such signals.

Moving average (MA) is one of the most popular methods for smoothing a series of data. It creates a series of average values of the consecutive subsets of the entire dataset. This is a special case of a finite impulse response filter (FIR), where the weights carry the same value of 1/N, where N is the length of the window. **Exponential moving average (EMA)** is a special case of moving average where weighting factors decrease exponentially [14].

Next group of methods is local regression. In **linear regression** each smoothed value is determined relative to the neighbouring data samples in a given window based on linear function models [15]. Analogously, **quadratic regression** smoothed each value based on quadratic model. These two methods have also **robust versions**, which are very helpful if smoothed data contains outliers. The robust procedure for linear and quadratic regression is given by [16]:

1. Calculate the residuals from the regression smoothing method,

2. Compute the robust weights for each data point in the window. These weights are defined by:

$$w_i = \begin{cases} \left(1 - \left(\frac{r_i}{6MAD}\right)^2\right)^2, & |r_i| \leq 6MAD \\ 0, & |r_i| \geq 6MAD, \end{cases} \tag{1}$$

   where r_i is the residual of ith data point and MAD is the median absolute deviation of the whole window, MAD=median(|r|). If r_i is greater than 6MAD, assigned weight is zero and this data point is excluded from the calculation in next iteration. If the r_i is small compared to 6MAD, then the weight is close to 1.

3. Repeat steps 1-2 for a total of five iterations,

4. Smooth data using the robust weights.

3 Application to real-life data

Machine investigated in this paper is a self-propelled wheel loader operated in one of the Polish copper ore underground mines for loading, hauling and dumping purposes (see Fig. 1). Total time of loader operations in production area lasts 4-shift work a day (6 hours per shift). Loaders operate in this mode 6 days a week with the exception of short intermissions for maintenance or repair tasks. Total distance of haulage – defined as route from mining face to dumping point – does not exceed 800 meters. For some longer distance routes, loaders are intended for filling cargo boxes of haul trucks. All sub-processes are performed in cyclic way.

Figure 1. Load-Haul-Dump machine (LHD)

As mentioned above, typical LHD machine work in production area consists of a few operation regimes:

- loading of the blasted ore in mining face,
- haulage to the dumping point (full bucket),
- unloading onto screen,
- returning to mining face (empty bucket).

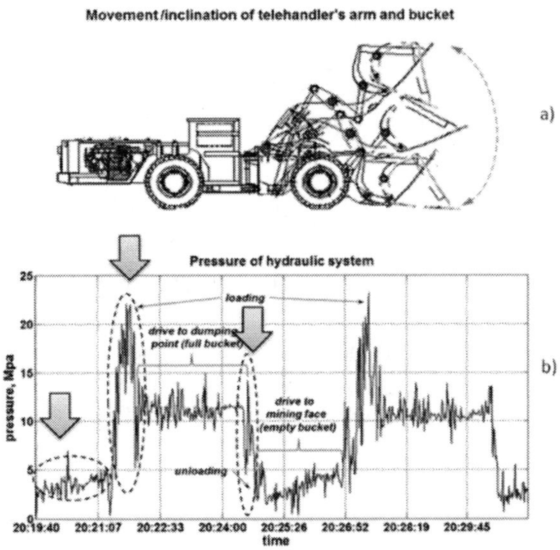

Figure 2. Swing angle of loader's arm and bucket (panel a) and variability of pressure of hydraulic system during different loader operations (panel b)

Each of these haulage sub-processes corresponds to different inclination of telehandler's arm as well as filling of the bucket (see Fig. 2a). These two features appear in the hydraulic pressure signal in form of specific pattern (see Fig. 2b). The high values of pressure with high fluctuations are responsible for loading process. Next, the signal is stabilising on around 12 MPa what indicates transport of the ore to the dumping point. Then, the rapid drop of pressure value is observed during unloading of bucket onto screen. After that, in last phase the LHD machine returns to mining face. Loader is driving the same route back with empty bucket and hydraulic pressure has settled low value.

3.1 Input data

Raw signal considered here consists of 280 minutes time series representing pressure in hydraulic system of loader under its typical load-haulage-dump operations (see Fig. 3). The signal has been acquired using bulit-in data acquisition system. On-board monitoring system records operating data with sampling period of 1 second. Data describes period of single work-shift of LHD machine.

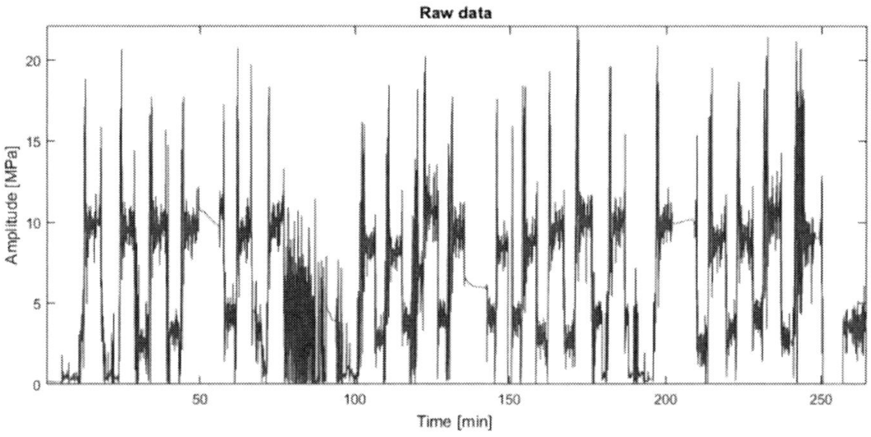

Figure 3. Raw data

3.2 Results

As a first step, presented input signal has been processed with all described methods. Since all methods use window length as a parameter, the same value of 61 samples has been provided not only for reliable comparison, but also due to the fact that it turned out to be optimal value for all of the methods individually.

Figure 4 presents results of smoothing with all methods using only a section where behavior of input signal was very much typical, operational cycles looked undisturbed and similar to each other. Such signal shape is expected when machine operates normally, without any disturbances. One can see a clear difference between results produced with tested smoothing methods.

Both moving average methods as well as both robust regression algorithms produced signals similar to a square wave, without very significant distinction of loading section in the beginning of the high state of the signal. They appear to be a good foundation for further segmentation. On the other hand both of basic regression methods provide more detail and precision for the loading section, and for the shape of entire signal.

Figure 5 presents results of smoothing with all methods, this time presented with a section containing non-typical events, e.g. fast oscillations between 75th and 100th minute, that may indicate other type of activity, extended loading action around 120th minute or stoppage around 140th minute. Such

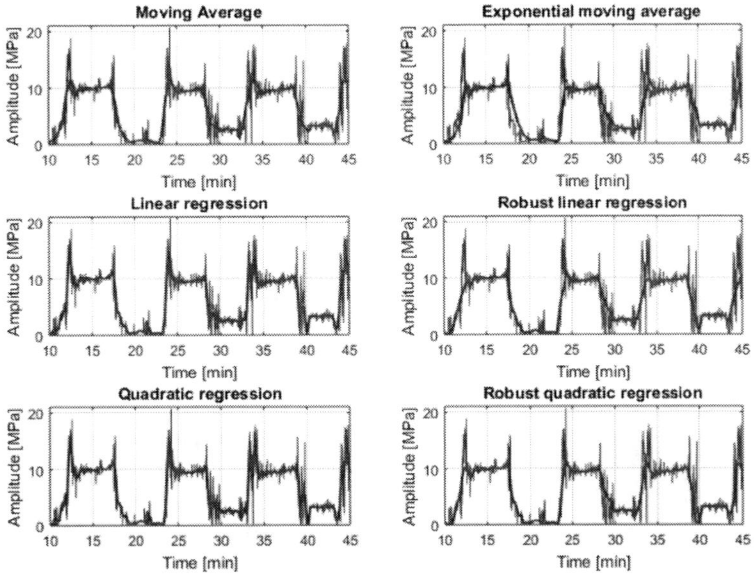

Figure 4. Signal section presenting typical behavior

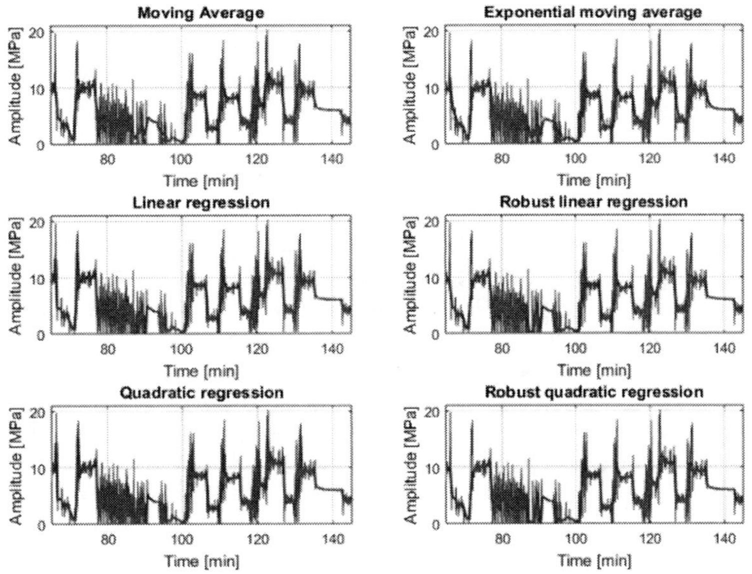

Figure 5. Signal section presenting anomalies and occasional alterations of typical behavior

cases can translate into unpredictable effects of threshold-based segmentation and are key aspects for smoothing performance evaluation.

Such segments especially visualize major differences between smoothing results. Again, basic regression methods appear to produce the most details, which can be misleading for segmentation procedure. Other methods produce smoother results with robust regression methods appearing to be the least jagged in details.

Figure 6 presents results of segmentation of smoothed signals. Segmentation is based on separating signal values above and below certain automatically calculated threshold, which is obtained as

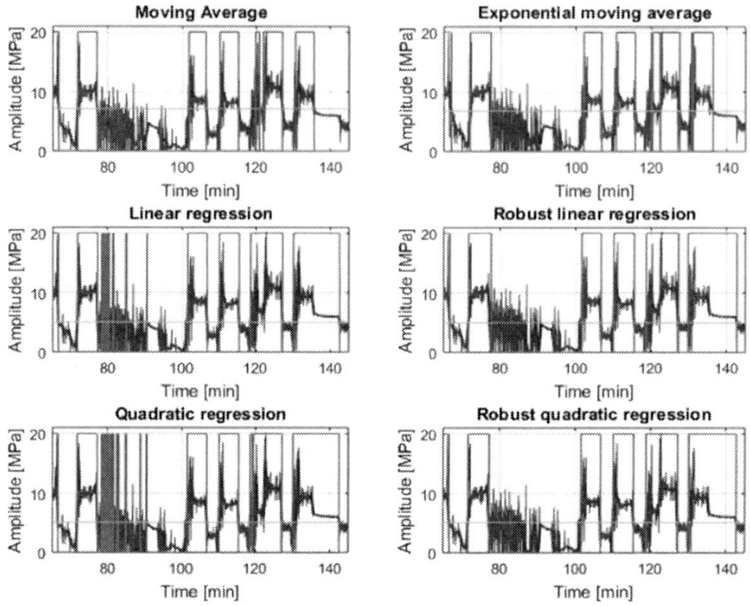

Figure 6. Non-typical signal section with segmentation results and threshold level

a central point of signal histogram (see Figure 7). It is identified as an average of two maim local maxima obtained after splitting the histogram in half according to amount of samples. Quantitative factor of smoothing quality is here expressed as a percentage of known cycles amount identified by segmentation.

As it can be seen in Table 1 basic linear regression methods are characterized with the worst quality, which is also visible in Figure 6. In those cases segmentation detects a lot of unnecessary cycles in the areas of unexpected events. Both moving average methods perform better, but they still detect some extra cycles, e.g. when the extended loading action occurs around 120th minute. They are also having problem with spanning the cycles in a correct way. When the stoppage occurs around 140th minute, MA methods close the cycle just before the stoppage, in contrast to the basic or robust regression methods, that capture this cycle correctly, according to the experts of LHD machines' operation.

Table 1. Results of segmentation based on denoised signals

Method	Number of cycles	Percentage
Known amount	20	100%
Moving Average	21	105%
Exp. Moving Average	23	115%
Linear Regression	29	145%
Robust Linear Regression	20	100%
Quadratic Regression	34	170%
Robust Quadratic Regression	20	100%

Finally, robust regression methods appear to perform as perfectly as it is possible for such type of data. They identify correct amount of cycles in the whole signal, and questionable cycles are captured in a correct way. Namely, cycle containing extended loading action should be captured as a whole,

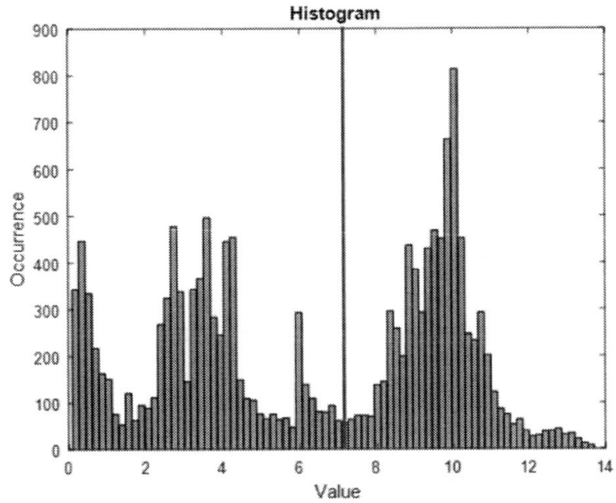

Figure 7. Example of histogram-based thresholding method. Threshold value is calculated as central point of histogram constructed from smoothed signal

along with the early phase of loading, as well as cycle containing stoppage should also be captured as a whole.

It is very important to note that one could easily implement security measures to avoid capturing too short cycles, which would eliminate excessive amount of cycles to the point when all of the methods would result in correct number of detected cycles. Although there are two points to be made in this matter:

1. As it can be seen in Figure 6 even after cleaning up short cycles, it is still important to properly capture correct cycles (see cycle with the stoppage),
2. The point of presented comparison is to verify which method can function properly without need to implement additional verification procedures. It is very important for final implementations in analytical systems to minimize the amount of additional procedures.

Taking those points into consideration, it can be summarized that robust regression methods are the best solutions for preprocessing similar types of signals. Additionally, since those methods focus on outliers analysis, it can be said that this aspect is very important while taking any smoothing actions.

4 Conclusions

Signals measured in underground mine reality are particularly susceptible to interference, which very often reduces the level of their informativeness, and this in turn might lead to wrong interpretation of the analysis results. The reasons for this are partly to be found in harsh mining conditions where machines operate in a cyclic way and their work consists of many sub-processes. In practice, there exist many ways to improve quality of measured signals commonly used at pre-processing level. In this paper, authors focused on review of smoothing methods for further segmentation of pressure signal in terms of further identification of work cycles of LHD machine. In fact, pressure signal measured at a bucket's hydraulic cylinder is very regular and at first glance seems like square wave clearly defined two main sub-processes of ore haulage (driving with full and empty bucket). Although, because of high fluctuations and noise level, it is not possible to detect a regime changes based on raw signal. Presented paper compares six smoothing methods in terms of readiness to segment smoothed signal produced by them. Extensive testing showed that robust regression methods perform the best in such application.

This work was financed by the Grant no. 0401/0128/17.

References

[1] A. Gustafson, H. Schunnesson, D. Galar, U. Kumar, International Journal of Mining, Reclamation and Environment **27**, 75 (2013)

[2] L.R. Cui, Y. Hayakawa, A. Gustafson, H. Schunnesson, D. Galar, U. Kumar, Journal of Quality in Maintenance Engineering **19**, 74 (2013)

[3] A. Gustafson, M. Lipsett, H. Schunnesson, D. Galar, U. Kumar, International Journal of Mining, Reclamation and Environment **28**, 342 (2014)

[4] A. Wyłomańska, R. Zimroz, in *Stochastic Models, Statistics and Their Applications* (Springer, 2015), pp. 469–478

[5] A. Wyłomańska, R. Zimroz, Diagnostyka **15** (2014)

[6] J. Gajda, A. Wyłomańska, R. Zimroz, Physica A: Statistical Mechanics and its Applications **464**, 123 (2016)

[7] J. Wodecki, P.K. Stefaniak, P. Sliwinski, M. Andrzejewski, SGEM2016 Conference Proceedings **2**, 157 (2016)

[8] P. Stefaniak, R. Zimroz, J. Obuchowski, P. Sliwinski, M. Andrzejewski, Procedia Earth and Planetary Science **15**, 797 (2015)

[9] M. Polak, P. Stefaniak, R. Zimroz, A. Wyłomańska, P. Śliwiński, M. Andrzejewski, *Identification of loading process based on hydraulic pressure signal*, in *The conference proceedings of 16th International multidisciplinary scientific geoconference SGEM 2016*, SGEM (STEF92 Technology Ltd., 2016), pp. 459–466

[10] P.K. Stefaniak, R. Zimroz, P. Sliwinski, M. Andrzejewski, A. Wyłomanska, in *Advances in Condition Monitoring of Machinery in Non-Stationary Operations* (Springer, 2016), pp. 197–210

[11] S.K. Mitra, Y. Kuo, *Digital signal processing: a computer-based approach*, Vol. 2 (McGraw-Hill New York, 2006)

[12] D. Valcarce, J. Parapar, Á. Barreiro, *Additive Smoothing for Relevance-Based Language Modelling of Recommender Systems*, in *Proceedings of the 4th Spanish Conference on Information Retrieval* (ACM, 2016), p. 9

[13] J. Friedman, T. Hastie, R. Tibshirani, *The elements of statistical learning*, Vol. 1 (Springer series in statistics Springer, Berlin, 2001)

[14] J.S. Hunter, J. Quality Technol. **18**, 203 (1986)

[15] W.S. Cleveland, C. Loader, in *Statistical theory and computational aspects of smoothing* (Springer, 1996), pp. 10–49

[16] W.S. Cleveland, Journal of the American statistical association **74**, 829 (1979)

Possibilities of using energy recovery in underground mines

Dariusz Obracaj[1], and *Sebastian* Sas[1,*]

[1]AGH University of Science and Technology, Faculty of Mining and Geo-engineering, 30 A. Mickiewicz Av., 30-059 Kraków, Poland

Abstract. In underground mines, there are many sources of energy that are often irrecoverably lost and which could be used in the energy structure of a mine. Methane contained in the ventilation air, the water from the dewatering of the mines and the exhaust air from the mine shafts are the most important sources of energy available to a mine. Among other sources of energy available in a mine, you can also distinguish waste energy from the process of the desalination of water or energy from the waste. The report reviewed the sources of energy available in a mine, assessed the amount of recoverable energy and indicated the potential for its use.

1 Introduction

The mining industry is one of the most energy-intensive industries in Poland [1]. High energy consumption translates into increased production costs, indirectly affecting the profitability of an underground mine. One of the energy-consuming energy systems of a mine is the installation of inlet air heating to the mine. This paper presents theoretical possibilities of extracting energy from the ventilation system, mine drainage, methane drainage and other sources of energy for heating the air introduced into a mine, analyzing theoretical possibilities of recovering waste heat from the ventilation shaft.

2 Heat recovery from ventilation air

A method of recovering heat and preheating the inlet air into a mine is the use of a cross-plate heat exchanger "GUPEX" (Fig. 1). Cold air passes through the heat exchanger taking away heat from the warm air removed from the mine. The heat exchanger may be part of the mine ventilation system, which requires an air flow of up to several million m^3/h. Thousands of kilowatts of heat are continuously transferred to the cold air entering the mine, which saves significant amounts of energy. The heat exchangers are made of corrosion-resistant material, which is of particular importance because of the high moisture content in the air. Heat exchanger plates are designed to cause a low-pressure drop and to minimize fan power consumption.

* Corresponding author: ssas@agh.edu.pl

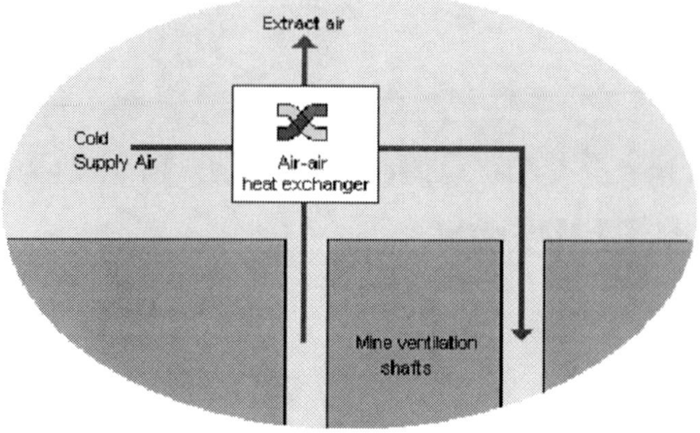

Fig. 1. Principle of operation of cross-plate exchanger for recovery of heat from the air removed from the mine [2].

3. Heat recovery from mine water

Polish literature lacks information on the use of energy derived from mine water as a source of energy. There are some references to the assessment of the heat recovery capacity of coal mines, in the "GIG" documentation [3], but these materials have not been published. In international journals, however, there is no shortage of examples of the use of energy from mine waters. The example from Stockholm [3] shows that even from water at 4°C, energy can be recovered by cooling it to 2°C.

An example of a mine water installation is located in England at the non-existent Caphouse mine located in Overton near Wakefield. The installation diagram is shown in Fig. 2.

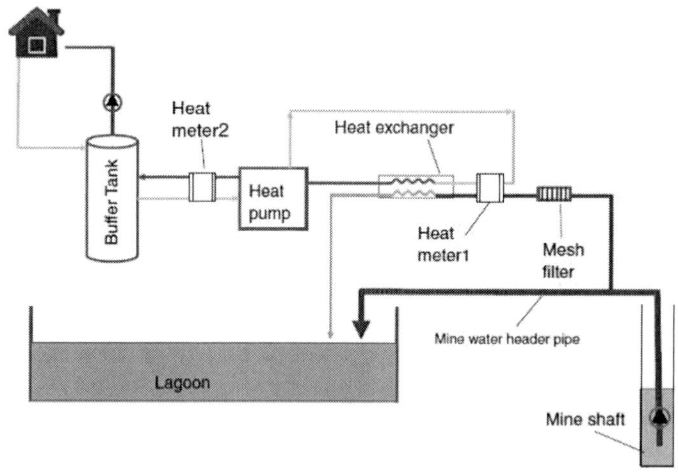

Fig. 2. Simplified diagram of heat recovery installations from mine water [4].

The heart of the installation is the heat pump which receives the energy transferred through the heat exchanger from the water pumped out of the mine. Due to the small size of the system, a part of the water discharge is transferred directly to the tank, while the remainder is directed to the heat exchanger where the heat exchange in the closed loop is carried out. Water circulates between the heat pump and the exchanger to which the water is to be drained. Due to the low quality of the mine water, characterized by high salinity,

3xidizing pyrite and the possibility of clogging of the exchangers in the system, two water filters are operated alternating during the maintenance periods of one of them, resulting in the possibility of the continuous operation. A heat pump used successfully in the summer with an average water temperature of 14.5°C, Δt = 5-6°C, achieved COP efficiency between 3.5 and 4.5 at Δt = 5°C [5].

4 Energy from cogeneration systems

Most of the captured methane is processed energetically in cogeneration systems in the Polish mining sector. These systems permit the generation of electricity and the recovery of the so waste heat that can be used to heat hot water, process steam, or to heat buildings.

Methane entrained in the drainage systems can also be used in existing gas boilers or co-fired in coal boilers.

The most economically justified system 3tilizing methane is now a tri-generation system that simultaneously produces three energy carriers in the form of power, heat and cooling (CHCP).

In Poland, the CHCP system works only in the "Pniówek" mine and a second trigeneration system is being prepared for work in the mine ZG Rudna Polska Miedz S.A. The installation is based on the combination of heat and power distribution system with a cooling system. The gas engine drives generators that produce electricity that is fed to the power system of the mine. Waste heat is used as drive energy in absorption chillers that provide cooling power for the central air conditioning system of an underground mine.

Less common solutions used in mines are, among others, gas turbines and microturbines that can be an alternative to gas engines. Their principle of operation is very close to each other. However, the microturbines are rarely used because of the relatively small output power [6].

5 Waste energy from other energy sources

Apart from the primary sources of waste energy mentioned in the above subsections, another source of heat that can be used as waste energy is the heat from the mine waste.

Coal waste sheds include at least a few percent of the carbonaceous and pyrite additives that cause them to slowly oxidize. The temperature inside the oxidizing dump can exceed the temperature of 600°C, where it has been tested at a depth of several meters at 250°C while the surface is overgrown with grass. Heat from the heap can be used for municipal purposes and the fired heap is used as a source of aggregate or clay shale or for recreational purposes. The proposed solutions to the problem of heap nuisance with the simultaneous reception of their thermal potential can be used in two main ways:

- burn the heap with simultaneous heat and then use it as a source of aggregate,
- burn out the heap with simultaneous heat reception and then afforestation and use as a recreational facility.

Heat from the heap can be exploited successively by passive heat transfer or by the active method using heap heater. Heat reception can take place using closed circuit heat exchangers, and heat transfer from the primary circuit to external heat exchangers can be held on the production head or the heater area near them [7]. A critical danger, however, which may arise during the migration of heat is the ignition of the heaps. Therefore, it is a useful further analysis of this method, to provide adequate insulation of the transported heat by insulating it from the rest of the heap.

Another important source of energy is the heat generated by the desalination of mine waters, which is a troublesome waste that can be economically exploited throughout the year.

Desalination involves the production of large amounts of clean waste water at 40°C. These waters cannot be disposed of by direct discharge into the open water because of their high temperature. Hot water could be used as the lower energy source for the heat pump or as a source of energy for heating the primary water supply, for example, hot water.

6 Cost and power needed to heat the inlet air

Heating the air supplied to mine shafts requires a large amount of thermal energy and generates considerable costs, the qualified majority as fixed costs (independent of production). The economic aspect should, therefore, be an incentive to modernize and streamline heating systems, particularly in the mining industry, which is characterized by high energy consumption. Figure 3 presents a graph of energy demand for inlet air heating from -20°C to 1°C. The chart shows that the peak demand for power in the case of a temperature of -20°C is about 17.2 MW for the largest inlet air stream of 35,000 m^3/min. Taking into account the statistical temperature distribution throughout the year, so much power demand is only required a few hours a year, plus these temperatures would not necessarily occur at the same time. The figure shows that the greatest deflection of the curve is observed in the number of hours of 80. For this value, the power demand is significantly smaller, because it is about 9 MW, which suggests that the peak demand theoretically could be an average of one hour each day during the whole heating season. It is therefore important to consider whether it is worthwhile to build a double-power installation just to provide air heating in such a short period, or to build a lower-power installation and an alternative system that would provide additional power at peak moments. Figure 4 shows the total Giga-hour energy demand (GWh) for the given air streams and may be the starting point for calculating the energy costs needed to heat inlet air. Figure 5 shows the instantaneous power requirements for the inlet air heating depending on the airflow and the temperature gradient between the expected temperature of 1°C and the outdoor air temperature in the mine [8-10].

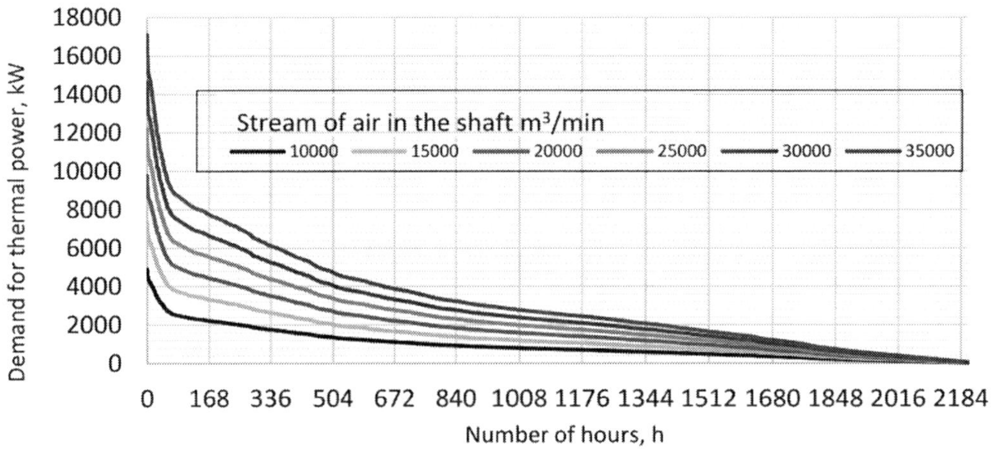

Fig. 3. Power demand curve for inlet air heating in the downcast shaft.

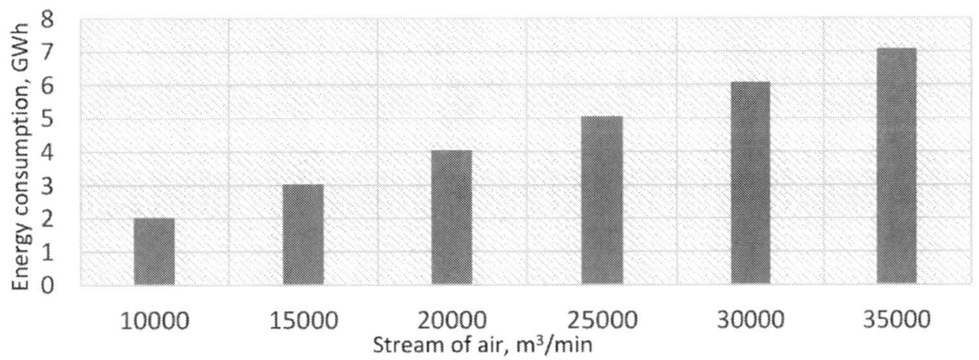

Fig. 4. Graph of GWh energy consumption for inlet air heating.

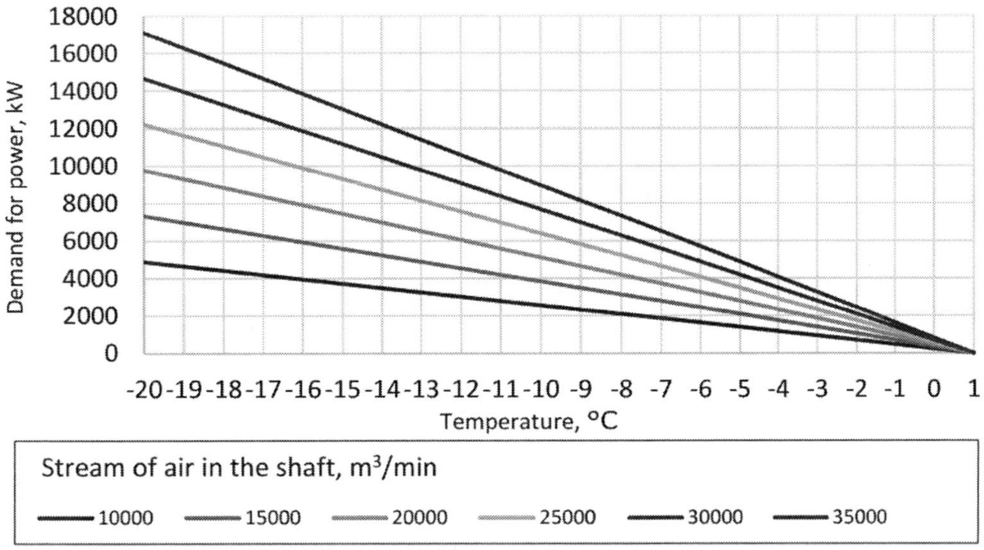

Fig. 5. Demand curve for the power needed to heat the inlet air.

7 Possibilities of using air exhausted by upcast shafts

The thermodynamic parameters of the exhaust air are shaped by many complex processes of heat and mass transfer. The heat transfer between the strata and the mining excavation is carried out explicitly by thermal conductivity, radiation, convection, and by evaporation or condensation (latent). With the air flow downward, the air is compressed, and when the ascent moves its expansion. Pressure reduction (without heat exchange) is accompanied by a decrease in temperature, while its increase increases the temperature. The values of heat transfer from rock to air depend on many factors such as:

- the depth of which the excavation is located, and the associated primary temperature of the rocks,
- the temperature of the flowing air and its change over time,
- geological construction of surrounding rocks, (physical properties of rocks, such as heat capacity, density, thermal conductivity coefficient),
- geometric dimensions of the shaft,
- period of ventilation,
- the flow and velocity of air flow,

- exchange of moisture and change of concentration of moisture contained in air,
- heat emitted from devices occurring in the windshield,
- the wall roughness caused by the unevenness of the shaft [8].

The air flowing through the excavation is generally at a different temperature than the surrounding rocks. The gradient of this value causes heat flow. Heat transfer in rocks is carried out by thermal conductivity. Rocks around the excavations to which atmospheric air enters are some kind of heat generator. In the summer, when the warm atmospheric air heats the rocks, the air cools, while in the cold, the cold air cools the rocks away from them, causing the seasonal fluctuations in the temperature of the mine air to decrease. These variations increase in the glazing, while in the excavations temperature changes are usually small. An example of an average expenditure and temperatures measured in the SW-VI-Jas exhaust shaft is shown in Figure 6.

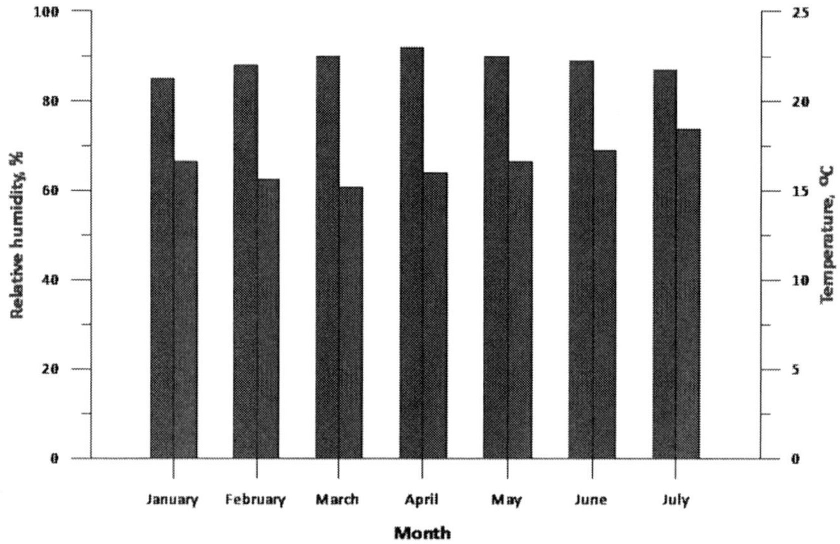

Fig. 6. Graph of mean temperature, humidity and in the upcast shaft in Jas-Mos mine [6].

The large energy potential of the mine removed from the mine in both summer and winter time can be used as the bottom source of energy by using the heat pump as an intermediate unit for the exchange of heat between the exhaust air and the mine.

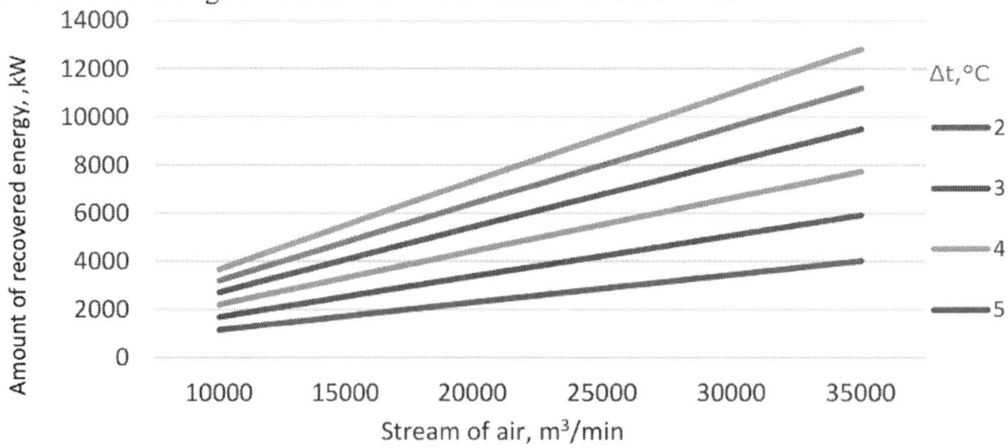

Fig. 7. Heat power obtained from the exhaust air from the upcast shaft as a function of airflow for variable differences of temperature.

Based on Figure 6, in the winter, the air stream has an oscillating temperature of 16°C, which allows for complete heat utilisation. The possibility of using waste heat could consist in mounting a heat exchanger in the ventilation drift in front of the main fan station or by using a bypass passage that would include an additional fan and a heat exchanger. The collected heat would be transferred to the upper source of heat which would be used to heat the water used to heat the inlet air [8, 9, 11]. Figure 7 presents a graph of the possibility of extracting heat from the exhaust air from the mine, depending on the gradient of temperature and airflow. Consistent with this figure, a power of 4 MW can be obtained for an air flow of 35,000 m^3/h and a difference of 3°C between the exhaust air and the exchanger wall temperature. With a temperature difference of 5°C, approximately 9.5 MW of power can be obtained. These values are significant considering the peak demand for 17 MW thermal power for heating inlet air at -20°C.

8 Conclusions

Actual demand for inlet air heating depends on current outdoor temperature conditions. From the ordered temperature chart for Katowice (Poland), the extreme air temperatures of less than -10°C are very rare throughout the heating season. They are only 26 hours according to the meteorological station in Katowice.

For heating the air enters the mine shaft, the heat power is about 12 MW for a volume flow rate of 35,000 m^3/h at the ambient temperature of -20°C. If the ambient temperature is -10°C, the power demand is approximately 6.2 MW.

You can use the heat from the mine exhaust air to heat the air to the coffin using a heat pump. For an air flow of 35,000 m^3/h, approximately 9.5 MW of power can be obtained assuming the temperature difference Δt = 5°C on the heat exchanger. For heating the circulating water, a heat pump can be used in the heat exchanger circuit.

The use of a heat pump as an indirect source of heat transfer between the exhaust air and the inlet air to the coals can be achieved by providing many savings in constant flow and CO_2 burning, which is undoubtedly a pro-ecological solution.

The problem to solve is the technical aspects of system design for heat exchange reliability. It is also necessary to minimize exchanger air resistance, which is crucial for the safety, failure-free and safe co-operation of main ventilation fans with the mining ventilation network.

References:

1. J. Kott, M. Kott, Z. Szalbierz, Management and Finance, R. **10**, nr 1, part 2, 585-593 (2012)
2. http://www.gupex.com/article.aspx?id=93224 (access: 15.06.2017)
3. E. Solik-Heliach, Research Reports Mining And Environment **2/2002**, 17-24 (2002)
4. A.P. Athresh, A. Al-Habaibeh, K. Parker, International Journal of Coal Geology, Vol **164**, 69-76 (2016)
5. E. Ostrowska, L. Ośródka, Materials of the 36th National PTG Congress, part II, Sosnowiec (1987)
6. N. Szlązak, M. Borowski, D. Obracaj, J. Swolkień, M. Korzec, *Methane drainge in coal mines* (AGH University of Science and Technology Press, Cracow, 2015)
7. A. Dybciak, Z. Heliasz, S. Ostaficzuk, Works of the Department of Basic Geology, Faculty of Earth Sciences, Silesian University, Sosnowiec, 119-128 (2000)
8. J. Wacławik, *Ventilation of mines t. 2*, Edition: 1 (AGH University of Science and

Technology Press, Cracow, 2010)

9. J. Pawiński, J. Roszkowski, J. Strzemiński, *Ventilation of mines* (Silesian National Publishing House. Katowice, 1995)
10. N. Szlązak, L. Kloc, *Influence of heating air in downcast shafts on the stability of fair streams in coal mine ventilation network* (AGH University of Science and Technology Press, Cracow, 2004)
11. M. Rubik, *Heat pump, Information center, Installation technique in construction* (Warsaw 2006)

Comparative analysis of sustainable value distribution for stakeholders in the mining industry

Sylwia Lorenc[1,*], *Arkadiusz* Kustra[1]

[1]AGH University, Department of Economy and Management in Industry, 30 A. Mickiewicz Av., 30-059 Kraków, Poland

Abstract. The objective of this article is the analysis and comparison of the value distribution process that takes place in enterprises for stakeholders. The following coal mining enterprises are subject of this work: JSW S.A., KGHM S.A., and LW Bogdanka S.A, for which the directions of value distribution in the years 2011-2016 were presented. The article defines the main groups of the partnerships' stakeholders, such as the owners, staff, suppliers, equity providers, the country budget and the local governments' budgets. The sustainable value benchmark was defined as the benefits received by the stakeholders through the allocation of financial surplus. The value defined for the abovementioned stakeholders was assessed according to the Free Cash Flow (FCF) methodology.

1 Introduction

Approaching a company's value in a way that is most beneficial to the shareholders' is no longer a current approach; especially considering the economy both at home and around the world. As Hop and Kudełko show, the current market conditions try to force companies to look for new ideas to achieve competitive advantage and gradually increase company's value [1]. Departing from traditional ways of value assessment, which are partly based on the shareholder theory, brought more attention to stakeholders of enterprises who either directly or indirectly contribute to the process of working out a company's economic success [2]. As the ideas of sustainable development and corporate social responsibility are developing, one can see a growing interest in building the sustainable development value of an enterprise based on meeting the needs of all the stakeholders of a given company (the stakeholder theory). It also stresses the need to adapt companies not only to legal requirements in this area, but also to carry out additional activities in the area of investing in human capital, environmental protection and stakeholder relations [3]. This is due to the existence of a contract theory according to, an enterprise consists of a set of contracts where the groups of stakeholders provide the enterprise with certain resources (skills, goods and services, cash and capital) and in return receive real benefits such as dividends, remuneration, earned interest, goods and services, taxes and financial donations [4].

* Corresponding author: slorenc@agh.edu.pl

Nowadays, there are several tools of value assessment for owners, both in science and in business. However, there is no offer of effective tools to measure sustainable value [5]. Creating a universal measurement tool is, therefore, a challenge to modern researchers. The difficulty in doing so lies mainly in the multidimensionality of the value assessment concept, as well as in the possibility of linking it with the idea of sustainable development, the dynamics of changes that take place in specific disciplines and in lack of clearly defined concepts in this area.

2 The model of sustainable value assessment based on the Free Cash Flow to Firm (FCFF) methodology

By meeting the needs of all its stakeholders, an enterprise creates sustainable value, at the same time ensuring that the objectives of sustainable development are achieved in the areas of economics, the environment, and society. Taking voluntarily into consideration the interests of communities and the environment, the expectations and needs of stakeholders, as well as the realization of value added for these groups makes a company socially responsible and seems to be one of the better ways of achieving market success [6].

According to Morley's assumptions, value added generated for stakeholders is the margin left from deducting current costs (the purchased goods and services) from the sales income [7]. Therefore, it constitutes a surplus that is distributed among beneficiaries, i.e., stakeholders. In return for tangible and intangible contribution to the organization's success, groups of individuals associated with the enterprise receive a reward in the form of 'social dividend', i.e., contribution to the staff, equity providers, shareholders, the Treasury, and others [8]. Thanks to their relationship with the enterprise, stakeholders benefit from remuneration and bonuses, pensions, loan interests and dividends, as well as taxes or profits from the long-term relationship with the business.

Defining the value that is distributable to a company's beneficiaries is made possible by using income methods, concepts of free cash flow, and operating and financial costs.

An appraisal of value added by means of cash flow calls for an application of the new concept of free cash flow from value added (FCF VA). It refers to the sum of cash flow for the respective groups of beneficiaries: owners (FCFE, i.e., Free Cash Flow to Equity), staff (FCFW, i.e., Free Cash Flow to Workers), equity providers (FCFD, i.e., Free Cash Flow to Debt), suppliers (FCFS, i.e., Free Cash Flow to Suppliers), municipal administration (FCFM, i.e., Free Cash Flow to Municipal Administration), and the state (FCFSt, i.e., Free Cash Flow to State). While for these groups the value distributed by the company constitutes a cash inflow benefit, for the company it is an expense related to cost [9].

Based on the above, free cash flow for owners was determined as a derivative of profit, depreciation, investments expenses, net working capital needs, and principal installments.

Cash flow to staff results from remuneration and other benefits.

Cash flow to banks comes from interest rates and principal installments payable to them.

Cash flow to suppliers results from product-related costs, power supply costs, and external services provided to the company.

Cash flow to counties, municipalities and the state is the derivative of taxation, fees, and income tax payable to them.

The methodology based on generic registry in the area of operational activity and used herein, together with value assessment for all stakeholders is legitimized by the necessity of putting total costs of business units in a bigger perspective. It seems reasonable to have a multidimensional perspective for companies' business activities and to consider all the costs in this regard. An overgeneralization of cost examination can result in insufficient data in the area of creating a real value for interest groups [10].

Sustainable value and the possibility of creating it become considerably important in the case of mining enterprises which have a major influence on different aspects of sustainable development. Sustainable development in mining requires not only a steady economic growth or social responsibility, but it also should protect natural resources and the environment by means of a reasonable exploitation of the resource or reduction of adverse effects that mining activities have on various forms of geological environment. Sustainable development in mining industry, thanks to a holistic approach to an enterprise and specific, assessed benefits to the society, can be a verification tool for a company's profitability with regard to liquidation or continuation of business activities, based on its current and prospective influence on immediate and more distant environment.

3 Analysis of Value Distribution

Analysis of value distribution was performed for the three largest domestic mining partnerships: Jastrzębska Spółka Węglowa S.A. (the largest producer of high quality coking coal), Lubelski Węgiel Bogdanka S.A. (provider of energy coal to power plants and heat and power stations, being one of the leading providers of black coal in Poland and proudly excelling in financial results and in efficiency of black coal exploitation, as well as in investments plans that aspire to explore new deposits), and KGHM Polska Miedź S.A. (one of the largest state-owned enterprises and a major producer of copper and fine silver in the world).

The abovementioned businesses are active in the area of sustainable development; for example, by following the social responsibility strategy. As listed companies, they make public their sustainability effects in CSRs and in integrated reports. Because of this, it is possible to create a positive image of the sector, which translates into value realization for shareholders and other stakeholders. Taking all this information into account, it seems that these enterprises can serve as benchmarks of good business practices; they may also constitute a starting point of this analysis.

A research method that combines both accrual and cash approaches allowed for the analysis of the distribution of the value generated by these enterprises from 2011 to 2016. The study was based on the official data made publically available in the financial reports issued by them.

The assessed and presented benefits that these partnerships produce were allocated to the respective groups of stakeholders. For the years that were examined, it was determined what the benefits amounted to financially as they reach the respective value beneficiaries.

The starting point of the analysis was the determination of free cash flow from value added FCFF VA, that is, the total value generated by the enterprises for their stakeholders from 2011 to 2016 (Fig. 1).

The analysis of the selected partnerships in terms of their capability to meet the financial needs of their stakeholders showed a varied pattern in value shaping.

In the case of LW Bogdanka S.A., a drop in total profits took place in the years 2013 to 2014 (6%). The highest dynamics of change was observed in the years 2011 to 2012 (growth of 110%), from 2014 to 2015 (growth of 24%), from 2012 to 2013 (growth of 10%). An increase more than twofold (110%) resulted from the shift in cash flow to owners (negative FCFE in 2011; positive FCFE in 2012), while the 24% growth was the consequence of the change in cash flow for the financing entities (FCFD). In the years of 2011 to 2016 total added value based on cash flow generated by the company for the selected stakeholders amounted to PLN 7, 746. 6 M.

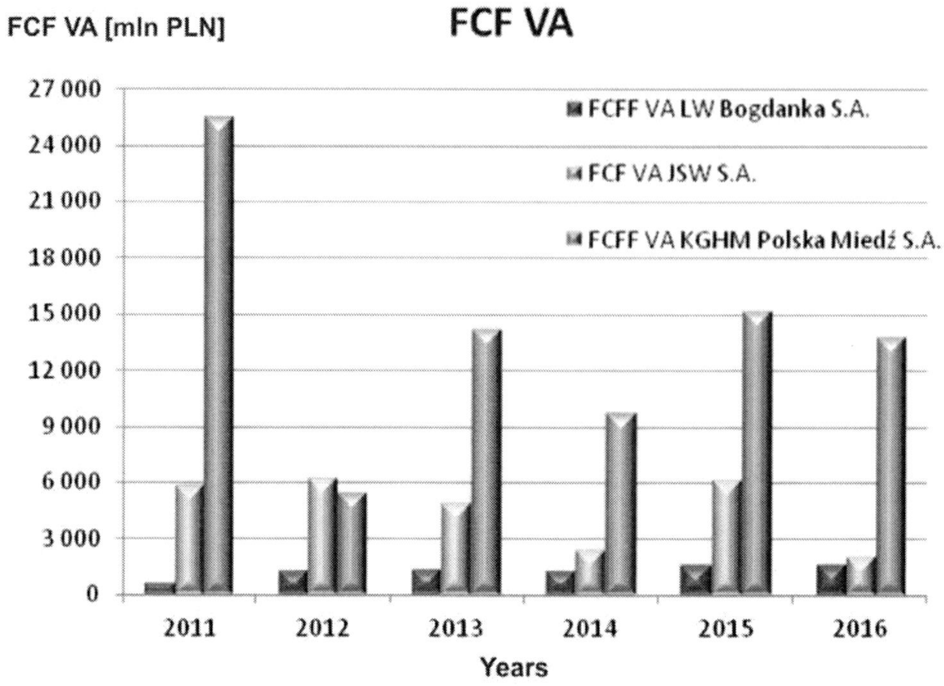

Fig. 1. FCF VA for all stakeholders generated by the analyzed companies in 2011-2016. Based on company data [11-13].

As far as JSW S.A. is concerned, there was a decrease in total cash flow generated by the company between 2012 and 2013 (21%), from 2013 to 2014 (51%) and from 2015 to 2016 (66%). The change was most dynamic between 2014 and 2015 (growth of 156%) and in the years 2011 to 2012 (growth of 7%). A more than double increase (156%) was caused mainly by the change in cash flow to creditors (a major negative FCFD in 2014; positive FCFD in 2015). On the other hand, an increase between 2011 and 2012 resulted from the reduction of negative values of cash flow for this particular interest group (FCFD). In the years between 2011 and 2016 total added value based on cash flow generated by JSW S.A. for stakeholders reached PLN 27, 606.3 M.

Nevertheless, KGHM Polska Miedź generated the highest value for its stakeholders, PLN 83, 959.0 M. The company saw the most dynamic change between 2012 and 2013 (increase of 161%) and between 2014 and 2015 (increase of 55%). A nearly triple increase (161%) was the effect of cash flow to ownership and banks (a major negative FCFE and FCFD in 2012; positive FCFE and FCFD in 2013), while the increase of 55% was due to reducing major negative cash flows to financing units FCFD and owners FCFE. A decrease in total value was observed between 2011 and 2012 (79%), from 2013 to 2014 (31%) and between 2015 and 2016 (9%) and was determined by the decline in FCFD and in FCCFE that reached negative values.

Figures 2-7 show the shaping of cash flow to those respective stakeholders who have their share in the value generated by the companies. The value is distributed in such ways as by remunerating and granting benefits to employees, paying taxes to municipalities and the state, earning interest by equity providers, paying dividends to ownership, and giving profits to goods and services providers.

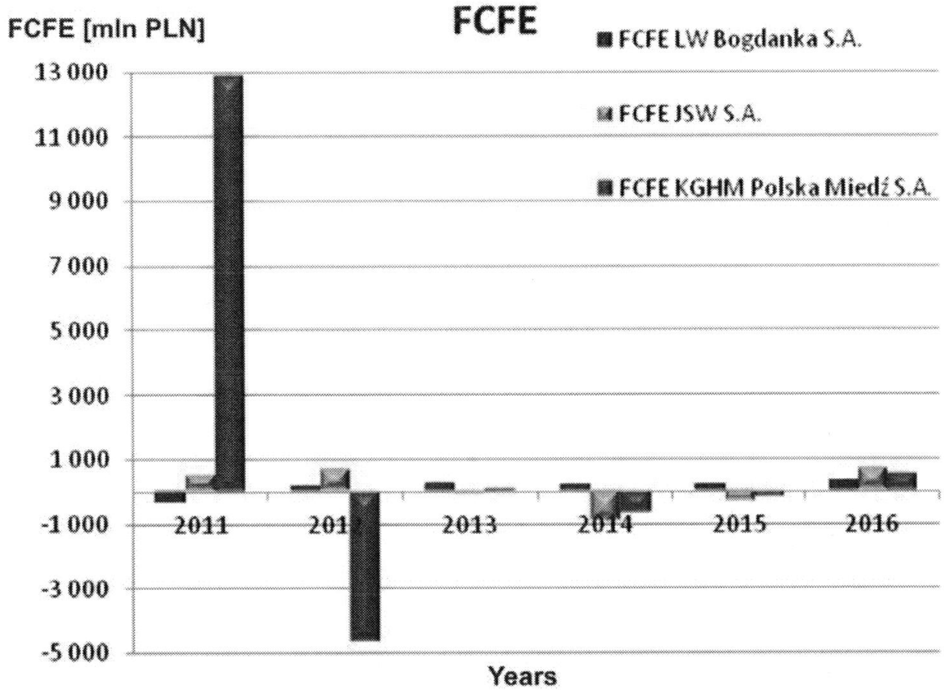

Fig. 2. FCF to equity in 2011-2016. Based on company data [11-13].

The analysis of the value created for the owners by the abovementioned units (Fig. 2) is reflected in the reported profits and losses for this interest group.

With regard to LW Bogdanka, the company generated positive cash flow between 2012 and 2016, with the exception of the year 2011, when, due to expenses incurred by investments (increase in fixed assets) and a bigger demand for net working capital, a negative value for this particular interest group was observed.

In the examination of business activity in JSW S.A., the years 2013, 2014, and 2015 show a negative cash flow to ownership. It indicates the necessity of providing the company with funds and, to a large extent, it is the result of major investment expenses and negative net profit. A positive value for this stakeholder can be seen in 2011, 2012 and 2016.

With respect to KGHM Polska Miedź S.A., a positive value for its owners was recorded in 2011, 2013 and 2016. The other years saw negative cash flow, the main reason of which was significant investments costs, negative net profit (net loss), and an increase in demand for net working capital.

Investments in fixed assets are one of the most common reasons that influence the level of cash flow to the owners. Investments freeze the accessible resources and lower FCFE. Nevertheless, taking into consideration the expenses in the long run, they are necessary to maintain and further generate current cash flow. Without investment costs, a steady development of companies is not possible.

Among the examined businesses, KGHM Polska Miedź S.A. provided its owners the highest value in 2011, that is, PLN 8, 045.8 M. The lowest value for this company was observed in 2012 and it amounted to PLN to 5, 007.4 M.

Equity providers constitute a group of stakeholders that, within the examined period of time, benefits the least from their relationship with the enterprises (Fig. 3).

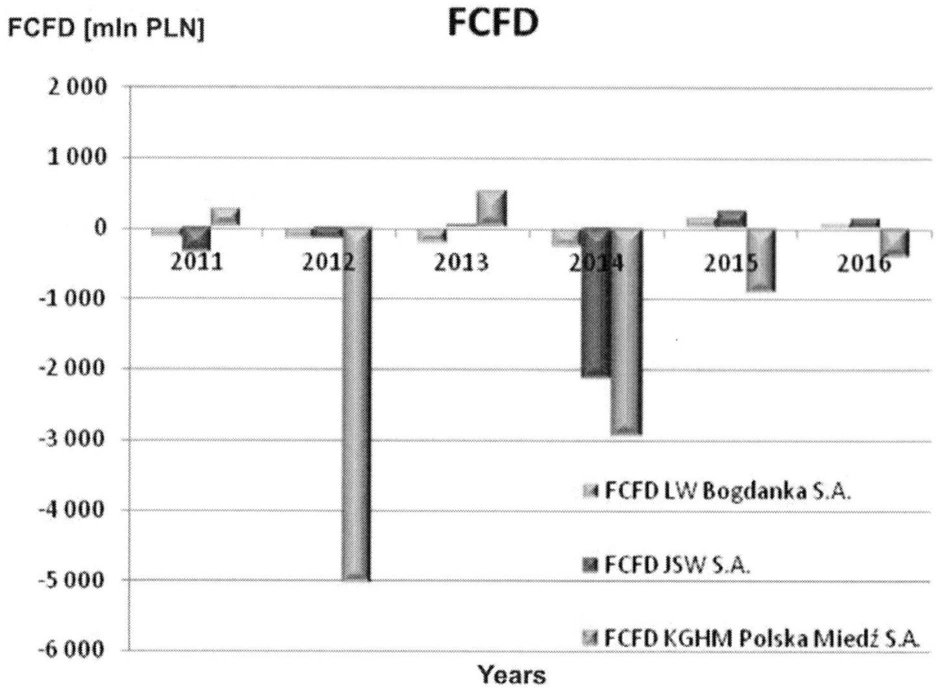

Fig. 3. FCF to debts in 2011-2016. Based on company data [11-13].

With regard to LW Bogdanka S.A., for the years 2015 and 2016, little value was recorded, namely, the generated profits were, respectively, PLN 170.5 M and PLN 79.7 M. Regardless of that, free cash flow to banks, for the majority of the examined years, shows negative values. In the period of 2011 to 2014 the banks did not benefit from their relationship with the enterprise.

Taking into account the results achieved by JSW S.A., a positive value for this stakeholder was reported in 2013, 2015, and in 2016, while during the other years the cash flow was negative. The year 2014 had the highest losses; that year saw the lowest value of PLN 2, 103.3 M.

KGHM Polska Miedź S.A. generated a positive value for equity providers only in 2011 and 2013. In the other years, the analysis shows strong negative values of cash flow for this group of stakeholders.

Negative values of cash flow to banks are the findings of analysis that is based on cash flow approach. Despite positive interest on loans that the banks receive on a regular basis and which the companies regularly pay off, there is a problem with high values of capital installments that are significantly higher than the interest. Negative values of capital installments are the consequence of an increase in foreign capital in the examined period of time. The implication is that, while the businesses take new loans, the banks benefit only from the interest payable to them. The exception applies to the years when foreign capital decreased, which further implies that businesses pay off capital installments, which in turn confirms an increased profit for equity providers.

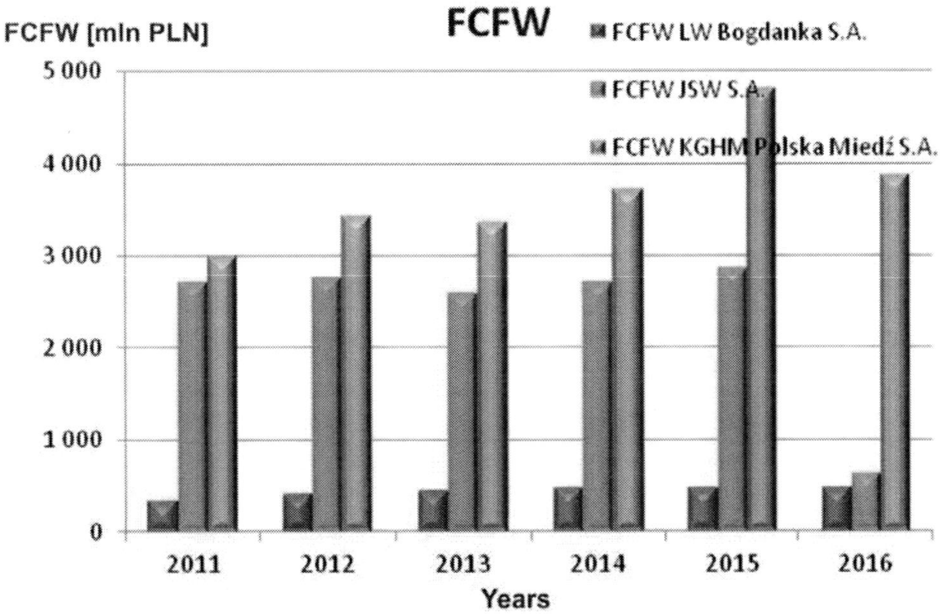

Fig. 4. FCF to workers in 2011-2016. Based on company data [11-13].

Employees are the group most closely related to businesses. It is confirmed by the fact that, through the relationship with the business they work for, employees are the main value beneficiaries (Fig. 4). Cash flow to this group fluctuates around the average amount of PLN 435.0 M for LW Bogdanka S.A., PLN 2, 380.9 M for JSW S.A., and PLN 3, 700.6 M for KGHM Polska Miedź S.A. The largest remuneration and benefits intended for distribution to staff was recorded in 2015. They amounted to PLN 4, 809.8 M and belonged to KGHM Polska Miedź S.A.

The suppliers are the second largest value beneficiaries of the examined businesses (Fig. 5). KGHM Poska Miedź S.A. generated the biggest profit for them. Such was the case in 2015 when the total amount was PLN 9, 580.8 M. In this respect, LW Bogdanka achieved the lowest score, namely, PLN 712.0 M on average reached the suppliers through the purchases of power resources and external services. Considering JSW S.A., the average values oscillated around PLN 2, 023.5 M, while KGHM Polska Miedź S.A. had the average value at the level of PLN 7, 871.6 M for its suppliers.

The businesses generate the lowest positive value for municipalities (Fig. 6). On average, funds payable to local governments' budgets as taxes and fees amount to PLN 27.5 M for LW Bogdanka, PLN 146.5 M for JSW S.A., and PLN 1, 446.0 M for KGHM Polska Miedź S.A. The biggest profit that this group of stakeholders earned was generated by KGHM Polska Miedź S.A. in 2015 and it equaled PLN 1, 985.9 M. LW Bogdanka S.A. had the lowest value of PLN 19.8 M.

The process of added value distribution in terms of profits earned by the Treasury (Fig. 7) indicates that this interest group had the highest income in 2011. In that year of the analysis, KGHM Polska Miedź S.A. paid the Treasury an income tax of PLN 2, 300.0 M. As opposed to this, the lowest value for this company was reported in 2015, that is, PLN 113.0 M (income tax return). The average liabilities that go in the Treasury vary around PLN 64.4 M each year for LW Bogdanka S.A., PLN 317.6 M for JSW S.A. and PLN 1, 043.3 M for KGHM for Polska Miedź S.A.

For all the three enterprises, a detailed analysis of value distribution for the abovementioned groups of stakeholders is presented in Figs. 8, 9 and 10.

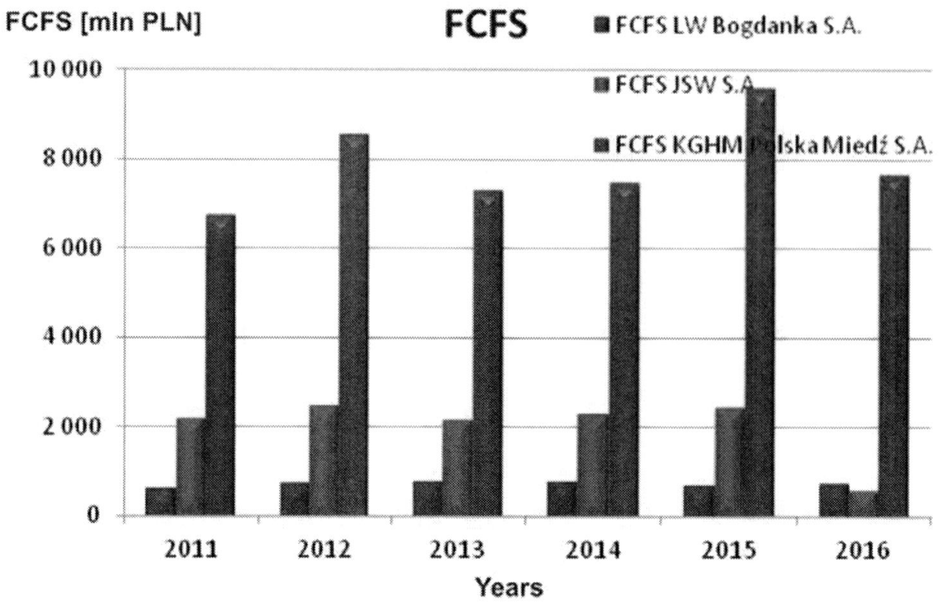

Fig. 5. FCF to suppliers in 2011-2016. Based on company data [11-13].

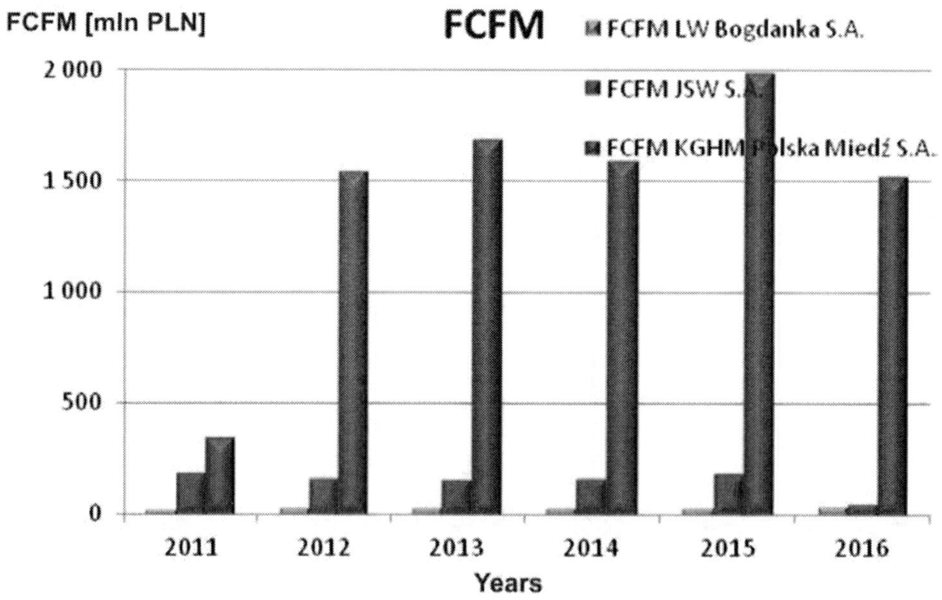

Fig. 6. FCF to municipal administration in 2011-2016. Based on company data [11-13].

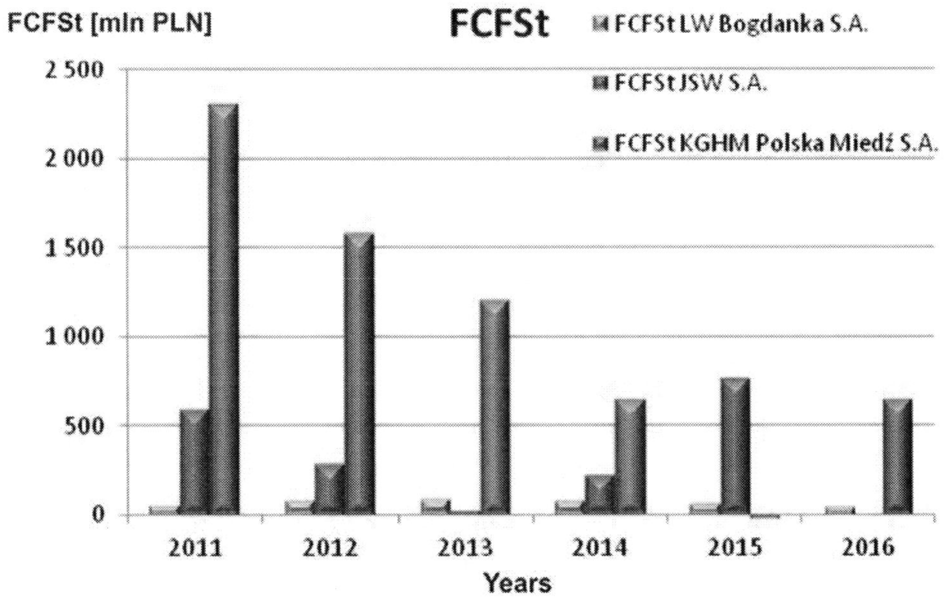

Fig. 7. FCF to state in 2011-2016. Based on company data [11-13].

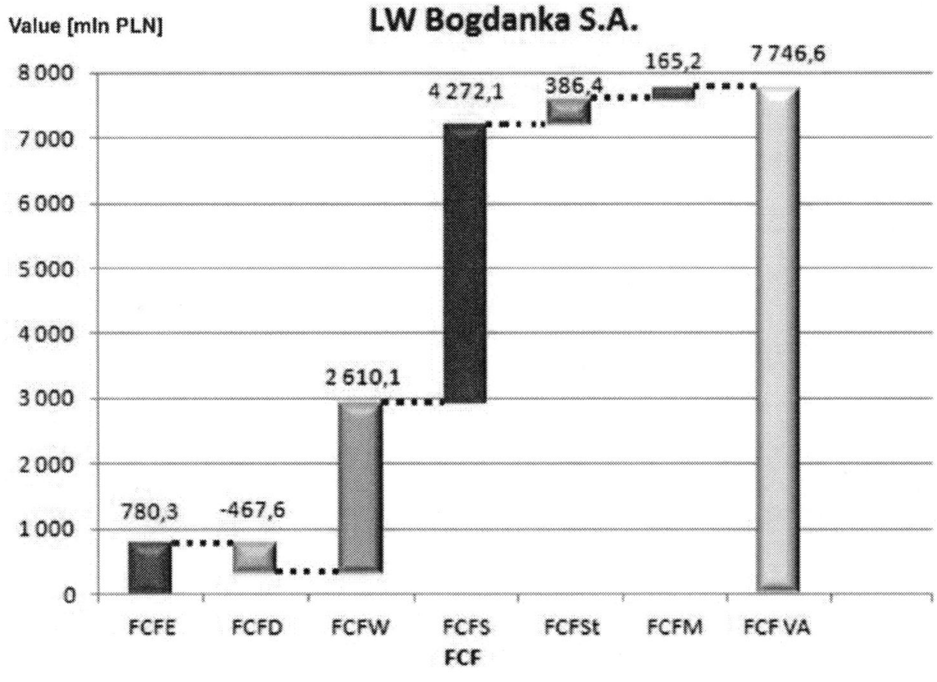

Fig. 8. Share of FCF for individual stakeholders in total FCF VA generated by LW Bogdanka S.A. Based on company data [11].

In terms of value distribution to all the stakeholders of LW Bogdanka S.A., the analysis shows that the suppliers had the biggest (55%) share in value consumption. For the years in question, the total sharable amount that was payable to this group was PLN 4, 272.1 M.

In terms of earned profits, the employees are the second most benefiting group with the amount of PLN 2, 610.1 M, which corresponds to a share of 34% of the total value. The trend in profit distribution is stable and suggests a steady growth. A minor decrease in remuneration and benefits for the staff was noted in 2015 and in 2016.

The owners are the third largest group of beneficiaries of the value generated by LW Bogdanka. In the years 2011 to 2016 they received PLN 780.3 M in total. Their share reaches 10%. With regard to this group, the biggest growth (154%) occurred in the years 2011 to 2012.

The share of the country budget fluctuates at around 5%. In the years that were examined, the Treasury's earnings from income tax amounted to PLN 165.2 M. The largest increase (44%), was in the years 2011 to 2012, while the biggest decrease (34%) was reported from the period of 2015 to 2016.

Municipalities received 2% of the total value from the company. As far as this interest group is concerned, there was a growth tendency in benefits such as taxes and fees that the businesses were charged in each year of the analysis. The years 2011 to 2012 had the biggest growth (34%).

The smallest share in the benefits generated by the company belongs to the banks (-6%). The negative value reported for this group comes from negatively generated cash flow.

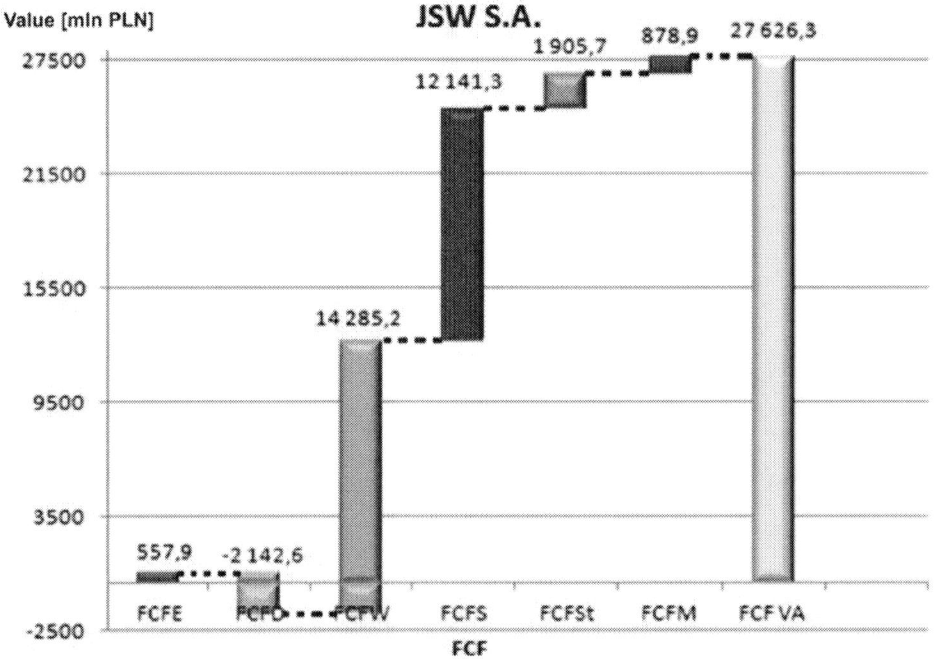

Fig. 9. Share of FCF for individual stakeholders in total FCF VA generated by JSW S.A. Based on company data [12].

The examination of JSW S.A. concluded that the largest share of the value generated by the company belonged to the staff (52%). Within the examined years, the employees received a total of PLN 14, 285.2 M. In the years 2011 to 2016 there was a stable trend in value sharing and it was evenly distributed. A fall in this trend took place in the last year of the analysis, which meant a drop in value of 77%. The most significant growth (5.3%) was between 2014 and 2015.

The second best-benefitting group of the company's value is the suppliers (44%). Thanks to the mutual relationship with JSW S.A., they gained a total benefit of PLN 12, 141.3 M. For this group of stakeholders, a drop in profits (15.4%) was recorded in the years 2012 to 2013 and between 2015 and 2016 (76%).

The country budget share is around 7%. The Treasury's gain from income tax between 2011 and 2015 reached PLN 1, 905.7 M. The biggest (104%) drop in value happened between 2011 and 2012, as well as between 2015 and 2016 (98%). Still, the highest value and, therefore, the largest tax payable to the state were in 2015. Between 2014 and 2015, the increase of value distribution for this interest group was 238%.

The average obligatory liabilities affecting the results that go to the municipalities' budgets through extractive activities of JSW S.A. account for 3% of the value generated by the company and correspond to PLN 878.9 M.

The lowest share in the benefits generated by the company belongs to the owners (2%) and the banks (-8%).

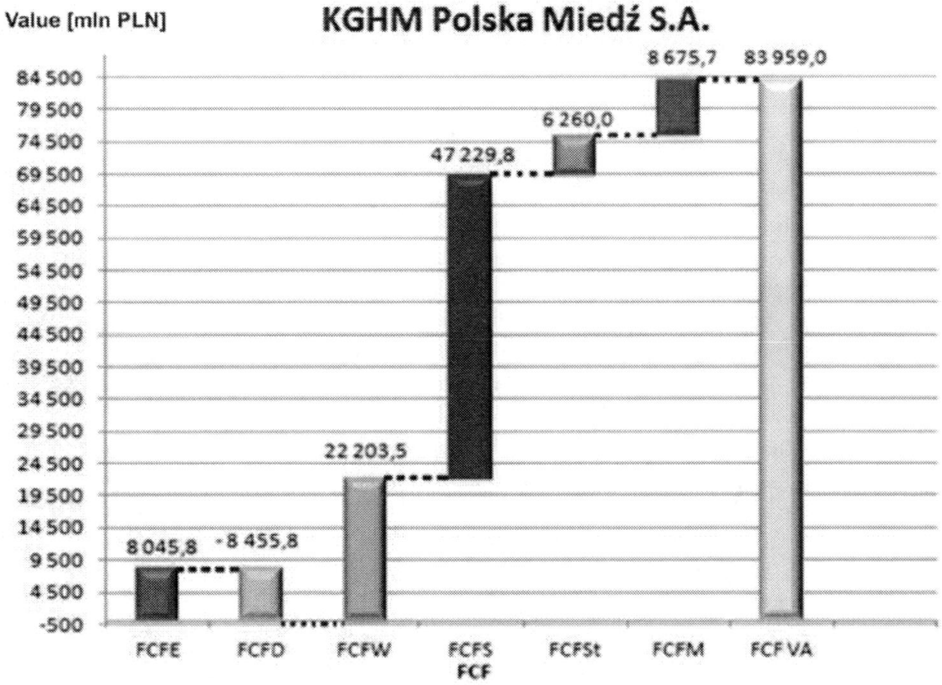

Fig. 10. Share of FCF for individual stakeholders in total FCF VA generated by KGHM Polska Miedź S.A. Based on company data [13].

The analysis of the value created by KGHM Polska Miedź S.A. shows that it is the suppliers who benefit the most from the relationship with the company (56% of the value). In return for providing the company with supplies and power, as well as external services, they received an equivalent of PLN 47, 229.8 M. The maximum increase (29%) was observed between 2014 and 2015, whereas the biggest decrease (20%) affected the years 2015 to 2016.

In terms of the value of the earned profits, the next interest group is represented by the employees. During the years that were subject to examination, the sum of remuneration and other benefits was PLN 22, 203.5 M. The employees' share in total value reaches 26%. The biggest decrease (19%) in payroll was noted between 2015 and 2016, while the years 2014 to 2015 saw the maximum increase of 29%.

Municipalities are the third largest beneficiary of the value. The value distributed for them reached PLN 8, 675.7 M, which corresponded to 10% of the distributed profits.

The owners' shares are at the level of 9.5%. All the profits that went to this group of the company's stakeholders were PLN 8 045.8 m. The years 2015 to 2016 were marked by the largest increase of 392%. Conversely, the largest decrease (643%) took place between 2013 and 2014, when positive values of cash flow changed to significantly negative ones.

Between 2011 and 2016, the country budget received a total of PLN 6, 260 M, which corresponded to approximately 7% of total profits.

The banks were the least benefiting group of stakeholders from the relationship with the company, with the negative value of PLN –8, 455.8 M.

4 Conclusions

Creating a long-term value for all stakeholders is an objective that each enterprise should strive to reach, at the same time ensuring the company's economic success and an improvement of social and ecological results.

Distribution of added value in mining enterprises comes from the aspiration of meeting the needs of all the interest groups. The specificity of the sector determines the necessity of making socially responsible decisions which will meet the financial needs of the enterprise and improve the economic situation of the society. The ongoing crisis in the coal mining sector and the unprofitability of modern enterprises related to that call for an examination that would demonstrate the consequences of shutdowns and discontinuation of business activities. The study presented above shows that the effects of such decisions will be visible and palpable not only to the owners, but also to the staff, suppliers, financing institutions, the country budget and other stakeholders.

The analysis accounts for the contribution of LW Bogdanka S.A., JSW S.A. and KGHM Polska Miedź S.A. to both private and public incomes; moreover, it illustrates how these companies' business activities can be beneficial to the society. The findings show that for the years 2011 to 2016, KGHM Polska Miedź S.A. managed to be the most beneficial to its stakeholders because of achieving the highest FCF VA. At the same time, the process of value distribution shows that it is the staff and the suppliers who are most closely connected to all the businesses covered in this analysis. They benefit most financially from the enterprises and from their relationships with them. The banks, however, were the least benefiting group between 2011 and 2016. They provide the businesses with capital for investment and current activities, and the only benefit for them comes from interest payable to them. Thus, this mechanism results in creating a negative value for this group.

In terms of the ability to generate added value, the examination of the enterprises presented above demonstrates that each of the interest groups has its own, bigger or smaller, share in the consumption of the value, at the same time benefitting from the mutual relationship with the company. Based on the findings, one can point out those interest groups that will suffer most from the wrong decisions taken at the management board level. In such a case, the staff and the suppliers will be most negatively affected, while the banks will be the least disadvantaged.

This information leads to the conclusion that this analysis, which can be a part of the model for sustainable value assessment, can also be a kind of account addressed to the society that raises awareness of advantages offered by mining enterprises. This model may also serve as one of the premises that facilitate making strategic decisions on a shutdown or continuation of extractive activities taken by socially responsible mining companies – ones that, in the pursuit of economic effectiveness, take into account the needs and expectations of the environment. This is due to the fact that the enterprises that belong to strategic sectors have a bigger influence on their immediate and indirect environments and the whole society than traditional business partnerships. In light of this, making strategic decisions should not be based on financial analyses only, but also on the findings of this work, which allow for assessing a given company in terms of benefits that it creates for the society. This would make it possible to have a comprehensive and holistic view of the current situation of mining enterprises both at home and abroad.

References

1. N. Hop, J. Kudełko, Dzieje górnictwa – element europejskiego dziedzictwa kultury, Oficyna Wydawnicza PWr, Wrocław (2013)
2. R.E Freeman, *Strategic Management: A stakeholder Approach*, Cambidge University Press, New York (2010)
3. J. Kulczycka, H. Wirth, Zeszyty Naukowe IGSMiE PAN, **79**, Kraków (2010)
4. M. Marcinkowska, ZN USz, **693** (2011)
5. A. Kijewska and A. Bluszcz and A. Sojda, Metalurgija, **4** (2015)
6. B. Kowal, A. Kustra, E3S Web of Conferences, **10**, (2016)
7. M.F. Morley, Accounting Rev, **3** (1979)
8. A. Riahi-Belkaouri, *Value Added Reporting –Lessons for the United States*, Quorum Books, New York (1992)
9. S. Lorenc, A. Kustra, IM, **2** (2016)
10. S. Lorenc, E3S Web of Conferences, **14** (2017)
11. http://www.ri.lw.com.pl/raporty-okresowe
12. https://www.jsw.pl/relacje-inwestorskie/raporty-gieldowe/raporty-okresowe/
13. http://kghm.com/pl/inwestorzy/prezentacje/raporty-finansowe

Use of IT platform in determination of efficiency of mining machines

Jarosław Brodny[1,*] and *Magdalena* Tutak[2]

[1]Silesian University of Technology, 26-28 Roosevelta St., 41-800 Zabrze, Poland
[2]Silesian University of Technology, 2 Akademicka St., 44-100 Gliwice, Poland

Abstract. Determination of effective use of mining devices has very significant meaning for mining enterprises. High costs of their purchase and tenancy cause that these enterprises tend to the best use of possessed technical potential. However, specifics of mining production causes that this process not always proceeds without interferences. Practical experiences show that determination of objective measure of utilization of machine in mining enterprise is not simple. In the paper a proposition for solution of this problem is presented. For this purpose an IT platform and overall efficiency model OEE were used. This model enables to evaluate the machine in a range of its availability performance and quality of product, and constitutes a quantitative tool of TPM strategy. Adapted to the specificity of mining branch the OEE model together with acquired data from industrial automatic system enabled to determine the partial indicators and overall efficiency of tested machines. Studies were performed for a set of machines directly use in coal exploitation process. They were: longwall-shearer and armoured face conveyor, and beam stage loader. Obtained results clearly indicate that degree of use of machines by mining enterprises are unsatisfactory. Use of IT platforms will significantly facilitate the process of registration, archiving and analytical processing of the acquired data. In the paper there is presented methodology of determination of partial indices and total OEE together with a practical example of its application for investigated machines set. Also IT platform was characterized for its construction, function and application.

1 Introduction

The dynamic development of Information and Communication Technologies (ICTs) has resulted in their increasing application in various areas of social and economic life. Nowadays, an increasing number of companies deploy cutting-edge IT tools to streamline their management processes. The practical application of these solutions improves the effectiveness of such companies, thus enhancing their competitiveness. The universal accessibility of advanced IT solutions and the growing competitiveness encourage other sectors to apply similar solutions. This is also true of the extractive industry and hard coal mining in particular. The growing competition on the energy market and the positive effects

* Corresponding author: jaroslaw.brodny@polsl.pl

of implementing IT tools into corporate management processes in other sectors have necessitated changes in mining companies as well. These companies are showing increasing interest in the application of IT tools to enhance their operational efficiency.

The area for which such tools are especially suitable is maintenance management. The use of state-of-the-art technologies, increasing automation of production processes and growing requirements as to the economic results are placing ever higher demands on maintenance departments [1, 2]. To address these demands, maintenance departments are forced to make wider use of the IT solutions offered. One of the primary goals set for these departments is to take full advantage of the technical resources owned by a given company. In the extractive industry, this mainly concerns the effective use of all types of machinery possessed by companies.

Determining the effectiveness of utilising mining machines is of the utmost importance for mining companies. The high costs of purchasing and leasing these machines make the companies strive to take as full advantage of their technical potential as possible. However, due to the specificity of mining production, this process does not always continue without disruptions. Practical experiences show that it is not easy to determine an objective measure for the utilisation of mining machines.

The relevant actions taken thus far in Poland have not had a significant impact. The effectiveness of utilising machines, especially those directly involved in mining the rock mass, is very low [3, 4]. At the same time, the methods applied to determine this effectiveness, based on dispatchers' subjective notes, do not guarantee objectivity of the resulting assessments.

The issue of assessing the effectiveness of machinery use in the extractive industry worldwide is addressed by few researchers. In their works, they use the OEE model [5, 6] and indicator based methods [7], as well as the reliability theory [7]. These works, however, are primarily focused on open-pit mines [5-7], whose specificity differs significantly from that of underground mines. The differences are particularly related to the conditions of the exploitation operations, including ventilation, existing hazards (methane, fire) and the deformative impact of the rock mass. The subject of machinery operation in underground mines has been discussed in publication [8], which presents the results of the effectiveness assessment for the operation of the longwall shearer in the Parvadeh coal mine, Iran. The analysis of these results unambiguously indicates that the operating conditions of underground mining machinery are very complicated. At the same time, such conditions make it difficult to achieve high effectiveness indicators. However, most of the data used for the analysis were acquired from reports by representatives of the mining supervision, which does not guarantee full objectivity of the data.

For this reason, the authors of this paper based their analyses on the data acquired from industrial automation systems, since they guarantee complete objectivity and reliability. At the same time, they provide great possibilities for analysing and interpreting the results obtained. It is indeed evident that the effectiveness of the machinery currently used in the hard coal mining industry in Poland is unsatisfactory.

The paper presents a proposal for how to solve this problem. To this end, use was made of an IT platform and the Overall Equipment Effectiveness (OEE) model [9, 10]. This model not only allows for assessing a given machine's operation in the areas of its availability, performance and quality of the product, but it also constitutes a quantitative tool of the TPM strategy [9-12]. Adapted to the specificities of the extractive industry, the OEE model – in combination with the data acquired from the industrial automation system – made it possible to determine the subindicators and the overall efficiency indicator of the machines under examination. The research was conducted for a machine combination directly involved in the process of coal exploitation, including a longwall shearer along with an armoured face conveyor and a beam stage loader. These machines are part of an automated longwall system.

The IT platform was used to determine the effectiveness of these machines. In the case at hand, the platform is understood to mean a set of interrelated components of IT techniques and technologies from the field of hardware, ICT and software infrastructure. This tool allows for practical application of the OEE methodology in the underground mining environment. The platform is primarily based on relational databases along with algorithms for data analysis. Accessible to the particular groups involved in the production and management processes, the platform makes it possible to generate results, as appropriate, and access the data. Generally, it may be presumed that the platform offers easy access to the information on the work of machinery for different decision-making levels in the mining company and the group of companies which includes the mine in question. At the same time, its structure is so universal that it can be used for a number of longwalls and mines.

The paper discusses the general structure of the IT platform and presents the results of its application for one of the longwalls in a hard coal mine. In the case at hand, the possibilities of the IT platform were used only partially. This is due to the fact that attention was focused on determining the effectiveness of utilising the machines directly involved in coal exploitation. However, it is expected to encompass a broader group of machines and users. In the authors' opinion, the platform developed should constitute an essential tool for supporting the management system of production processes, with the use of the OEE methodology.

2 Description of the IT platform

The dynamic changes in the area of maintenance, increased reliability of production equipment and machinery performance may be addressed by developing new methods and techniques that support the operations of maintenance departments. These changes also affect the mining industry. The globalisation leading to widespread competitiveness may be addressed by using increasingly sophisticated solutions that support corporate management processes. The Polish coal mining industry has a well ordered internal structure, which also encompasses the management of technical resources and company maintenance. However, in practice, it appears that the resources possessed are not capable of fully optimising the production process. Therefore, the new challenges faced by the mining industry necessitate the use of cutting-edge information and communication technologies (ICTs) in the management of mining companies.

In practice, such companies have already implemented certain IT tools with a view to supporting the management of resources and safety as well as further training of employees. However, the area of machinery operation lacks an efficient and reliable tool for objective (independent from the employee's subjective feelings) determination of the utilisation effectiveness of this machinery.

Such possibilities are offered by the IT platform which provides quick, flexible and unlimited access to data resources and information on the operation of the machinery under examination. This information is obtained for the purpose of managing the operational process and supporting business, educational and research processes. It can be assumed that platform, which has been designed and is currently being deployed, is a tool that allows for using IT technologies to support knowledge management in a production company, thereby facilitating effective flow of information coming from various sources amongst recipients for efficient and knowledge-based management of the production process [13-18].

The platform in question is therefore a solution which constitutes an integrated set of tools, including:

- administering, data-feeds and archiving measurement data (DW – Data Warehouse);
- administering the processing of measurement data to the form and scope relevant for determining a tree of the OEE indicators;
- securing the information layer on the production process on the longwalls for which the methodology adopted will be used;
- administering the calculation of OEE indicators;
- securing the information layer for analyses related to the methodology applied;
- administering the presentation and reporting layers.

The main component of the tool is the Data Warehouse (DW) composed of a series of relational databases. They contain input data on the operational parameters of the machinery examined, computation algorithms and result-based databases. Particular importance must be given to input data. It is their quality and the accuracy of the analytical process that determine the quality of the results obtained. In order to reduce the problems related to data acquisition and ensure systemic processing mechanisms, further layers of data were introduced for subsequent transformations:

- the layer of source (raw) data (offline data, file databases, online data);
- the layer of standardised source data (DW, layer I);
- the layer of selected, standardised source data (DW, layer II);
- the layer of processed data based on the model of time buckets necessary to determine the OEE indicator (DW, layer III).

The OpenEye system was used for the management of the entire IT mechanism and the data acquired in particular. This solution uses elements of the supervisory control and data acquisition (SCADA) system. It is made up of a series of tools for quick data analysis and visualisation of the results obtained. The OpenEye system contains built-in visualisation products and cooperates with any relational databases, including Oracle, MS SQL and PostgreSQL.

The structure of the platform allows for direct cooperation with the SZYK2 management support system commonly used in Polish mines. Referential integrity with the KPT module (Production-Technical Complex) of this system was implemented. This module is responsible for supporting the administration of business processes related to production as well as its preparation and maintenance. This allows for direct acquisition of the data on the parameters of the machines in operation, the mining and geological conditions, the geometry of mine workings as well as much more information and the mining operations in progress.

The whole system also includes a series of components from an industrial automation system (sensors, controllers, relays, etc.), an optical fibre transmission network and an entire data archiving and processing system. The main parts of the entire system include engines for data acquisition, archiving and analytical processing, for reporting and visualisation, as well as an engine for exchanging information, opinions and inserting comments.

Figure 1 presents the diagram of the acquisition process for the data on machinery operating parameters. It is one of the fundamental parts of the entire IT system, demonstrating the manner of creating databases for the particular layers of data. A particularly important role in this process is played by the removal of holes in the data acquired and the qualification of the data for further processing. As a result of this action, data are supplied to the part of the data warehouse which contains the database of processed data. These data give rise to further analytical processing.

Additionally, the system has been equipped with an engine for registering the causes of downtimes for the particular machines. The module of the IT platform for registering unscheduled downtimes of machines may be viewed in Figure 2 (original view with description in the Polish language).

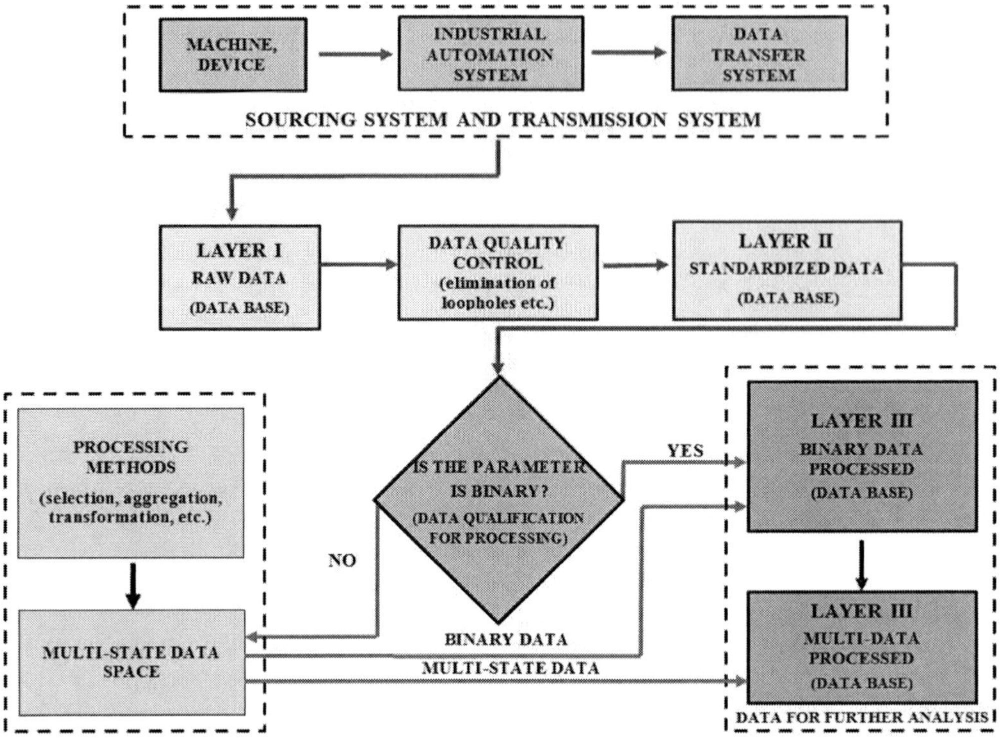

Fig. 1. The diagram of the acquisition process for the data on machinery operating parameters.

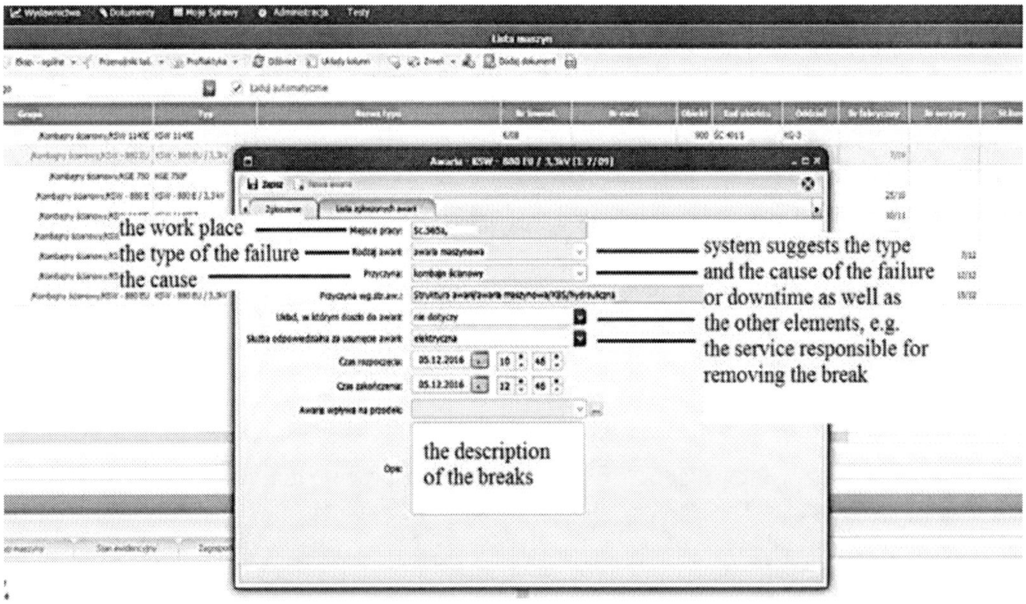

Fig. 2. The module of the IT platform for registering unscheduled downtimes of machines.

It should also be highlighted that one of the important objectives behind the implementation of the solution presented is the possibility of exchanging views and comments for all the individuals involved in mining exploitation. This applies to employees and dispatchers as well as the management of the group of companies.

3 Description of the IT platform

The solution in question was used to assess the operational effectiveness of the machines integrated into an automated longwall system for three longwalls in one of the hard coal mines. The publication demonstrates its practical application for one of these longwalls. The results presented encompass merely a small part of what was obtained from this longwall.

The longwall examined was 139 metres long and approximately 3.0 metres high, with the longwall panel length amounting to approximately 750 metres. The longwall's longitudinal inclination was 1o, while its transverse inclination – 6o. The longwall was located in seam 510 at the depth of 665 metres. The only major risk in the area of the longwall was the methane hazard, which had no greater influence on the exploitation process. It can therefore be assumed that the exploitation conditions were acceptable.

In the case of the mine under examination, exploitation is carried out in a unidirectional mining system. The longwall shearer's operating cycle consists of the sumping phase, coal-getting and reverse (manoeuvring) motion. The average advance speed of the shearer during coal-getting operations is approx. 5-6 m/min., and within the range of 16-20 m/min. during a reverse motion.

The effectiveness of the machinery set under examination was determined by means of the Overall Equipment Effectiveness (OEE) model. Due to the specificity of the mining industry and the underground exploitation of hard coal in particular, the manner of calculating the particular indicators of the OEE method was modified. These modifications covered all three indicators [3, 4]. The availability (A) for each of the machines examined was determined as the ratio of the actual working time (T_r) of a given machine to the productive time (T_w) of its work. The latter is also referred to as the operational time and constitutes part of the normative time (T_n) less scheduled downtimes (T_p). On the other hand, the actual working time of a machine is calculated as the difference between the productive (operational) time and the duration of unscheduled downtimes (T_u). The dependency which makes it possible to determine the availability indicator has been presented in the following equation (1):

$$A = \frac{T_r}{T_w} = \frac{T_w - T_u}{T_n - T_p} \tag{1}$$

The most crucial component of the availability indicator is the actual working time of a machine. The methodology at hand used discreet methods to register the operational parameters of the machines examined. In the case of the longwall shearer, these included the amperage values of the currents consumed by the shearer's motors, the speed of advance and the shearer's location in the longwall specified in relation to the powered support section. In the case of the armoured face conveyor, the registration encompassed the amperage values of the currents consumed by the motors of its drives, while in the case of the beam stage loader – the amperage of the current consumed by its propulsion engine.

The actual working times of these machines (T_r) were calculated based on the registered behaviours. It was assumed that the difference between the actual time (T_r) and the productive (operational) time (T_w) constituted unscheduled downtimes for a given machine. The algorithms used made it possible to determine these times and the duration of the particular downtimes.

The productivity indicator (E) was defined as the ratio of the actual quantity of the mined rock mass in tonnes (E_r) to the quantity of the mined rock mass in tonnes as assumed in the technical plan (E_n). This indicator thus specifies the ratio of actual productivity to that assumed in the technical plan for the given longwall.

$$E = \frac{E_r}{E_n} \tag{2}$$

The productivity indicator was determined based on the registration of the actual productivity of each working shift, conducted by dispatchers. Additionally, accurate measurements were taken of the exploited area every ten days. On that basis, with account being taken of the assumed densities of the mined rock, the actual mass of the mined rock was determined. No significant deviations were noted in this regard during the examinations.

The most difficult to actually determine in underground conditions is the excavated coal quality indicator. The mines in question have no systems for continuous control of the excavated coal material quality. Once this indicator had been determined, use was made of the information obtained from the mine's preparation plant and the excavated coal quality control department. Based on the examination results of the excavated coal quality, carried out once every ten days, and the coal size assessed during each working shift, the excavated coal quality indicator (Q) was determined.

This indicator was determined as the ratio of the actual excavated coal quality (Q_r) to that assumed in the technical plan for the given longwall (Q_p), with both of these factors taking into account the quantity of the gangue in the excavated coal material and the coal size (3).

$$Q = \frac{Q_r}{Q_p} \tag{3}$$

The value of the total OEE indicator was calculated as the product of the particular subindicators (4):

$$OEE = A \times E \times Q \times 100\% \tag{4}$$

The dependencies presented were used to determine the values of the subindicators as well as the total OEE indicator for the particular machines and the entire machine combination for a single working shift (Fig. 3).

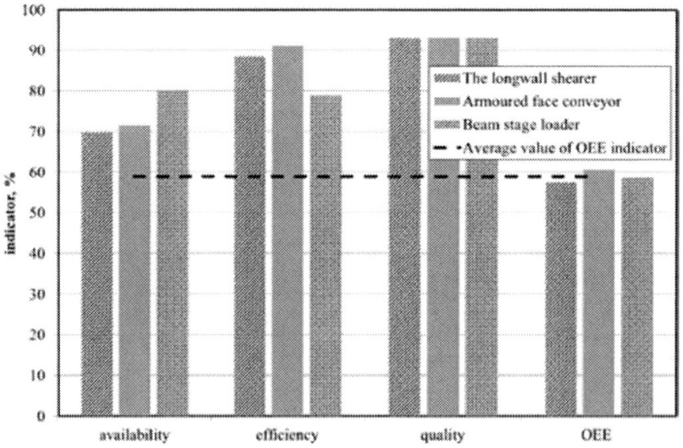

Fig. 3. The values of the subindicators as well as the total OEE indicator for the particular machines and the entire machine combination for a single working shift.

Figure 4 presents the values of the overall equipment effectiveness (OEE) indicators for the particular machines under examination and the average value for the entire machine combination for 30 working shifts (for one measurement decade). Figure 5 presents average values of unscheduled downtimes per one working shift for the particular machines. After determining these times, it was assumed that a working shift lasted 6 hours, or 360 minutes.

Based on the data registered, it is also possible to analyse the unscheduled downtimes of the machines examined. This analysis may encompass the structure of these downtimes and their causes. The results obtained may constitute a very important source of information for users and machine producers alike.

Figure 6 presents a summary of the quantity of unscheduled downtimes for the armoured face conveyor and the beam stage loader registered by the industrial automation system. The summary also takes into account the structure of these downtimes. Their duration was divided into six sections. The summary covers the entire period under examination, i.e. 30 working shifts.

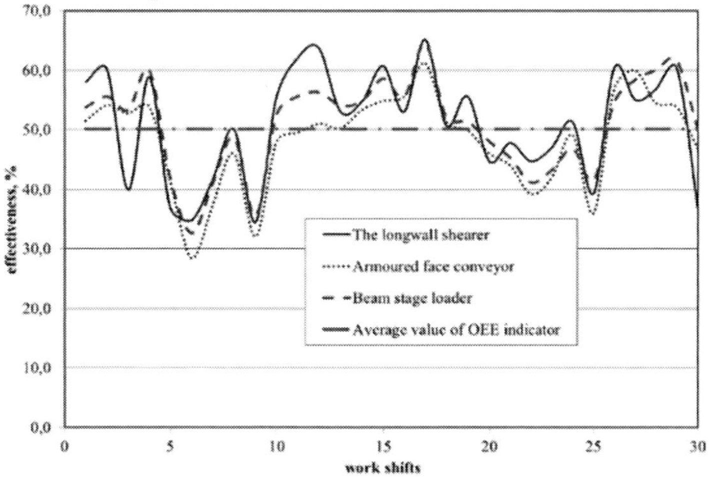

Fig. 4. The values of the overall equipment effectiveness (OEE) indicators for the particular machines under examination and the average value for the entire machine combination for 30 working shifts (for one measurement decade).

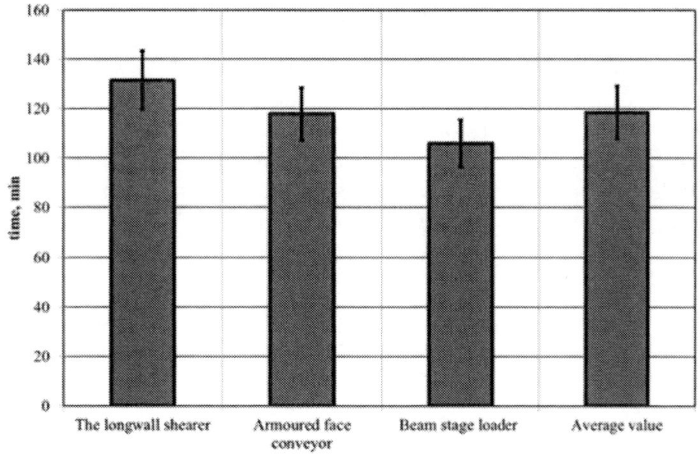

Fig. 5. The average values of unscheduled downtimes per one working shift for the particular machines.

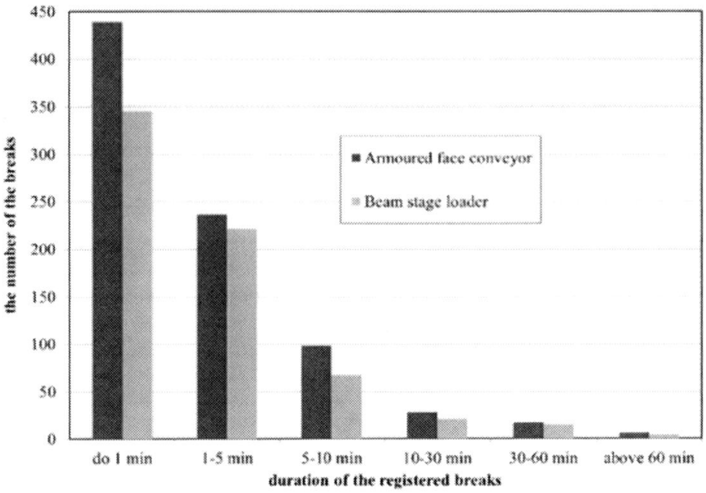

Fig. 5. The quantity of unscheduled downtimes for the armoured face conveyor and the beam stage loader registered by the industrial automation system.

4 Conclusions

The IT platform presented and discussed in the publication is intended to support the maintenance management process in a mining company. It is one of the main areas offering considerable reserves for optimising the hard coal production process in Polish mines. In the authors' opinion, the IT solution in question may significantly facilitate this process.

The paper presents only some of the analytical and cognitive possibilities offered by the platform. First and foremost, it allows for direct information exchange (both online and offline) between all the individuals and institutions involved in the production process that uses the machines under examination.

Another significant aspect is the fact that, in the area of identifying the causes of unscheduled downtimes, the personnel in charge of inputting this information remain anonymous. This makes it possible to acquire tacit knowledge, which, so far, has been practically impossible to acquire due to employees' solidarity. Using the industrial automation system to acquire data on the operational parameters of the machines allowed for complete verification of the information registered by the dispatchers. Ever since the system was implemented, there has been a significant improvement in the quantity and quality of the information provided by the dispatchers.

Currently, the system is used for analysing the operation of the machines in the automated systems of two longwalls. Thanks to full integration with the management support system, the measurement data, along with additional information, are transferred to the platform in the online system. The system provides quick access to information, both within the mine (for the particular decision-making levels) and for external units (the management of the group of companies, the manufacturers of the machines, scientific bodies, etc.). It may therefore be assumed that the IT platform allows for collecting and processing information on the work of the machines (their operation), and exchanging such information between all individuals having access to this process.

The methodology presented in the paper for acquiring and analysing data on machine operation as well as the results obtained indicate that the actions taken offer extensive possibilities for examining and assessing the operation of these machines. It should subsequently allow for improving the effectiveness of their use and produce positive effects on the company's economic performance. It should be noted that the improvement is also dependent on certain external factors, which may be of paramount importance in the case of underground exploitation.

The detailed analysis of the results obtained in terms of the effectiveness of using the machines under examination demonstrates a great potential for improving the indicators determined. The obtained values of these indicators are unsatisfactory. Within the data presented in the paper (30 working shifts) and the entire longwall under exploitation (8 months), none of the working shifts exceeded the effectiveness value of 66%. During the period covered in the paper, the highest value of the effectiveness indicator recorded was 65.19%. The average value of this indicator for that period amounted to 50,14±7,64%. Despite the specificity of mining exploitation already mentioned, the results obtained are unsatisfactory. In the subsequent stages of work on this issue, it seems advisable to undertake a deeper analysis of the causes behind such a low utilisation of the machines of the longwall system. After all, it appears that the mining industry offers considerable reserves in the area of utilising the technical resources possessed.

To conclude, it may be stated that the application of the OEE model, in combination with state-of-the-art IT solutions, makes it possible to objectively monitor and assess the operating status of mining machinery. The information acquired may give rise to actions undertaken with a view to improving these indicators and optimising the process of coal exploitation. Due to editorial requirements (i.e. limited page-count of a publication), the paper discusses only selected parts of the system's operation. However, the authors anticipate further publications in this area.

This article is the result of the research project ***No. PBS3/B6/25/2015 „ Aplication of the Overall Equipment Effectiveness method to improve the effectiveness of the mechanized longwall systems work in the coal exploitation process*** " realized in 2015-2017, financed by NCBiR.

References

1. P.C. Cozzucoli, J. Qual. Stati. Rel., Vol. **2009**, (2009)
2. K. Kandananond, Qual. Stati. Rel., Vol. **2010**, (2010)
3. J. Brodny, M. Tutak, M. Michalak, *Beyond Databases, Architectures and Structures. Towards Efficient Solutions for Data Analysis and Knowledge Representation*. DOI: 10.1007/978-3-319-58274-0_35 (2017)
4. K. Stecuła, J. Brodny, M. Tutak, *GeoConference SGEM, Conference Proceedings*, Book1 Vol. 2, 65-72, DOI: 10.5593/SGEM2016/B12/S03.009 (2016)
5. M. Mohammadi, P. Rai, S. Gupta, Arch. Min. Sci. **62**, 1, 105-120, DOI 10.1515/amsc-2017-0008 (2017)
6. S. Elevli, B. Elevli, Acta Mont. Slov. Vol. **15**, číslo 2, 95-101 (2010)
7. M. Mohammadi, P. Rai, S. Gupta, Acta Mont. Slov. Vol. 21, no **1**, 1-8 (2016)
8. S.H. Hoseinie, B. Ghodrati, U. Kumar, *Sixth International Symposium High Performance Mining*, (2014)
9. W. Mazurek, *Wskaźnik OEE – Teoria i praktyka*, wydanie II Neuron (2014)
10. S. Nakajima, *Introduction to TPM*, Productivity Press (1988)
11. B. Gajdzik, Log., nr **3** (2014)
12. Japan Institute of Plant Maintenance: *TPM dla każdego operatora*. The Productivity Press Development Team (2012)
13. R.M. Grant, Strat. Manag. J., Nr **17**, 109–122 (1996)
14. T. Gwiazda, *Wstęp do informatyki w zarządzaniu*, Wyd. Wydz. Zarz. UW (2004)

15. Z.J. Klonowski, *Systemy informatyczne zarządzania przedsiębiorstwem*, Oficyna Wydawnicza Politechniki Wrocławskiej (2004)
16. M. Miłosz, *Eksploatacja systemów informatycznych – od teorii do praktyki* (2004)
17. P. Rave, Int. J. of R. Devel, Vol. **42**, no 6, (2008)
18. R.M. Stair, G. W. Reynolds, *Principles of information systems*, Cengage Learning (2009)

Analysis of conditions and the concept of multidirectional revitalization of the dolomite quarry in Siewierz

Katarzyna Pawełczyk[1,*]

[1]AGH University of Science and Technology, Faculty of Mining and Geoengineering, Department of Surface Mining, 30 A. Mickiewicz Av., 30-059 Kraków, Poland

Abstract. The development of mining of rock raw materials in Poland has significantly contributed to economic growth and the quality of life of local communities. However, mining activity, besides a number of positive effects, also implies broadly understood changes in the environment and the formation of brownfields. Reclamation and redevelopment of post-industrial areas, and especially post-mining areas, is currently a huge environmental and socio-economic challenge. Revitalization of post-mining areas is particularly important for small towns and municipalities, where mining was one of the main pillars of development and prosperity. An example of such a municipality is Siewierz in Silesian Voivodship. In the paper analysis the conditions have been conducted and the concept of revitalization of dolomite quarry Górnicze Zakłady Dolomitowe S.A. has been proposed. The AHP method has been used to analyse and select the optimal method of revitalization. As a result of the analysis, the concept of multidirectional revitalization with socially, economically and environmentally beneficial functions has been created, maximizing the potential of the quarry.

1 Introduction

Development of mining industry which could have been observed in the previous century, undoubtedly contributed to economic growth so from economic point of view was very beneficial and desired phenomena. Opening of new mine gave not only the possibility of gaining valuable raw materials, but also provided new work places and, what goes with it, raising living conditions and financial situation of local community. It concerns especially small towns, where deposits of natural minerals were a huge potential for growth. However, it is worth remembering that with widely understood social-economic growth of areas where mining activity took place, degradation of natural environment progressed. Negative influence of extraction of minerals appears through many phenomena, occurring in different environmental components.

Reclamation, through entering it into law making ceased to be treated as additional activity, going beyond entrepreneur's responsibility. Nowadays it is one of the steps of

* Corresponding author: pawelczyk@agh.edu.pl

mining activity and according to the bill from 9th of June 2011 Geological and Mining law to its realization rules of the bill from 3rd of February 1995 about protection of agricultural and forest lands are applied. Moreover, suitable and thought-through development of post-mining area may significantly raise its attractiveness, at the same time improving the image of mining in the society [1]. Because of the scale of interference into natural environment, the process of revitalization is long, complex and demands introduction of many interdisciplinary analysis. However, it is worth underlining that thorough analysis of conditioning and coming to correct conclusions, and consequently formation of suitable concept of redevelopment, gives huge possibilities of creating new role of the area which attractiveness may reach far beyond local community [2].

Revitalization of post mining areas is extremely significant It is unacceptable to leave areas degraded as a result of exploitation without giving them new, useful, from the local communities point of view, functions which will allow for effective use of the potential to build a new land value according to rules of sustainable development. It's worth being mindful of that a mine may be a huge chance for development for the community, not only during undergoing extractive activity but also after its end. Perfect example of it may be brown coal mine "Konin", the ground of which has higher soil quality class after performing reclamation than before taking it over for mining purposes [2], reclamated sulphur mine „Machów", excavation of which was turned into water reservoir Tarnobrzeskie Lake very popular and liked by local community or finally hard coal mine „Kleofas" in Katowice, on place of which commercial center Silesia City Center was created – its design, architecture and incorporation of mine elements refers to mining tradition of the region.

2 Subject and aim of the papers

The subject of this paper is the quarry of Triassic and Devonian dolomite exploited by Górnicze Zakłady Dolomitowe S.A. located in Siewierz in Silesian Voivodship. The aim of the paper is to choose optimal reclamation direction of the mine as well as working out the concept of target redevelopment of the area based on all relevant factors that have been analyzed.

3 Research method

In order to determine optimal way of reclamation and redevelopment of the quarry of Triassic and Devonian dolomite exploited by Górnicze Zakłady Dolomitowe S.A. following elements of post exploration area development design method were applied [3]:

- systematization of factors and criteria which have an influence on the choice of the direction of reclamation and their analysis;
- determination main and complementary functions for particular parts of the area based on factor analysis;
- hierarchization of factors and the choice of optimal concept of area's redevelopment;
- establish of rules of realization of chosen concept and functioning of revitalized area.

The choice of optimal direction of revitalization is based on selected elements of AHP method, i.e. analytic hierarchy process. Postulates of the method is simplification of decision-making process in case when the choice must be done based on many criteria and with the help of immeasurable parameters. It makes difficult and complex problems come down to sequences of paired comparison [3] as it is shown on figure 1. In the case of this article, all factors have been compared in pairs, then as a result of this comparison, the most important factors have been identified and their hierarchy has been created. The next step was to compare the previously created concepts of revitalization (based on a detailed analysis of factors) with respect to the most important factors and selection a concept that would meet the criteria resulting from the adopted hierarchy in the highest degree.

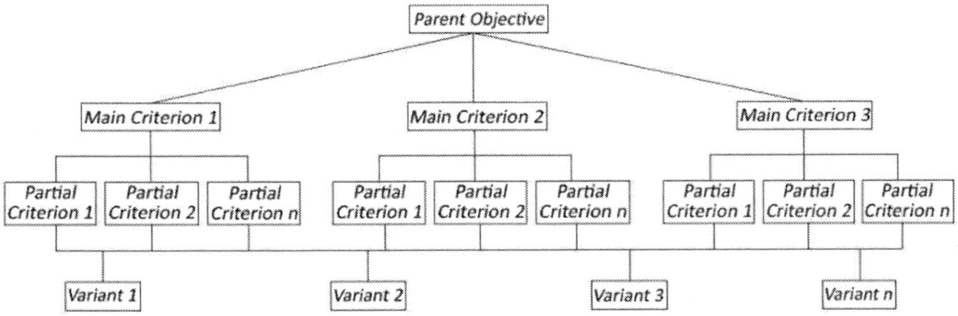

Fig. 1. Diagram of general hierarchic structure [3].

4 Analysis of conditions

In this chapter characteristics of the quarry of dolomite in Siwierz has been discussed on the base of documentation provided by the Górnicze Zakłady Dolomitowe S.A. as well as on-site inspection.

4.1. Spatial factors

The quarry of dolomite Zakłady Dolomitowe S.A.is located in the north end of Siewierz in Silesian Voivodship, about 35 kilometers east-north of Katowice and about 40 kilometers south of Częstochowa. Significant spatial factor is location of the mine in the external zone of upper Silesian conurbation, which makes polycentric settlement form, consociating 54 boroughs, including 2 boroughs of Lesser Poland Voivodship. Siewierz borough borders also with Silesian Metropolis [4]. The quarry is located directly by the express road E75 of international status connecting Cieszyn and Gdańsk. Other significant communication routes are located nearby as well. Such as highway A1 (Pyrzowice – Gorzyczki), trunk road 78 (Chałupki – Chmielnik), trunk road 86 (Podwarpie – Tychy) or trunk road 94 (Zgorzelec – Korczowa), connecting west with east of the country. Many communication routes in the nearest neighborhood of the mine makes it easy to get there not only from Silesian area, but also from all over the country. Moreover, public transport stops are located nearby.

The nearest surroundings of the mine are:

- on the north – wasteland area with buildings of Dziewki village;
- on the east – wasteland, industrial areas and the village of Kazimierzówka with dense residential buildings and singular orchards;
- on the south – wasteland areas, meadows and forests;
- on the west – wasteland areas, behind which there is express road E75 and further buildings of Brudzowice village [5]

Layout of neighboring villages may be described as linear settlement with housing on each side of local roads, therefore the area isn't high-densely housed and significant part of it consists of agricultural wasteland and small forests. Area located around the mine was shown on ortophotomap in the figure 2.

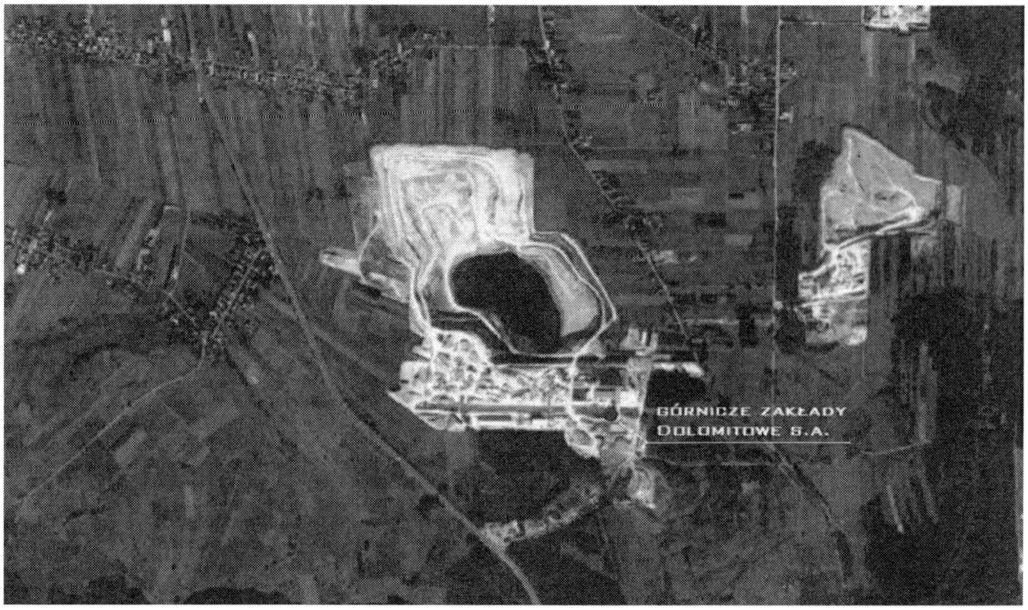

Fig. 2. The surroundings of the dolomite mine. Source: https://www.google.pl/maps.

4.2 Geological-engineering factors

Geological conditions were characterized on the base of geological documentation of dolomite deposit "Brudzowice" [6]. Dominant formations in geology of the area are Devonian, Triassic and Quaternary formations. Deposit is located on the north-east outskirts of Upper Silesian Coal Basin within anticline Dziewki-Brudzowice. A part of the deposit, including Devonian dolomites is located in central part of tectonic edge, forming erosive hill characteristic of uneven surface and steep slopes. South and north part of the hill develops into abrasion platform with gradual slopes.

Devonian dolomites are developed in the form of collarian carbonate layout, which is characteristic of grey, dark grey and violent color, crystallinity and huge hardness. The Triassic on the described area is represented by motley sandstone, conchoidal limestone and keupers. The overburden is built of quaternary formations, which consist of: soil, sands and rock waste clay contains dolomite crumbs. Thickness of the layer is about 5 meters. Exploitation is done in mine working on seven levels, beginning from +340 meters above sea level and ending on +280 meters above sea level. The lowest level is flooded with water.

Exploitation is done by wall system on following levels in east, west and north direction. At a distance of 6 m from the border of the plot, a protective shaft is formed from the overburden removed. The surface of the excavation is about 88,7 ha and average depth is around 65 m [5]. Current development of the mine divides its area into four parts. The first of them is partially flooded excavation with technological roads, the second is external dump, the third is processing plant together with raw material and processed products warehouses and the fourth is administration buildings with the board of Górnicze Zakłady Dolomitowe S.A.

4.3 Hydrological factors

The quarry is located in catchment of Vistula river between two rivers - Czarna Przemsza and Brynica. The northern part of the mining area is a watershed for the rivers. Described area is poor in surface natural watercourses. In the nearest neighborhood of the mine there is Jordan stream which is right tributary of Czarna Przemsza. The stream is about 800 meters north-east away from the mine and it is a place of draining water from mine dewatering [7].

In the direct neighborhood of the quarry there is no other aquatic reservoir. However, in 25 km radius there are other big reservoirs of recreational and economic character, among others: Pogoria I, Pogoria II, Pogoria III i Pogoria IV, which are post exploitation excavations of sand mines. According to the division of the National Hydrogeological Service of the Polish territory into groundwater bodies (JCWPd), the area is located in the Central Vistula Subregion, ie JCWPd No. 112. The quarry is located in the northern part of the Main Underground Water Reservoir No. 454 Olkusz - Zawiercie and in the southern part part of the Main Underground Water Reservoir No. 337 Lubliniec - Myszków. Average depths of water intakes in the areas of reservoirs range from 100 m to 135 m. However, the average depth of groundwater deposition in the mining area varies within about 2 m in the southern part to about 10 m in the northern part of the area [7]. Currently dewatering is performed up to +280 meters above the sea level.

4.4. Formal-law factors

The area being the subject of the study is covered by both the study of conditions and directions of spatial development as well as the local spatial development plan for the mining area. Those documents imply the direction and way of developing the lands on which the mine is located. The study of conditions and directions of spatial development of the town and borough of Siewierz determines water direction of excavation reclamation. However, areas adjacent to the excavation located in the Górnicze Zakłady Dolomitowe are destined to afforestation. The area where the buildings are located are designed for commercial and production activity [8].

Local spatial development plan for mining area predicts water direction of target redevelopment of the excavation. For the dumps forest reclamation direction is decided. The buildings, roads and squares could be used for economic purposes. Created reservoir is to have a role of storage reservoir without a possibility to use it for recreational purposes. The ground mass is to be used for formation of target land form. Moreover, isolating green zones were designed as well as open spaces destined for meadows, pastures and crops [9].

The area of the quarry is not under any form of environmental protection in the understanding of the bill from 16th of April 004 about environmental protection (tekst jedn. z 2016 r. Dz.U. z 2015 poz. 1651) and it does not belong to European ecological net Nature 2000. On the area of the mine there is no monumental objects included in the Registry of Monuments of Silesia Voivodship.

4.5. Environmental factors

The mine is located on the lands greatly covered by forests, mainly coniferous forests. Moreover, in the river valleys there are ash-alder forests as well as currant alder and peaty alder forests. However, on the calcic base surface turfs are developed. Important element of natural world are also meadows [7]. On the south and north slopes of the excavation young trees such as: birch, spruce, pine and willow are located. Moreover, by the office and warehouse decorative species of shrubs and trees were introduced such as: hedgerows or larches. Surrounding of the mine consists of coniferous and birch tree stand, crop fields and wastelands of meadow character.

Fauna and flora on the quarry area is very poor because of the opencast dolomite exploitation. The nearest surrounding of the quarry has anthropogenic character as well because of communication infrastructure and housing. Current development of the area makes it quite difficult to create suitable conditions for animal habitat. The significant part of soil in the mine area was extracted because of making deposit available and building technical infrastructure necessary for functioning of the mine. Extracted soil is stored in the form of protection banks, giving a possibility of using it again during the reclamation process.

In order to protect stored soil and soil located around the mine much work is performed if comes to suitable waste and wastewater management as well as taking care of technical state of machines and devices so as to mineralized their negative influence on the environment [7].

4.6. Cultural-historical factors

Siewierz borough is one of the oldest settlement in Poland and has very rich history. Someday it was a capital of principality and developed trade and industry center located on the crossroad of many significant trade routes. Buyers traditions reach Middle Ages times. There are many historical and cultural heritage objects in the town such as for example: The Castle of Silesian Princes, the main square, chapels and sacral objects. All those monuments are located at least 3 kilometers away from the mine [10].

All devices, machines and buildings on the mining area cannot be qualified as monuments even in common undertaking of this word because of relatively young age. Nevertheless because of direct neighborhood of Upper Silesia, where mining culture and tradition are strongly developed and rooted, all material goods connected with mining are perfectly written into industrial heritage of the region.

Historical character of the borough and great number of objects of recreational-cultural character creates a question about the point of creating another area of high touristic potential. It is worth remembering though that because of long exploitation period – the mine started its activity in 1954, this area is characterized by mining identity, which is worth underlining and cultivating.

4.7. Social factors

Demographic structure on the studied area is very varied. According to date of GUS from 2014 [11], 12284 people lived in the area of town-village borough Siewierz. Women are the majority among residents. The level of unemployment is lower than in the county. Especially the employment of women is on much better level according to statistics from 2015, what may be connected with majority of services in the local market and small part of industry in the borough's economy. Economy of Siewierz is in much better position comparing to neighboring boroughs [8].

Because of village-town character of the borough, various social needs need to be taken into consideration. Expectations of Siewierz town residents will be different to nearby villages residents. The quarry is located on the outskirts of the town in the direct neighborhood of Brudzowice and Dziewki village, residents of which are employed in the mine. Agricultural activity decreases year by year, therefore for this social group it is greatly important and essential to keep and create new work places. The possibility of undertaking work in small distance from the place of residence is also beneficial, what is worth taking into consideration on the level of giving new function to post mining area.

4.8. Economic factors

Economic factors play greatly significant role in the process of reclamation and revitalization. Mining entrepreneur is responsible for reclamation of the quarry according the law. Górnicze Zakłady Dolomitowe S.A. have a license for mineral exploitation given by the Minister of Environmental Protection, Natural Resources of Minister of Forestry. Therefore, according to the bill from 9th of June 2011 Geological and Mining Law (tekst jedn. z 2016 r. Dz.U. z 2015 poz. 196) have the duty to create Mine Liquidation Fund and gathering there the means designed for liquidation of the mine and reclamation of the grounds. Because of the water-retaining destination of the reservoir [5], protection and adjusting of banks and areas around the excavation will demand less work, and so less expense, than in the case of recreational use. Introduction of greenery will not demand brining special soil, because good quality soil which was taken off during making deposit available is storage by the excavation and is destined for reclamation purposes. This aspect also will decrease expenses necessary to perform the reclamation work.

New functions of the area should be designed in such way to make it possible that they will finance themselves. What is more, it is necessary to use significant potential in the form of facilities and buildings located in the borders of Górnicze Zakłady Dolomitowe S.A., which now are in good state. Their adaptation will cause decrease of reclamation costs, and may also bring potential private investors, since adaptation will be connected with lower costs than building new facilities.

5 The choice of reclamation direction

The analysis of particular factors influencing the choice of reclamation direction has shown preferences and restrictions concerning future redevelopment of the area. According to postulates of AHP methods the main aim is to work on the optimal concept of the reclamation of the dolomite quarry in Siewierz. Detailed analysis of all the factors makes possible multidirectional development of dolomite quarry emerge, such as:

- Option 1: Water direction – water-retaining for the excavation, forest-landscape for the areas around the excavation and didactic for area of processing facilities, warehouses and administration buildings.
- Option 2: Water direction – water-retaining for the excavation, forest-landscape for the areas around the excavation and business-service for the area of processing facilities, warehouses and administration buildings.
- Option 3: Water direction– water-retaining for the excavation, forest-landscape for the areas around the excavation and economic-industrial for the area of processing facilities, warehouses and administration buildings.

Optimal development of the lands of processing facilities, warehouses and administration buildings emerges at least three mentioned above possibilities. According to the AHP method after comparing all the factors in pairs done by the author, the most important factors (with the highest priority) based on previous analysis are decided to be: economic, formal-law, social and environmental. Local law acts forbid recreational use of reservoir and determine target destination of the lands located in the borders of Górnicze Zakłady Dolomitowe S.A. Water retaining redevelopment of excavation and afforestation of the surrounding areas comes from urban planning documents of the borough. Prohibition for recreational usage of the reservoir may be connected with high concentration of lakes located in nearby neighborhood of described area what is connected with the lack of the point of creating next place for recreational purposes. Moreover, there is already many recreational lands in Siewierz. Because of economic factors it is demanded to keep the investment profitable as it will allow to attract the investors and keep the investment. Good prospering of introduced investments relates directly to social factors by creating new work places. Obviously other factors must be also taken into consideration, such as for example environmental. Analysis of previously described options of development in the face of assumed hierarchy of factors deciding about optimal direction of revitalization leads to the choice of option 2. The process was shown in the figure 3.

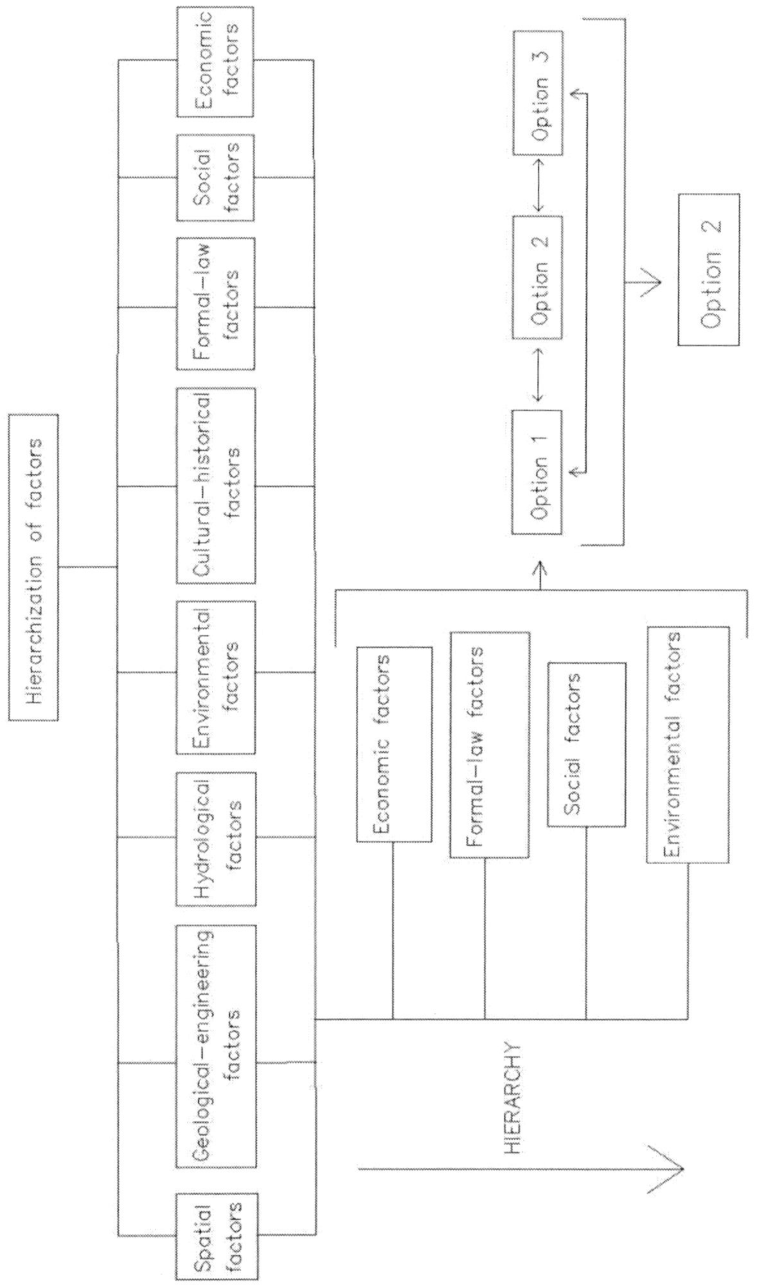

Fig. 3. The use of elements of AHP method to select the optimal concept.

Water direction – water-retaining for the excavation, forest-landscape for the areas around the excavation and business-service for the area of processing facilities, warehouses and administration buildings seems to best meet basic postulates of the process of reclamation, as this direction will improve the state of environment, influence positively on economic growth and will bring back spatial order as well as will significantly improve the quality of life of local community. What is more there is a possibility of protection of industrial heritage in the form of machines and mining devices, through compassing them into business park. Such activity will allow underling character, identity and history of the place whole adapting it to new, socially and economically useful functions at the same time.

6 The concept of the reclamation of dolomite quarry in Siewierz

The new function of the quarry and the surrounding lands assumes creation of the service-business park with business incubators directed toward development and innovation with keeping minding-industrial tradition and identity. All the buildings of the current mine will be adapted and adjusted to the needs of modern business. Apart from the office buildings there will be training and conference rooms. There will also be a possibility of locating research center cooperating with the industry and business. Joining the science and business as well as creation of good and friendly environment for the development of new companies will filly use significant potential of this place. What is more, such type of investment in the place of past quarry will perfectly write into the trend of the region's development with strong accent on industrial and mining tradition which in Silesia is strongly advertised. The area of the mine including the surroundings of the excavation was divided into four zones mutually dependent and interlocking, as shown in the figure 4.

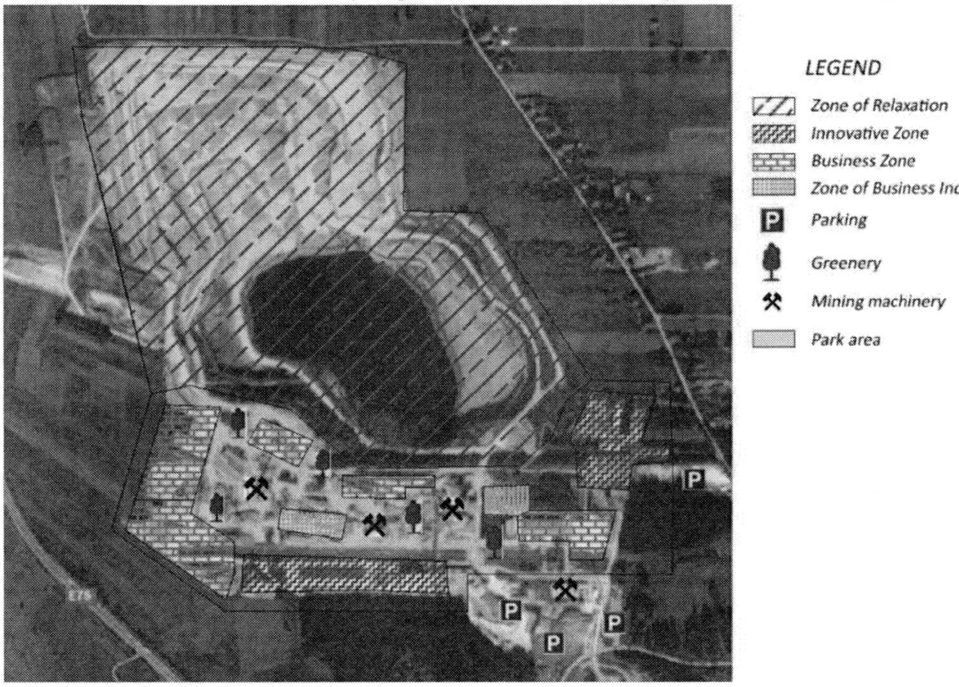

Fig. 4. Development of the quarry area.

All the zones were designed depending on its function, it is necessary to mention the connection thanks to which they are to complete each other. Moreover, they were composed into park, which is a mutual place unifying particular zones. Zone of relaxation surrounds the excavation and is located on the areas destined for greenery. Because of forest-landscape direction of reclamation of those lands there is not much plans for building recreational infrastructure. This zone from the side of excavation will be fenced and protected by wooden barriers, matching the forest surroundings. There will be path planned and at the points of the biggest landscape value there will be some benches. The zone is created with the thought of business park workers who will be able to breathe in fresh air and clam down during their break and inspire by surrounding nature. Moreover, this zone will be arranged in such way so its landscape value will be visible from the office windows what will make the work more pleasant and create friendly working environment. It will be possible thanks to the layer growing i.e. from the building side lower and medium-height greenery is designed and gradually way from the buildings it will move to higher greenery. Essential element of the landscape will be water-retaining reservoir created in the excavation therefore it is important not to cover it with trees and plants. On the area of the whole zone and surrounding areas there will be a silence rule, not to scare animals and keep forest character of the place. Innovation zone is destined for research centers and centers cooperating with universities which will provide qualified staff and transfer of knowledge and with companies which will show needs and practical demand for innovative solutions and technologies. In this zone it will be possible to locate various kinds of laboratories and server rooms as well as auditorium and interactive training rooms, thanks to what it will be possible to exchange knowledge of various levels. Location of Innovative Zone is to provide easy access to it for new firms from business incubators and huge companies from business zone. Business zone are facilities designed for businesses with strong position in the market. There are office facilities as well as such which can be adapted for other purposes, according to needs of the given company. Buildings of this zone are located in four places creating with the buildings of innovative zone the frame of the area where currently administration buildings and processing facilities are located. Buildings located closer to the excavation will be smaller comparing to those in the south of the complex. Such step is to keep spatial order and providing access to green landscape also from the office windows not located directly next to excavation and forest area. The Zone Of Business Incubators are objects designed for young business and small developing firms, which according to idea of business incubators will be provided with support and help of specialists from various fields in making first steps in the market. In the offices of the Zone variously-sized offices will be located, so as to young businessman could rent space adjusted to their needs. There will also be rooms of coworking type where it will be possible to rent a desk with the access to office devices and fast internet. Those facilities are located between the buildings of Business and Innovation Zones, with the aim of cooperation and integration and kind of care over developing business.

Between the buildings of the zones mining machines which remained after the quarry activity will be arranged into the space for keep the memory of history of the place before reclamation. One of the machines will be also located by the entry parking. Such objects are to relate to the past and identity. Moreover, between buildings and machines it is planned to introduce lower greenery and singular trees, which will complete the landscape.

The space between the zones is planned as car free, therefore all parking area at the main entrance from the south side and one by the Innovative Zone on the east part of the area. There will be a map of the complex on each parking making it easy to know where to go. The area of the business park will be space which symbolizes development and unlimited possibilities and freedom of activity. Important element will be putting accent on to integration with nature and relation to mining tradition. It is to be friendly and pro creativity place.

7 Summary

Revitalization of post-industrial areas, including post mining is a very important issue. Quite often huge potential in post-industrial areas is wasted. Therefore, revitalization process is crucial, which may be a chance of regaining old greatness through giving new, socially, environmentally and economical useful functions. The whole area of the dolomite quarry in Siewierz was divided into functional zones. Such step allows to use all benefits and values of diverse space as well as will meet the demands of local spatial planning documents.

Analysis of external and internal conditions showed possibilities and potential as well as restrictions of the described area. On this base three possibilities of future redevelopment of quarry came up. During analytical hierarchical process, the most important criteria of choosing optimal direction of reclamation - economic and formal-law factor - were emerged. Comparison of particular redevelopment possibilities with regard to factors hierarchy showed that the best concept is to be service-business direction with water-retaining development of the excavation and forestry development of the area directly neighboring with it.

Such multidirectional way of revitalization of Górnicze Zakłady Dolomitowe S.A. brings with itself a chance for economic and social growth of Siewierz borough and creating it into modern business-science and people –friendly place and also it will improve the state of the environment as well as creation of spatial order. Multidirectional revitalization will therefore guarantee a synergy effect. Moreover, realizing the reclamation project of the quarry, one may think of the usage of wasteland located in the neighborhood creating complex and integrated program of revitalization of neighboring boroughs which thanks to improving its attractiveness could influence creation of new and worthy point of the Silesian conurbation map.

References

1. Z. Kasztelewicz, M. Ptak, Mining Sience, **132**, 165-175 (2011)
2. Z. Kassztelewicz, S. Hajdo, S. Sypniowski, Problemy Ekologii, **14**, 22-32 (2010)
3. A. Ostręga, *Sposoby zagospodarowania wyrobisk i terenów po eksploatacji złóż surowców węglanowych na przykładzie Krzemionek Podgórskich w Krakowie* (2004)
4. R. Krzysztofik, *Encyklopedia Województwa Śląskiego* (2014)
5. *Plan ruchu odkrywkowego zakładu górniczego* (2015)
6. *Projekt Zagospodarowania Złoża dolomitów triasowych i dewońskich „BRUDZOWICE"* (2015)
7. Raport o oddziaływaniu na środowisko dla przedsięwzięcia: „Odkrywkowa eksploatacja dolomitów triasowych i dewońskich ze złoża „Brudzowice" (2012)
8. Uchwała Nr XX/158/2000 Rady Miejskiej w Siewierzu z dnia 25 kwietnia 2000r. w sprawie: studium uwarunkowań i kierunków zagospodarowania przestrzennego gminy Siewierz (2000)
9. Uchwała Nr XXVIII/206/2004 Rady Miejskiej w Siewierzu z dnia 30 września 2004r. w sprawie: miejscowego planu zagospodarowania przestrzennego dla terenu górniczego kopalni dolomitu ze złoża "Brudzowice" oraz terenu przyległego w Gminie Siewierz (2004)
10. https://www.siewierz.pl
11. *Statystyczne Vademecum Samorządowca* (2015)

Analysis of hard coal quality for narrow size fraction under 20 mm

Tomasz Niedoba[1], and *Paulina* Pięta[1,*]

[1]AGH - University of Science and Technology, Faculty of Mining and Geoengineering, Department of Environmental Engineering and Mineral Processing, 30 A. Mickiewicz Av., 30-059 Kraków, Poland

Abstract. The paper presents the results of an analysis of hard coal quality diversion in narrow size fraction by using taxonomic methods. Raw material samples were collected in selected mines of Upper Silesian Industrial Region and they were classified according to the Polish classification as types 31, 34.2 and 35. Then, each size fraction was characterized in terms of the following properties: density, ash content, calorific content, volatile content, total sulfur content and analytical moisture. As a result of the analysis it can be stated that the best quality in the entire range of the tested size fractions was the 34.2 coking coal type. At the same time, in terms of price parameters, high quality of raw material characterised the following size fractions: 0-6.3 mm of 31 energetic coal type and 0-3.15 mm of 35 coking coal type. The methods of grouping (Ward's method) and agglomeration (k-means method) have shown that the size fraction below 10 mm was characterized by higher quality in all the analyzed hard coal types. However, the selected taxonomic methods do not make it possible to identify individual size fraction or hard coal types based on chosen parameters.

1 The market of small and medium-sized hard coal products in Poland

The IEC 2016 results indicate that fossil fuels remain dominant primary energy sources in the world as they will provide about 78% of total energy consumption in 2040. Globally, coal is still the second most frequently used energy source in the world, after oil and before natural gas, renewable energy, nuclear energy and lignite energy. Despite all trends, coal remains the main fuel in the power sector in Poland, both among small customers, as well as industrial plants.

In Polish hard coal processing plants, the enrichment of raw material in the full range of size is abandoned. One reason for this approach is prices of final products and market fluctuations. It is obvious that the use of higher quality parameter coal products have economic and environmental justification. Comparing unit prices of coal products for the same granulation and different calorific values, the higher cost of better quality assortments

* Corresponding author: ppieta@agh.edu.pl

is only apparent. The real difference in gaining the same amount of energy from enriched coal and raw coal with the same granulation oscillates around 14% (in case of fine products). However, the additional costs are incurred during a purchase of lower quality coal, for example transport cost, which will constitute a disadvantage.

The pro-ecological actions of the government and the EU have resulted in a high demand for high-quality coal, called peas for commercial purposes, supplying retort boilers to private consumers, which has resulted in higher price and limited supply of the coal product since 2007. Facing this problem, the Institute of Chemical Processing of Coal has developed a technology that allows efficient burning of the fine coal in the retort boilers. The production of coal under 20 mm is maintained at a constant high level and the cost of fine is 10-15% lower than the peas [1-4]. However, the quality of coal types depends not only on maintaining the technological regime and a compliance with the developed control systems in coal processing plants, but it is largely determined by the natural characteristics of raw material, deciding on the possibilities of its using.

2 Characteristic of raw materials

The industrial suitability of fine-grained coal products for power and coking plants depends on many factors, but the most important ones are: calorific value, ash content or sulfur content. The basis for the rational use of coal in the combustion and coking process is knowledge of its qualitative characteristics and its correct assessment [1, 4].

Coal consists of combustible organic matter and ballast, which is a moisture and mineral substance. These are analytical components, the content of which is respectively within the following limits: 50-98%, 1-18%, 2-30%. The organic and mineral substances come in many different chemical compounds and therefore comprehensive characterization can cover more than 40 individual determinations, which entail considerable financial effort and they are labor-intensive.

Heterogeneous coal construction is closely linked to the formation of the raw material. This includes such factors as: type of plant material, time, temperature, pressure, humidity and type of microorganisms. What serves as confirmation is the significant quality differentiation of hard coal in Polish coal basins, mines or even stratum exploited in a single mining plant [2, 5].

The commercial and industrial assessment of hard coal requires the determination of certain analytical parameters characterizing the energy source in terms of its thermal resistance, ballast quantity and calorific value. In order to examine these characteristics a technical analysis is carried out under which the moisture content, ash and volatiles, as well as the heat of combustion and calorific value are determined. These properties constitute an important element in settlements between mining plants and consumers [5, 6].

Quantitative and qualitative characteristics of the basic components of coal and its physicochemical properties are an important element for selecting an optimal processing and use method. As the range of end users is wide and the properties of the raw material are different, there exists i.a. the Polish classification of hard coal which systematizes particular information (PN-ISO 11760: 2007P). However, in the era of the development of modern and highly efficient combustion technology of the discussed solid fuel and the deterioration of the quality of deposits, it is necessary to analyze thoroughly the properties of fine coal fractions, whose market share is predominant.

3 Materials and methods

The paper presents the results of analyzes of differentiation of hard coal properties in narrow size fraction. Raw material samples were collected in selected mines of Upper Silesian Industrial Region and they were classified according to Polish classification as types: 31, 34.2 and 35. As a result of the sieve analysis, 9 size fraction were obtained: 0-1, 1-3.15, 3.15-6.3, 6.3-8, 8. -10, 10-12.5, 12.5-14, 14-16, 16-20 [mm] for each sample. They were further characterized by the following characteristics: density (ρ), ash content (A), calorific value (Q), volatile content (V), total sulfur content (S) and analytical moisture (W).

The complex nature of hard coal is described by the 7-dimensional vector of properties of the grained material $\overline{W}$ (1), in which measurable and non-measurable variables are included [7-10]:

$$\overline{W} = [coal\ type, d, \rho, A, Q, V, S, W] \tag{1}$$

The application of multidimensional characterization of the raw material allows numerous relationships and a more detailed description of the properties to reach better understanding of the topic. Problems of mineral processing were considered according to the multidimensional description in the following work [7, 11-16].

The taxonomic methods were used to order and classify the experimental data concerning size fractions of various types of hard coal. It is not a very common way of consideration the nature of the mineral processing issues, but cluster analysis has been used in following work successfully [17-24].

The application of multidimensional comparative analysis requires the formation of a Z (2) unitarized data matrix. The process of unitarization is an extremely important element from the point of view of elimination of the influence of scale on the results of grouping and agglomeration. Based on these values, the taxonomic distance matrix D (square matrix of n×n) is defined (4). There is a large number of metrics that use a different approach to transform the description of phenomena from n-dimensional space into a straight line. The Euclidean distance d_{ij} (3) was used in the paper as a measure of the variability of the hard coal pairs studied. Another element of cluster analysis is the selection of the appropriate hierarchical method, which is the grouping algorithm of the analyzed cases. For the purpose of the analysis of hard coal quality and the creation of a dendrogram, the Ward method (5) was used, where the sum of squared deviations within the assumed groups is minimized. A detailed description of the steps involved in the implementation of different taxonomic methods can be found in the following work [11, 18, 25].

$$Z = \begin{bmatrix} z_{11} & z_{12} & \cdots & z_{1l} \\ z_{21} & z_{21} & \cdots & z_{2l} \\ \cdots & \cdots & \cdots & \cdots \\ \cdots & \cdots & \cdots & \cdots \\ z_{n1} & z_{n2} & \cdots & z_{nl} \end{bmatrix} \tag{2}$$

$$d_{ij} = \sqrt{\sum_{p=1}^{n} \left(z_{pi} - z_{pj}\right)^2} \tag{3}$$

$$D = \begin{bmatrix} 0 & d_{12} & \cdots & d_{1l} \\ d_{21} & 0 & \cdots & d_{2l} \\ \cdots & \cdots & \ddots & \cdots \\ d_{l1} & d_{l2} & \cdots & 0 \end{bmatrix}_{[n\times n]} \quad (4)$$

$$ESS = \sum_{i=1}^{k} \left(x_{ij} - \bar{x}_j\right)^2 \quad (5)$$

where: n – number of examined cases, i.e. coal types and size fractions,
l – number of random variable, i.e. number of parameters,
z_{ij} – unitarized value of x_j property, which was calculated according to formula: $z_i = \frac{x_i}{x_{max}-x_{min}}$,
x_i – the value of a single property measurement,
x_{max}, x_{min} – the maximum and minimum value of the property,
x_{ik} - the value of the property for the each cluster at the mean value of the characteristic $\bar{x}$,
k - the number of elements in a cluster.

4 Results and discussion

In order to present the chosen question a qualitative analysis of selected coal types in a narrow range of particle size less than 20 mm was performed. A cluster analysis was used to investigate the structure of hard coal quality parameters. Tab. 1 summarizes data as a starting point for further assessing the variability of size fraction properties.

Table 1. The matrix of unitarized value of qualitative coal parameters.

Coal type (d)	ρ	Q	A	S	V	W
31 (1)	3.037974	1.232133	0.444946	0.250241	1.750197	1.191265
31 (3.15)	2.849456	1.353137	0.297351	0.208513	1.856297	1.121108
31 (6.3)	3.145346	1.308598	0.330727	0.263389	1.878694	0.930461
31 (8)	3.244557	1.077405	0.695272	0.179093	1.431426	0.788933
31 (10)	3.552371	0.838955	0.95161	0.187945	1.284642	0.74155
31 (12.5)	3.486440	0.861240	0.938284	0.217675	1.301985	0.731613
31 (14)	3.542248	0.757164	1.025184	0.164913	1.297361	0.65390
31 (16)	3.607795	0.773379	1.031595	0.302379	1.257185	0.67845
31 (20)	3.736203	0.664083	1.154617	0.168892	1.179073	0.680875
34.2 (1)	2.736203	1.664083	0.154617	0.107378	1.796995	0.191265
34.2 (3.15)	3.070923	1.393565	0.432640	0.126284	1.430637	0.236474

34.2 (6.3)	2.748028	1.599577	0.223936	0.090345	1.670455	0.311218
34.2 (8)	2.808994	1.540958	0.280232	0.087022	1.612480	0.329088
34.2 (10)	2.841770	1.521321	0.324632	0.088438	1.576281	0.400133
34.2 (12.5)	2.849267	1.514370	0.320033	0.07742	1.57281	0.298834
34.2 (14)	3.368400	1.055552	0.727415	0.115825	1.277294	0.251008
34.2 (16)	2.955492	1.644815	0.330215	0.089411	1.466929	0.290437
34.2 (20)	2.764356	1.426877	0.41377	0.091261	1.466126	0.355175
35 (1)	2.884867	1.498100	0.352919	0.100936	1.192387	0.312277
35 (3.15)	3.047743	1.340802	0.510616	0.100471	1.121604	0.310905
35 (6.3)	3.329298	1.141705	0.714939	0.109130	1.002651	0.404173
35 (8)	3.502603	0.986963	0.879483	0.103406	0.939860	0.388168
35 (10)	3.407665	1.058374	0.805628	0.095228	1.022796	0.416712
35 (12.5)	3.515951	0.973397	0.897462	0.092415	0.93192	0.389232
35 (14)	3.734799	0.798653	1.074937	0.074777	0.885933	0.417079
35 (16)	3.711644	0.799705	1.079979	0.077208	0.881488	0.386538
35 (20)	3.735637	0.787784	1.104560	0.074470	0.878694	0.392187

The most direct method of calculating the distance between the coal pairs in a multidimensional space is a determination of Euclidean distance (3). On the basis of this method, agglomeration of Ward method was conducted (5). The algorithm allowed to find coal types that were the closest in terms of size fractions quality. A graphical interpretation of this method is the dendrogram (Fig. 1). On the other hand, the stages of drawing up the agglomeration graph are shown in Fig. 2.

The ratio for determining the quality of the each coal type is the taxonomic distance (6):

$$DE = d_j = \sqrt{\sum_{i=1}^{n}\left(1 - s_{ji}\right)^2} \tag{6}$$

where: $s_{ji} = \frac{z_{ji}}{z_{jmax}}, i = 1, \ldots, l; j = 1, \ldots, n.$

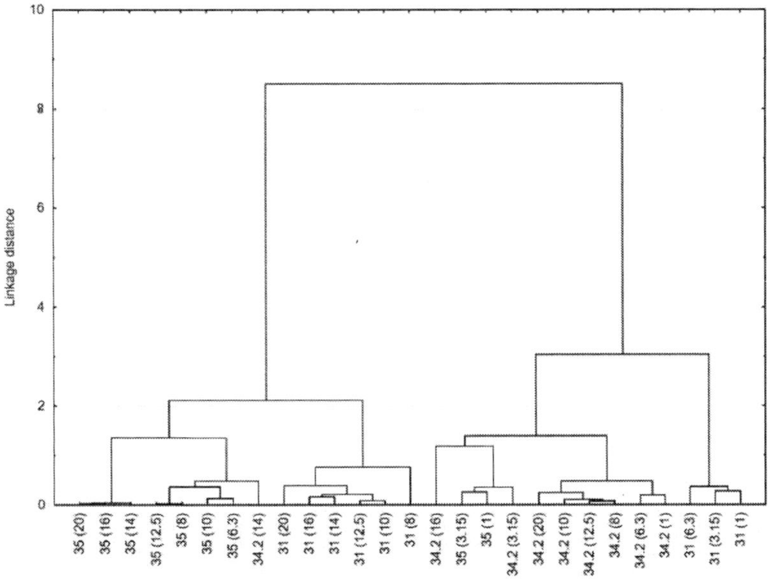

Fig. 1. Dendrogram of examined size fractions and coal types (Ward method).

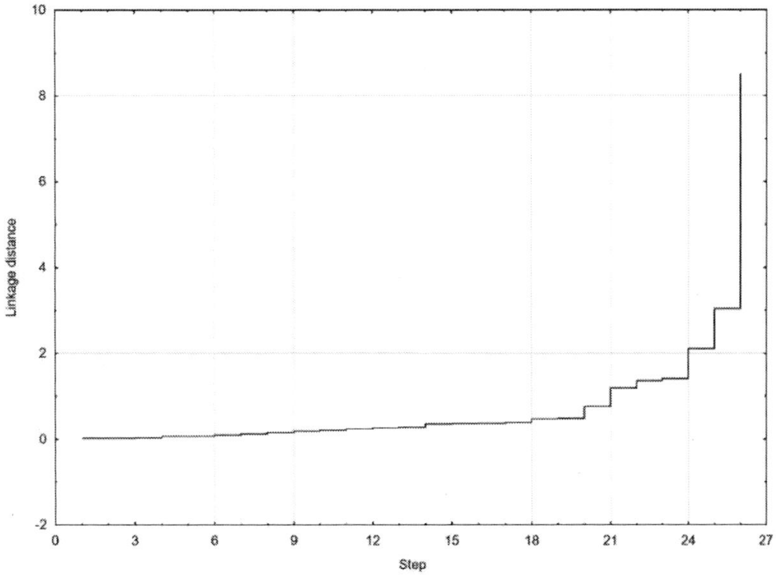

Fig. 2. Steps of dendrogram creation.

Table 2. Summary data to assess the quality of the size fraction and coal types.

Coal type (d)	1/ρ	Q	1/A	1/S	V	1/W
31 (1)	0.329167	1.232133	2.247464	3.996154	1.750197	0.839444
31 (3.15)	0.350944	1.353138	3.363025	4.795864	1.856297	0.891975
31 (6.3)	0.31793	1.308599	3.02364	3.796666	1.878694	1.074736

31 (8)	0.308208	1.077405	1.438286	5.583685	1.431426	1.267535
31 (10)	0.281502	0.838955	1.050841	5.320704	1.284642	1.348511
31 (12.5)	0.286826	0.86124	1.065776	4.594012	1.301985	1.366843
31 (14)	0.282307	0.757164	0.975434	6.063814	1.297361	1.529076
31 (16)	0.277178	0.773379	0.969373	3.307105	1.257185	1.473929
31 (20)	0.267651	0.664083	0.866088	5.920932	1.179073	1.468699
34.2 (1)	0.36547	1.664083	6.467604	9.312868	1.796995	5.228359
34.2 (3.15)	0.325691	1.393565	2.31139	7.918671	1.430637	4.228801
34.2 (6.3)	0.363897	1.599578	4.465554	11.06864	1.670455	3.21318
34.2 (8)	0.355999	1.540958	3.568476	11.49137	1.61248	3.038702
34.2 (10)	0.351893	1.521322	3.080409	11.30732	1.576281	2.49917
34.2 (12.5)	0.350967	1.51437	3.124675	12.91507	1.57281	3.346343
34.2 (14)	0.296877	1.055552	1.374731	8.633729	1.277295	3.983941
34.2 (16)	0.338353	1.644815	3.028325	11.18425	1.466929	3.443088
34.2 (20)	0.361748	1.426877	2.416802	10.95753	1.466126	2.815512
35 (1)	0.346636	1.4981	2.833512	9.907242	1.192387	3.202282
35 (3.15)	0.328112	1.340802	1.958421	9.953124	1.121604	3.21642
35 (6.3)	0.300364	1.141705	1.39872	9.163358	1.002651	2.474187
35 (8)	0.285502	0.986964	1.137032	9.670653	0.93986	2.576205
35 (10)	0.293456	1.058374	1.241267	10.50112	1.022796	2.399739
35 (12.5)	0.284418	0.973397	1.114253	10.8207	0.931929	2.569164
35 (14)	0.267752	0.798653	0.930287	13.37308	0.885934	2.397627
35 (16)	0.269422	0.799705	0.925944	12.95202	0.881488	2.587067
35 (20)	0.267692	0.787784	0.905338	13.42814	0.878694	2.549803

The results of the analysis of the price parameters of the raw materials are presented in Table 3.

Table 3. Evaluation of hard coal quality based on Euclidean distance.

Coal type (d)	DE	Coal type (d)	DE	Coal type (d)	DE
31 (1)	1.306034	**34.2 (1)**	0.309537	**35 (1)**	0.825045
31 (3.15)	1.169706	**34.2 (3.15)**	0.844396	**35 (3.15)**	0.954841
31 (6.3)	1.221354	**34.2 (6.3)**	0.537645	**35 (6.3)**	1.157722
31 (8)	1.313538	**34.2 (8)**	0.650659	**35 (8)**	1.215767
31 (10)	1.419609	**34.2 (10)**	0.778689	**35 (10)**	1.171355
31 (12.5)	1.431495	**34.2 (12.5)**	0.659067	**35 (12.5)**	1.207031
31 (14)	1.402514	**34.2 (14)**	1.037122	**35 (14)**	1.283496
31 (16)	1.503578	**34.2 (16)**	0.693524	**35 (16)**	1.269353
31 (20)	1.466693	**34.2 (20)**	0.841281	**35 (20)**	1.2784

* shaded fields mean coals with the best costogenic parameters

The best quality characterizes the coking coal (34.2) with particle size of 0-12.5 mm and 14-20 mm, and the second coking coal (35) in narrower granulation (0-3.15 mm). For these cases the index values are the lowest.

Subsequently, the k-means grouping was performed, in which the algorithm assigned the cases as described in Tab. 4. Fig. 3 shows the distribution of mean values in both groups.

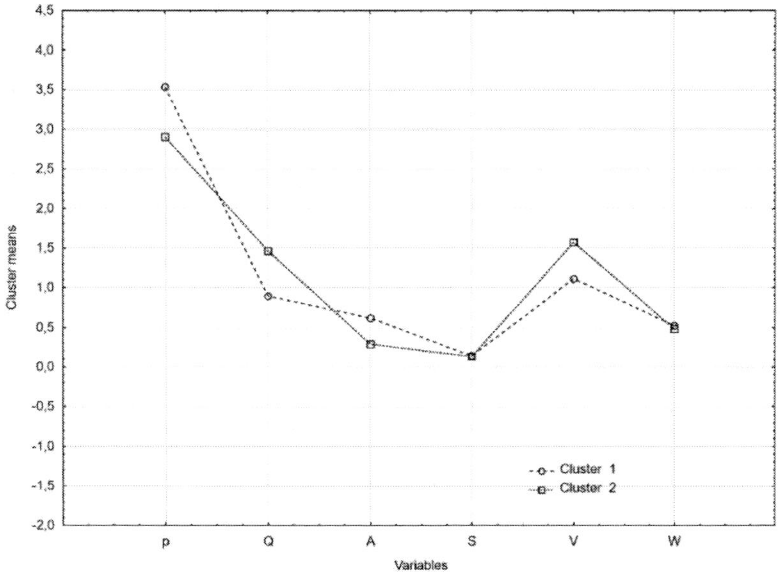

Fig. 3. Distribution of mean values of parameters in cluster assigned k-means method.

Table 4. Summary cases grouped in two clusters by k-means method.

Cluster 1	Claster 2
31 (8) 31 (10) 31 (12.5) 31 (14) 31 (16) 31 (20) 34.2 (14) 35 (6.3) 35 (8) 35 (10) 35 (12.5) 35 (14) 35 (16) 35 (20)	31 (1) 31 (3.15) 31 (6.3) 34.2 (1) 34.2 (3.15) 34.2 (6.3) 34.2 (8) 34.2 (10) 34.2 (12.5) 34.2 (16) 34.2 (20) 35 (1) 35 (3.15)

Cluster no. 1 includes the following size fractions: 8-20 mm (31 coal type), 12.5-14 mm (34.2 coal type), 6.3-20 mm (35 coal type), which exhibit poorer quality as compared to the other cases. In cluster area no. 2 there are: energetic coal 31 type with granulation 0-6.3 mm, a coking coal 34.2 type with granulation 0-12.5 mm and 14-20 mm, and a coking coal 35 type with particle size 0-3.15 mm. Based on Fig. 3, it was found that group no. 1 had higher mean density values and ash content, while group no. 2 was characterized by higher calorific content and volatile content. Parameters such as moisture and sulfur content did not significantly affect the differentiation between the coals. The following parameters were most important in grouping the chosen method: density, calorific value, ash content and volatiles. These conclusions were obtained through an analysis of variance (Tab. 5), which is an extremely useful tool for investigating the impact of certain factors of the phenomenon under consideration.

Table 5. Analysis of variance.

Variables	Test F	Probability level p	
p	121,8886	0,000000	The verification of assumptions: $p<0.05$
Q	108,5164	0,000000	
A	20,1889	0,000139	
S	0,1844	0,671288	
V	30,4180	0,000010	
W	0,1435	0,708032	

Conducting further qualitative analysis of the distinguished clusters of coals and size fractions a more precise segregation of the raw material was possible. In cluster no. 1 with poorer quality parameters, 3 subgroups were diversified, mainly in terms of ash content, caloric content and volatiles content. Among them the best quality was in the size fractions 6.3-8 mm (31), 12-14 mm (34.2) and 3.15-10 mm (35), slightly worse parameters were recorded in 35 coking coal with granulation 6.3-20 mm, whereas energetic coal with granulation 8-20 mm was characterized by the weakest quality. In cluster no. 2 also an internal division of cases into 3 subgroups was possible. According to the results of assessment based on Euclidean distance, the first and the highest quality subgroup consisted of 34.2 coking coal practically throughout the whole grain size range, excluding one size fraction (12-14 mm), and 35 coking coal (1-3.15 mm). Other types of coal were segregated into two separate internal clusters. The weakest quality was energetic coal with a granulation 0-3.15 mm, while the size fraction 3.15-6.3 mm of the same raw material was characterized by a slightly higher quality. Three parameters proved to have the greatest influence on the division within the cluster no. 2: moisture content, sulfur content, calorific value.

5 Conclusions

The applied methods of cluster analysis enabled detailed examination of hard coal quality in narrow size fractions, as well as their classification. However, the selected taxonomic methods do not make it possible to identify individual size fractions or types of hard coal based on chosen parameters. As a result of the analysis it can be stated that the 34.2 coking coal in the whole range of grain can be characterized by the best quality, at the same time in terms of price parameters the size fractions 0-3.15 mm of 35 coking coal type and 0-6.3 mm of 31 energetic coal matched. The methods of grouping and agglomeration have shown that in all the analyzed hard coal types the size fractions below 10 mm were distinguished by higher quality.

The good quality of fine coal is one of the reasons why enrichment operations in the full range of grain size have not been introduced in some Polish hard coal preparation plants. However, more and more price parameters of hard coal are deteriorating, and the expanding operation includes coking coal decks, to which the technological systems of processing plants are not adapted largely. In contrast, in the structure of production of commercial sorts, a fine coal accounted for a significant part, about 80-90%. An accurate technical analysis of hard coal with a granulation 0-20 mm is an important element due to the continuing structure of coal production and the need to adapt domestic furnaces and boilers for solid fuels produced.

References

1. W. Wawrzynkiewicz, Journal of the Polish Mineral Engineering Society, **4**(1), 27-37 (2003)
2. T. Olejnik, A. Surowiak, T. Gawenda, T. Niedoba, T. Tumidajski, AGH Journal of Mining and Geoengineering, **4**(1), 207-216 (2010)
3. J. Zawistowski, S. Janiszewski, Rynek Instalacji, **7**(8), 52-56 (2010)
4. J. Grabowska, Logistyka, **2**, 579-588 (2012)
5. A. Karcz, *Coke engineering Vol. I*, (AGH, Cracow, 1991)
6. A. Surowiak, Journal of the Polish Mineral Engineering Society, **32**, 153-158 (2013)
7. T. Tumidajski, MRM, **23**(6), 111-123 (2010)
8. T. Niedoba, A. Surowiak, Proceedings of the XXVI International Mineral Processing Congress, Vol. **1**, 3844-3854 (2012)
9. T. Niedoba, Bulletin of the Mineral and Energy Economy Research Institute of the Polish Academy of Sciences, Cracow (2013)
10. T. Niedoba, Physicochem Probl Mi, **49(1)**, 175-188 (2013)

11. C. Aldrich, *Exploratory Analysis of Metallurgical Process Data with Neural Networks and Related Methods. Process Metallurgy 12* (Elsevier Science B.V., Netherlands, 2002)
12. T. Niedoba, *Wielowymiarowe charakterystyki zmiennych losowych w opisie materiałów uziarnionych i procesów ich rozdziału* (Wydawnictwo Instytutu Gospodarki Surowcami Mineralnymi i Energią PAN, Kraków, 2013)
13. T. Niedoba, D. Jamróz, Arch. Min. Sci., **58(4)**, 1317-1331 (2013)
14. F. Nakhaei, M. Irannajad, A. Sam, A. Jamalzadeh, Physicochem. Probl. Mi., **52 (1)**, 252-267 (2015)
15. T. Niedoba, A. Surowiak, P. Pięta, J. Pol. Miner. Eng. Soc., **17**(1), 15-22 (2016)
16. T. Niedoba, P. Pięta, A. Surowiak, D. Jamróz, MEC2016, E3S Web of Conferences, 8, 01036 (2016)
17. K. Pulkkinen, R. Ylinen, S.-L. ämsä-Jounela, M. Järvensivu, XVIII International Mineral Processing Congress, Sydney, Parkville, Australia, 325-334 (1993)
18. D.W. Ginsberg, W.J. Whiten, Trans. Instn. Min. Metall., **100**, 139-146 (1991)
19. D.W. Ginsberg, W.J. Whiten, Int. J. Miner. Process., **36**, 63-79 (1992)
20. D.W. Ginsberg, W.J. Whiten, Miner. Eng., **6(1)**, 31-40 (1993)
21. S. Laine, H. Lappalainen, S.-L. Jämsä-Jounela, Miner. Eng., **8(6)**, 637-648 (1995)
22. J.R. Whiteley, J.F. Davis, Comput. Chem. Eng., **16**(4), 329-346 (1992)
23. J.R. Whiteley, J.F. Davis, Comput. Chem. Eng., **18**(7), 637-661 (1994)
24. C. Aldrich,.CEC, **177**(1), 121-137 (2000)
25. A. Stanisz, *The accessible statistics course using STATISTICA PL on examples from medicine. Multidimensional analysis* (StatSoft, Cracow, 2007)

Methane and carbon dioxide emission measurements in selected area of the Polish Western Carpathians

Marzena Mojżysz[1], *Adrianna* Góra[1,*], *Henryk* Sechman[1], *Piotr* Guzy[1], and *Anna* Twaróg[1]

[1]AGH University of Science and Technology, Faculty of Geology, Geophysics and Environmental Protection, Department of Fossil Fuels, 30 A. Mickiewicz Av., 30-059 Kraków, Poland

Abstract. The paper presents measurement results of methane and carbon dioxide emission in Western Carpathians. The investigations were carried out with application of static chamber method in 10 stations. An emission and dynamics of changes within methane and carbon dioxide concentrations were examined by measuring the emission directly from the land surface as well as after drilling an additional hole. In total, 40 gas samples were taken, which were then subjected to chromatographic analysis. The measurements of methane and carbon dioxide emission have not shown significant emission values of these gases. The average value of methane emission from soil to the atmosphere at selected measurement sites was -0.451 $mg*m^{-2}*d^{-1}$, while the average methane emission with the use of the hole, equalled to -0.361 $mg*m^{-2}*d^{-1}$. The changes of carbon dioxide concetration were opposite. Higher emission of this gas was showed due to its natural escaping to the atmosphere with an average concetration of 1829,3 $mg*m^{-2}*d^{-1}$. Conducted studies show that methane and carbon dioxide emission are low in the studied area. It does not confirm the elevated values of light hydrocarbons and associated components which were previously recorded in the soil gas samples.

1 Introduction

The surface geochemical survey are among unconventional measurements methods used mostly for hydrocarbon prospection and environmental protection. For many years now, the Outer Carpathians have been an area of the studies on the effects of hydrocarbons penetration from deep-seated accumulations and the zones of their generation to the land surface. Migration of hydrocarbons from deep accumulations towards the surface is controlled by diffusion and effusion. The effects of these processes are referred to as microseepage [1, 2]. Such phenomena still exist in the Carpathians [3]. In the past, hydrocarbon's seepages were used as direct indicators of the occurrence of subsurface accumulations of oil and gas. The hydrocarbon seepages are typical of petroleum basins of complicated tectonics and are observed mostly in the elevated, axial zones of fold structures, and in dislocation zones [3, 4].

* Corresponding author: agora@agh.edu.pl

The following paper aims to estimate the quantify of methane and carbon dioxide emissions from the soil to the atmosphere in the selected area of the Polish Western Carpathians.

The basic materials and results were presented in MSc. Eng. project prepared by one of the authors [5].

2 Location of study area

The study area is located in the western part of the Outer Carpathians. The measuring points were arranged within the range of the occurrence of three litho-stratigraphic units: Magura, Fore-Magura and Silesian units. These units differ from each other in terms of the development, formations age and the tectonic style.

For gas sampling, ten sites were selected to carry out emission measurement (Fig. 1). They were located along the geochemical profile where soil gas samples had been taken [6]. Methane concentrations in soil gas samples ranged from 2,05 to 37,96 ppm, whereas carbon dioxide from 0,091 to 0,842 vol.%. At locations where inside soil gas samples elevated levels of methane and/or carbon dioxide were detected, emission measurements were made. The second criterion for the selection of emission measuring points were a 100 ml volume of the soil gas samples taken during the geochemical sampling, which indicated the good permeability of the soil at the gas sampling depth of 1.2 m.

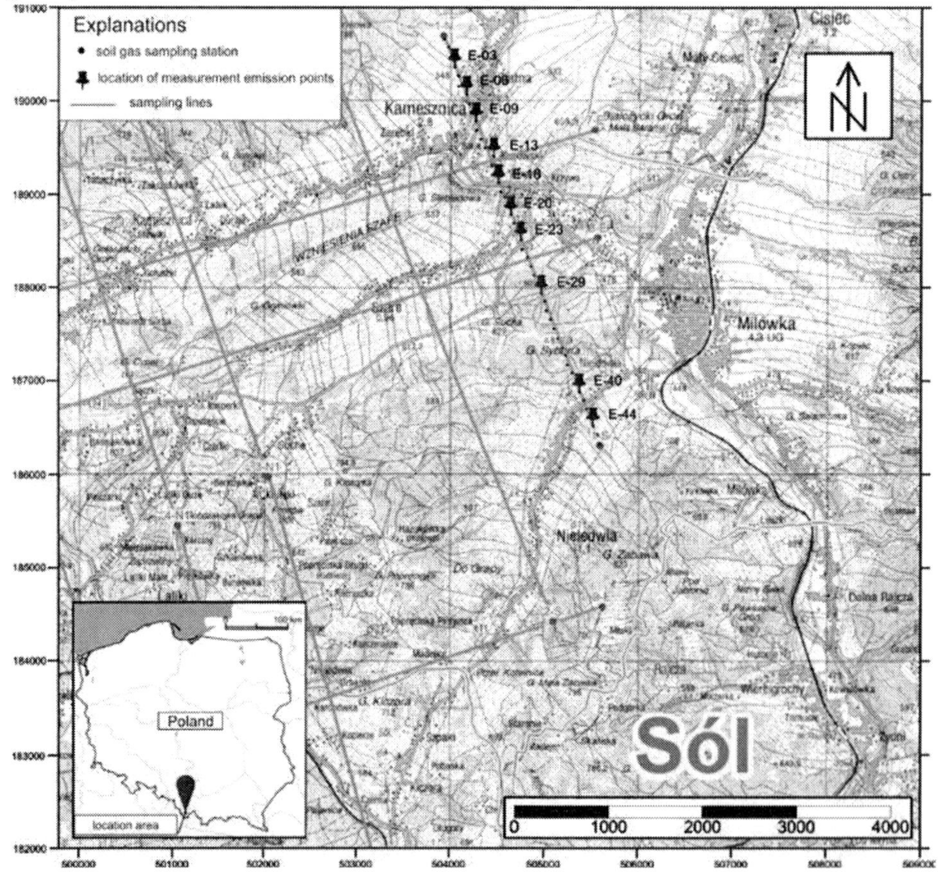

Fig. 1. Location of the soil gas and emission sample stations.

https://doi.org/10.1051/e3sconf/20182900017

3 Methodology

3.1 Field sampling

In the field, the measurement of emission was performed in two variants: natural emission (on the surface of terrain) and with excluding of near-surface zone (with an additional hole). The patented static chamber method was applied [7]. This method is commonly used in the near-surface geochemical survey for determination the gas release from Earth to the atmosphere.

The hole simulates a disruption of the continuity of the layers of rock, which allows for the examination of uncontrolled flow of gas into a closed space. In the case of the measurement of the emission on the land surface without the hole, it is possible to estimate only the amount of methane that has not undergone bacterial destruction and being released into the atmosphere [1]. It is assumed that – depending on the environmental conditions – the largest activity of methanotrophic bacteria occurs up to the depth of approximately 0.8 m [8]. This stratum is considered by some researchers to be a sort of "bacterial filter", that may significantly reduce or completely eliminate the release of methane into the atmosphere [9].

In general, the method applied in the field involves a collection of gas samples from the inner space of closed and sealed chamber adjacent to the land surface in predefined time intervals. The static chamber method does not involve air flow into the chamber. The chamber itself is a vessel of volume 10.8 dm^3 made of stainless steel and it is provided with a sealed opening to allow a sample of gas from the interior volume of the chamber (Fig. 2). The area directly covered by the chamber is 7.1 dm^2.

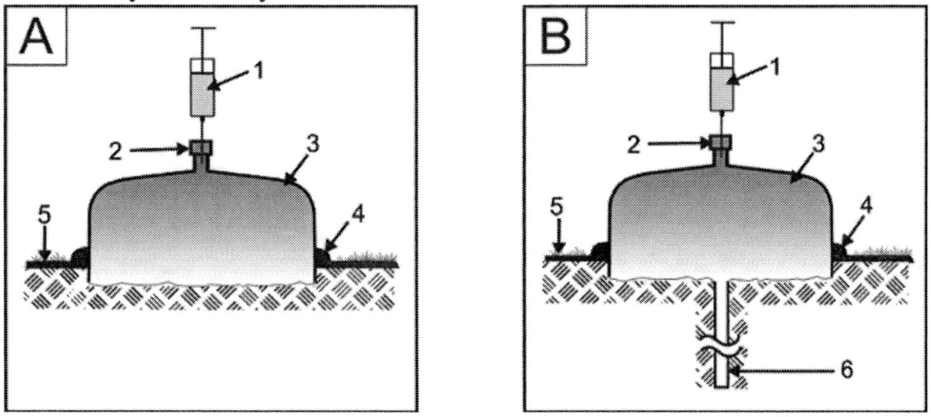

Fig. 2. Measurement methods for natural emission (A) and with the use of a well (B): 1 – gas tight syringe, 2 – rubber diaphragm, 3 – cap of the static chamber, 4 – seal, 5 – land surface, 6 – well.

Before the installation of the chamber, a hole was drilled up to the depth of approximately 0.5 m using a tube with a 40 mm diameter (Fig. 2). The side surface of the 100 cm x 3.14 cm cylinder created by the hole formed an additional emission surface, which was included in the calculations. Additionally, contact between chamber edge and soil was sealed. Afterwards, at each site, gas was sampled immediately after chamber setting, then after 10, 20 and 30 minutes. During the sampling, the atmospheric pressure was measured together with soil temperature at 10 cm depth. Samples of 50 ml were pumped to special glass vessels filled with brine. Totally, 40 gas samples were collected from 10 sampling sites.

3.2 Analytical procedure

Gas samples collected from static chamber method were taken to the laboratory and analyzed using gas chromatography. The analyses of gas samples were focused on detection of methane and its higher homologues (ethane, propane, i-butane, n-butane) as well as on gaseous alkenes (ethylene, propylene, 1-butene) and carbon dioxide. Analyses were carried out at the Department of Fossil Fuels, AGH University of Science and Technology in Kraków with a FISSONS Instruments GC 8160 gas chromatograph equipped with a FID and CARLO ERBA Instruments GC 6300 gas chromatograph equipped with a thermal-conductivity detector (TCD).

First instruments used a metal column (internal diameter 4 mm, length 1.3 m) filled with Activated Alumina (mesh 100/120). The carrier gas (He) flow rate was 60ml/min. The programmed column temperatures were: 80°C for 3 min, 80-200°C increment at 30°C /min rate and 200°C for 3 min. The FID working temperature was 270°C, the injection chamber temperature was 120°C and the volume of each injected sample was 2 ml.

The analyses of CO_2 were carried out with a TCD detector. The following analytical conditions were applied: metal columns (internal diameter 4 mm, length 1.5 m) filled with the HaySep; carrier gas (Ar) flow rate 30ml/min; constant column temperature: 65°C, TCD detector temperature 60°C, TCD filament temperature 260°C and volume of sample was 2 ml injected with an automatic valve.

Both the FID and TCD readings were processed with the Peak Simple integration program. The gas chromatograph was calibrated with „Scott II" gas standard distributed by the Supelco and Alltech. Precision was estimated at 5% of measured values.

3.3 Methodology of emission calculation

The quantities of methane and carbon dioxide emissions were calculated using the formula established during the previous geochemical surveys [10].

$$E^{CH_4} = \frac{V_p}{A} \cdot \frac{M_W \cdot T_0 \cdot p}{M_V \cdot p_0 \cdot (T_0 + t_{gl})} \cdot n^{CH_4} \quad (1)$$

$$E^{CO_2} = \frac{V_p}{A} \cdot \frac{M_W \cdot T_0 \cdot p}{M_V \cdot p_0 \cdot (T_0 + t_{gl})} \cdot n^{CO_2} \quad (2)$$

where:
E^{CH4} – emission of CH_4 ($g*dm^{-2}*min^{-1}$);
E^{CO2} – emission of CO_2 ($g*dm^{-2}*min^{-1}$);
M_W – molecular weight (for CH_4: M_W = 16 g, for CO_2: M_W = 44 g);
M_V – volume of 1 moles of gas in normal conditions ($p_0 = 1013*10^2$ Pa, T_0 = 273 °K) (M_V = 22,4 dm^3);
t_{gl} – soil temperature (°C), (T_0 = 273 °K);
p – atmospheric pressure (Pa);
V_P/A – chamber parameter (V_p – volume [dm^3], A – surface of emission dm^2);
n^{CH4}=dc/dt for $t \rightarrow 0$ (CH_4 concentrations rise rate in the chamber ($ppm*min^{-1}$) obtained from the concentration rise function);
n^{CO2}=dc/dt for $t \rightarrow 0$ (CO_2 concentrations rise rate in the chamber ($vol.\% *min^{-1}$) obtained from the concentration rise function).

The values of the soil temperature and atmospheric pressure in the relations (1) and (2) were used as averaged values from the measurements taken during sampling. This was due to the small variability range of the measured values.

The calculated values of the changes in methane concentrations were given in ppm per hour [ppm/h] and carbon dioxide in vol.% per hour [vol.%/h]. These values were summarized in tables (Table 1,2). The tables also contain the methane and carbon dioxide emission (E^{CH4}, E^{CO2}). Their values were given in milligrams per square meter per day [mg*m^{-2}*d^{-1}].

4 Results and discussion

4.1 Results of natural emission

The dynamics of the changes in the concentrations of methane naturally escape from near-surface zone to the atmosphere (n^{CH4}) is determined by the range from -0.846 to 0.204 ppm/h. The decreasing trends were observed at all locations (Tab. 1). The natural methane emission values are negligible and range from -1.469 to 0.354 mg*m^{-2}*d^{-1}. Only two measuring points recorded positive values. (Tab. 1). The positive emission indicates that methane is continuously released to the atmosphere, while the negative emission are the result of bacterial destruction of this gas in the near-surface zone.

The dynamics of the changes in carbon dioxide concentrations varies from -0.101 to 0.120 vol.%/h. In 20% of sampling sites, the decreasing trends were observed (Tab.1). The natural carbon dioxide emission values are between -4812.8 and 5658.7 mg*m^{-2}*d^{-1}. Two measuring points recorded negative values, which may be related to an inflow of air into the chamber. However, positive emission values are predominant, represents a transfer from soil to the atmosphere.

Table 1. Values of changes of methane and carbon dioxide concentration and natural emission calculated on the basis of measurements with application of the static chamber method.

Item number	Number of site	METHANE		CARBON DIOXIDE	
		Change of concentration n^{CH_4} (ppm / h)	Value of methane emission E^{CH_4} (mg*m^{-2}*d^{-1})	Change of concentration n^{CO_2} (vol.% / h)	Value of carbon dioxide emission E^{CO_2} (mg*m^{-2}*d^{-1})
1	E-3N	-0.846	-1.469	-0.101	-4812.8
2	E-6N	-0.204	-0.355	0.080	3845.4
3	E-9N	0.204	0.354	-0.011	-543.5
4	E-13N	0.168	0.288	0.120	5658.7
5	E-16N	-0.204	-0.355	0.053	2554.1
6	E-20N	-0.336	-0.579	0.053	2501.0
7	E-23N	-0.216	-0.370	0.004	168.2
8	E-29N	-0.708	-1.216	0.040	1871.1
9	E-40N	-0.426	-0.731	0.064	3030.3
10	E-44N	-0.042	-0.072	0.086	4020.4

4.2 Results of emission with an additional hole

The dynamics of concentrations changes of methane (n^{CH4}) after making an additional hole, is expressed by the range from -0.750 to 0.390 ppm/h. In 60% of sampling points, decreasing trends were found (Tab. 2). The methane emission values calculated based on the static chamber method with application of the hole are very similar to the natural emission values and is illustrated by values from -1.133 to 0.316 mg*m^{-2}*d^{-1}.

The dynamics of the changes in carbon dioxide concentrations from the drilled hole ranges from -0,083 to 0,111 vol.%/h. Carbon dioxide showed emission values from –3432,5 to 4657,9 mg*m^{-2}*d^{-1}. The reduction of the emission values after making the hole may be due to the fact that carbon dioxide, which is heavier than the air, may fall gradually during the measurement onto the bottom of the drilled well.

The calculated values of natural emission and with excluding of near-surface zone are comparable to the values obtained in different areas by other researchers, including in the Carpathians [2, 10, 11].

Table 2. Values of changes of methane and carbon dioxide concentration and emission with an additional well calculated on the basis of measurements with application of the static chamber method.

Item number	Number of site	METHANE		CARBON DIOXIDE	
		Change of concentration n^{CH_4} (ppm / h)	Value of methane emission E^{CH_4} (mg*m^{-2}*d^{-1})	Change of concentration n^{CO_2} (vol.% / h)	Value of carbon dioxide emission E^{CO_2} (mg*m^{-2}*d^{-1})
1	E-3W	-0.072	-0.110	-0.080	-3342.8
2	E-6W	-0.372	-0.568	0.111	4657.9
3	E-9W	0.390	0.032	0.025	55.3
4	E-13W	0.210	0.316	0.060	2482.3
5	E-16W	0.060	0.092	-0.054	-2266.0
6	E-20W	-0.750	-1.133	0.043	1795.3
7	E-23W	0.246	0.020	-0.043	-94.6
8	E-29W	-0.582	-0.877	-0.083	-3432.5
9	E-40W	-0.474	-0.714	0.050	2062.3
10	E-44W	-0.444	-0.664	-0.014	-592.0

5 Conclusion

The sampling sites were situated at locations where previously in soil-gas samples have been detected elevated levels of methane and/or carbon dioxide as well as the presence of higher homologues of methane, denoting to their deep-seated origin. Totally, ten points were selected for sampling. Emission measurements using the method of static chambers were conducted (PL 206259 B1 patent [7]). Changes in the concentrations of methane and carbon dioxide released naturally to the land surface, as well as from the depth of 0.5 m were measured. Due to the fact that the measurements were taken within a short period of time the seasonal changes could not have had any influence on the established emission value.

The natural methane emission measurements ranged from -1.469 to 0.354 mg*m^{-2}*d^{-1}. The performed emission measurements indicate that the study area is dominated by negative CH_4 emission values. This may be evidence of microbial assimilation process of hydrocarbons occurring in the soil. Similar values are observed after making a hole ranging from -1.133 to 0.316 mg*m^{-2}*d^{-}1.The changes of the natural carbon dioxide emissions ranged from -4812.8 to 5658.7 mg*m^{-2}*d^{-1}. The variability range for the emissions of carbon dioxide with an additional hole is smaller and is illustrated by values from -3432.5 to 4657.9 mg*m^{-2}*d^{-1}. However, the drilling of the well increased the percentage share of the measuring points with a negative CO_2 value. The obtained results indicate a small intensity of the inflow of methane and carbon dioxide into the near-surface zone. Moreover, negligible emission values, that not confirm elevated methane and carbon dioxide concentrations recorded in soil gas samples may be due to a too shallow hole. The emission was measured over a hole, which was drilled up only to the depth of approximately 0.5 m, but the samples of the soil gas were collected at a depth of 1.2 m. Therefore, we believe that the hole should be drilled below the active exchange zone (at least 1 m).

Conducted studies show that methane and carbon dioxide emission from soil to the atmosphere are low in the studied area. It does not confirm the elevated values of light hydrocarbons and associated components which were previously recorded in the soil gas samples. This indicates low activity of gases migrating from the depths. The negative values of methane emission may also be evidence that processes of methanotrophic methane oxidation are active in the near-surface zone.

The research was funded as a part of the Project of the National Centre for Research and Development Blue Gas (BG2/ShaleCarp/14).

References

1. G. Etiope, R.W. Klusman, Chemosphere, **49**, 777-789 (2002)
2. G. Etiope, Atmos. Environ., **43**, 1430-1443 (2009)
3. J. Kuśmierek, G. Machowski, Pr. Inst. Nafty i Gazu, **150**, 247-250 (2008)
4. D.F. Saunders, K.R. Burson, C.K. Thompson, AAPG Bull, **83**, 1 (1999)
5. M. Mojżysz, Master thesis. Arch. Fac. of Geology Geophysics and Environmental Protection, AGH University of Science and Technology (2017)
6. M. Dzieniewicz, H. Sechman, no. 184080, WUP, **08/08** (2002)
7. A. Korus, M.J. Kotarba, M. Dzieniewicz, H. Sechman, no. 206259 B1, WUP, **07/10** (2010)
8. W.J.H. Kunicki-Goldfinger, *Życie bakterii* (PWN, 1994)
9. R.W. Klusman, AAPG Bull., **89**, 8 (2005)
10. H. Sechman, M. Dzieniewicz, Geology., **35**, 129-153 (2008)
11. J. Tang, Z. Bao, W. Xiang, Q. Gou, J Environ Sci, **20**, 1055-1062 (2008)

Petrophysical parameters of the porous space of the “tight” type sandstones of the Skole Unit - Preliminary analysis

Michał Maruta[1,*], and *Vitalij* Kułynycz[1]

[1]AGH University of Science and Technology, Faculty of Drilling, Oil and Gas, 30 Mickiewicza Av., 30-059 Krakow, Poland

Abstract. The scientific goal of the paper is the physical characteristics of pore space of the Inocereamian Sandstones located in the Skole Unit as a part of the Outer Carpathians – The Carpathian Flysch. Rock samples were tested using mercury porosimeter. Using this method, cumulative curves of effective porosity were obtained, as well as the pore geometry distribution and pore surface area distribution. geometry and distribution. In the article the authors determine the physical parameters of the pore space for 30 samples, such as porosity, permeability, size and distribution of pore diameter, specific surface area and geometry of a pore space. Preliminary analysis of rock samples is to answer the question of the existence of sandstones capable of forming "tight" type deposits of natural gas and determining their reservoir parameters.

1 Introduction

The "tight" type deposits are formed by the accumulation of hydrocarbons in reservoir rocks with very poor filtration parameters. These types of deposits can achieve high porosities. On the other hand, the microporous development of the parameters of the pore space gives a permeability less than 0.1 mD. This microporous development is the only factor that distinguishes the "tight" deposit from conventional deposits [1].

Porosity, permeability and pore space parameters analysis are extremely important for gas and oil-bearing rocks, allowing the estimation of accumulated hydrocarbons in the rock, while properly measured parameters of the pore space formation also assess the gas or oil production of a rock formation [1, 2].

In this study the application of the mercury porosimetry technique in determination of different reservoir parameters have been investigated. Mercury injection capillary pressure (MICP) data have been widely used to characterize reservoir rocks, evaluating sealing capacity for traps and to explain the locations of hydrocarbon accumulations and transition zones. This method can determine a broader pore size distribution more quickly and accurately than other methods. It can be used to characterize pores ranging from 0.003 μm to 360 μm using a single theoretical model. The data produced (mercury intrusion volume

* Corresponding author: maruta@agh.edu.pl

at various pressures) can be used to calculate numerous sample characteristics such as pore size distributions, total pore volume, total pore surface area, median pore diameter, and sample densities (bulk and skeletal) [3].

2 Materials and methods

2.1 Geological background

From the geological point of view the Polish part of the Outer Carpathians is built from many nappes, folds and scales overthrusted overthrown usually in the North direction. These formations were deposited in the presence of turbidity currents with a variable density in a deep-sea environment. Alternately lay shoals of conglomerates, sandstones, siltstones and claystones formed geological layers called the Flysch [4, 5].

The study area is located within the Skole unit (Fig. 1), which stretches from the Romanian Carpathians through the cities of Skole and around Przemyśl, Rzeszów to Bochnia, where fading hidden under the Silesian unit. The width of the belt on the eastern edge of the unit is 40 km and narrows toward the west [4-6].

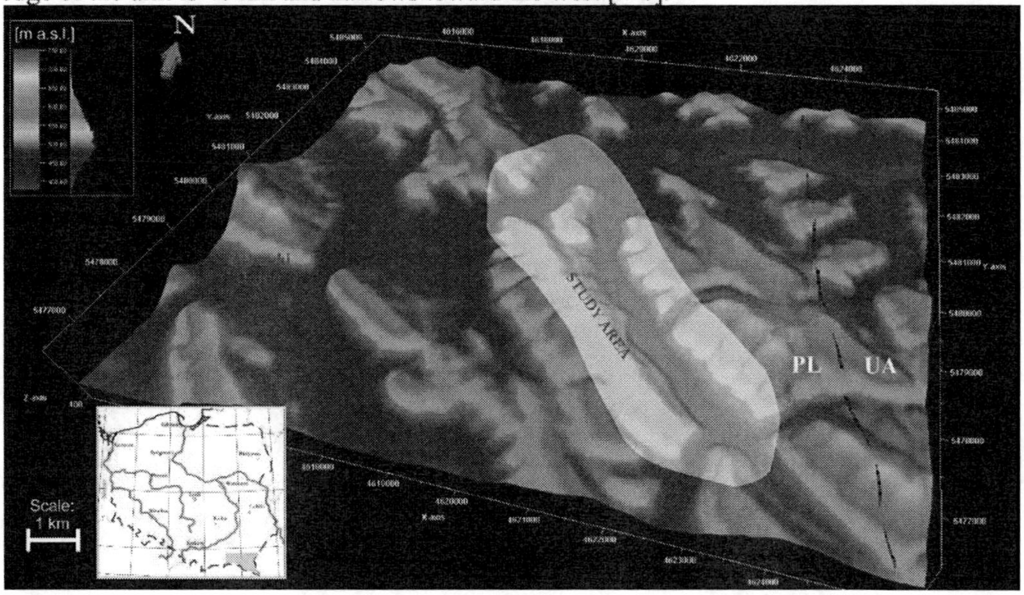

Fig. 1. Study area location.

2.2 Methodology

30 samples of the Inoceramian sandstones of the Skole Unit, as part of the Outer Carpathians, were investigated. The permeability was measured on the basis of the known Swanson model based on the threshold pore diameter. Parameter measurements of pore space of sandstone rock samples will be taken by AutoPore II 9220 firm Micromeritics (Fig. 2). The apparatus used for this process control computer injection of mercury from a pressure less than ambient to 413.4 MPa, allowing penetration of pores or fractures with a diameter or aperture of 0.003 μm to 360 μm. These tests are based on a cylindrical pore space model, in which the pore space is treated as a bundle of cylindrical capillaries conductive fluid reservoir.

Based on the results of porosimetry measurements were determined the following values, characterized pore space geometry of analyzed sample [1].

The bulk density. Density relating to the total bulk volume of the sample. Bulk volume of the sample is determined by mercury pycnometer.

Fig. 2. Porosimeter AutoPore IV 9520 (Micromeritics) [7 , modified].

Porosity count of the porosimetry. This porosity is different than the open porosity in that the count of nonwetting liquid volume which has been pressed into the sample. This volume does not include all the sub-micropores, whose diameter is too small for mercury penetration. This value will be so dynamic porosity. Porosity coefficient in porosimetry is calculated by dividing the total volume of mercury penetrated into the sample by the bulk volume:

$$\phi_d = \frac{V_{Hg}}{V_b} \tag{1}$$

where:

ϕ_d – porosity coefficient [%],

V_{Hg} – the total mercury volume penetrated into the sample [cm^3],

V_{obj} – the bulk volume of the reservoir rock [cm^3].

The average pore diameter used to assess the quality of the reservoir rock. This value is calculated as a weighted average of the pores weight, and not the percentage of pore space.

Percentage distribution of the pore diameter in the pore space is determined by counting the fractional volume of mercury penetrated to the sample. To determine the relationship pores - fractures is possible to determine the percentage of pores and fractures determining the possible hydrocarbon migration and accumulation.

The measurement consists of injecting mercury into the sample at the desired pressure and measuring the pressure and the volume of mercury which injected into a sample at this pressure. The cumulative curve is obtained according to the applied pressure – the volume of mercury. Knowing the sample weight and the setting during the porosimetric measurement its volume and a bulk volume of the rock skeleton by converting the results obtained Washburn formula gives the distribution of pore diameter in a given sample and a partial volume, the porosity of the sample, the skeletal density and apparent.

The total area of the pore space. This is the total pore surface area per unit volume (weight) of test rocks and a measure of the quantity of the resistance of the porous medium flowing fluid. It is calculated based on the cylindrical model of Washbourna [1]:

$$\sum \Delta A = \frac{\sum P_c \cdot V}{\sigma \cdot cos\theta} \tag{2}$$

where:

A – surface area [m^2/m^3],

ΔV – partial volume corresponding to the pressure P_c [m^3],

σ – surface tension [N/m],

θ – contact angle [°],

P_c – capillary pressure [Pa].

The irreducible water content. It is the sum of the residual water content in the sample and the micropore volume of such small diameter in which mercury cannot penetrate. The value is associated with the coefficient of dynamic porosity.

3 Results and discussion

The values of mercury porosimetry properties were obtained for each of the core samples. It was measured: the compactness, pore intrusion volumes, pore surface areas, pore specific surface areas, average pore diameters, bulk and skeletal densities, effective porosities.

As a result of the research conducted by the authors, cumulative curves of effective porosity and distribution of pore geometry were obtained. Figure 3 shows the cumulative porosity curves for 30 samples taken from sandstone outcrops of the selected area. These curves consist of the image of heterogeneous rocks with a porous-fracture type reservoir rock space of the mono-modal character. The shape of the cumulative curves indicates that the highest effective porosity is found in samples BD4 and BD5. This is confirmed in the pore geometry distribution graph (Fig. 4), where the pore system for these samples is in the range of 0.07 to 0.16 μm. For the rest of the samples, the pore system is in the range of 0.03 to 0.05 μm in diameter, indicating low effective porosity. Within the micropore range, single isolated peaks are visible in the geometry graph, which corresponds to microfractures with the aperture to several tenths of μm.

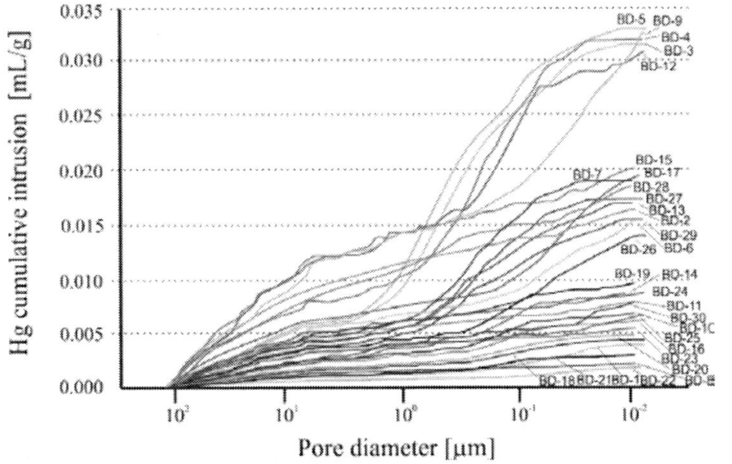

Fig. 3. The cumulative curves of the effective porosity.

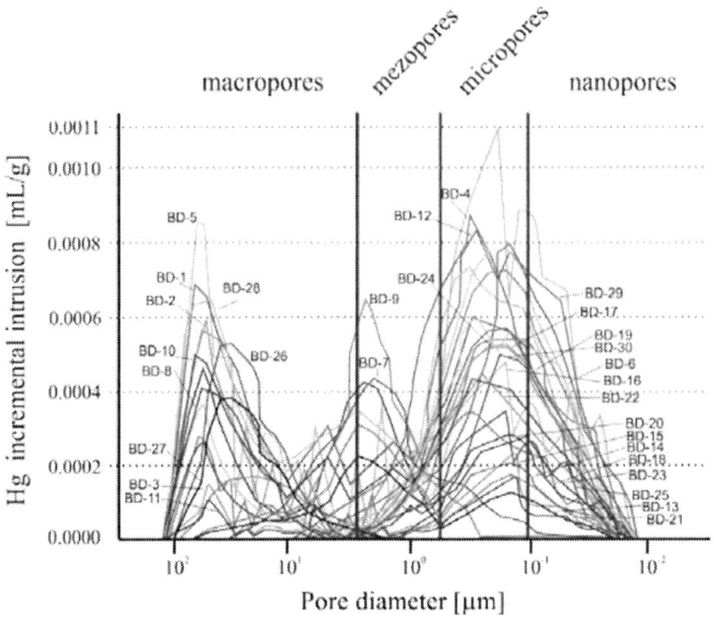

Fig.4 Pore geometry distribution.

It is worth adding that the cumulative curves of effective porosity of samples number BD3, BD4, BD5 flatten at the final fragment (Fig. 4), which shows that these samples are fully saturated with mercury at the applied pressure range. For the remaining samples, such flotation does not occur and indicates incomplete mercury saturation and the possibility of continuing the mercury loading process in the pores using a pressure higher than the maximum pressure that can be used. This indicates that in these samples there may be smaller pores out the measuring range of the porosimetry method. Looking still on the figure it is worth noting that the curves of cumulative effective porosity of samples are flattening in the final fragment which proves that sample is full saturated with mercury at applied pressure range and also indicates the reliability of data results. For other samples, such flattening does not exist and it proves incomplete saturation of its mercury to continuing the process of injecting mercury into the pores by applying a pressure higher than the maximum pressure, possible to use. This demonstrates that in these samples can occur pores far smaller, which are outside the measurement range of porosimetry method.

The juxtaposition of cumulative curves for samples indicates that the higher the mercury intrusion, and therefore a higher apparent porosity, due to the fact that in the porosimetry studies is greater availability of mercury into the pores of tested samples.

The results of the permeability were obtained using the Swanson model, which is based on the threshold diameter. The results indicated that the samples were characterized by low permeability values up to 0.00307 mD for the „DB10" sample.

Based on literature data [8, 9] the „tight" type reservoir rock have a permeability of 0.0001 - 0.1 mD, a close to the „tight" type reservoir rock having a permeability of 0.1 – 1 mD and a conventional reservoir rock of 1 – 1000 mD. The permeability analysis shows that most samples exhibit the characteristics of the „tight" type reservoir rock.

Authors obtained the statistical description of the mercury porosimetry measured data. They are summarized in Figure 5 and Table 1.

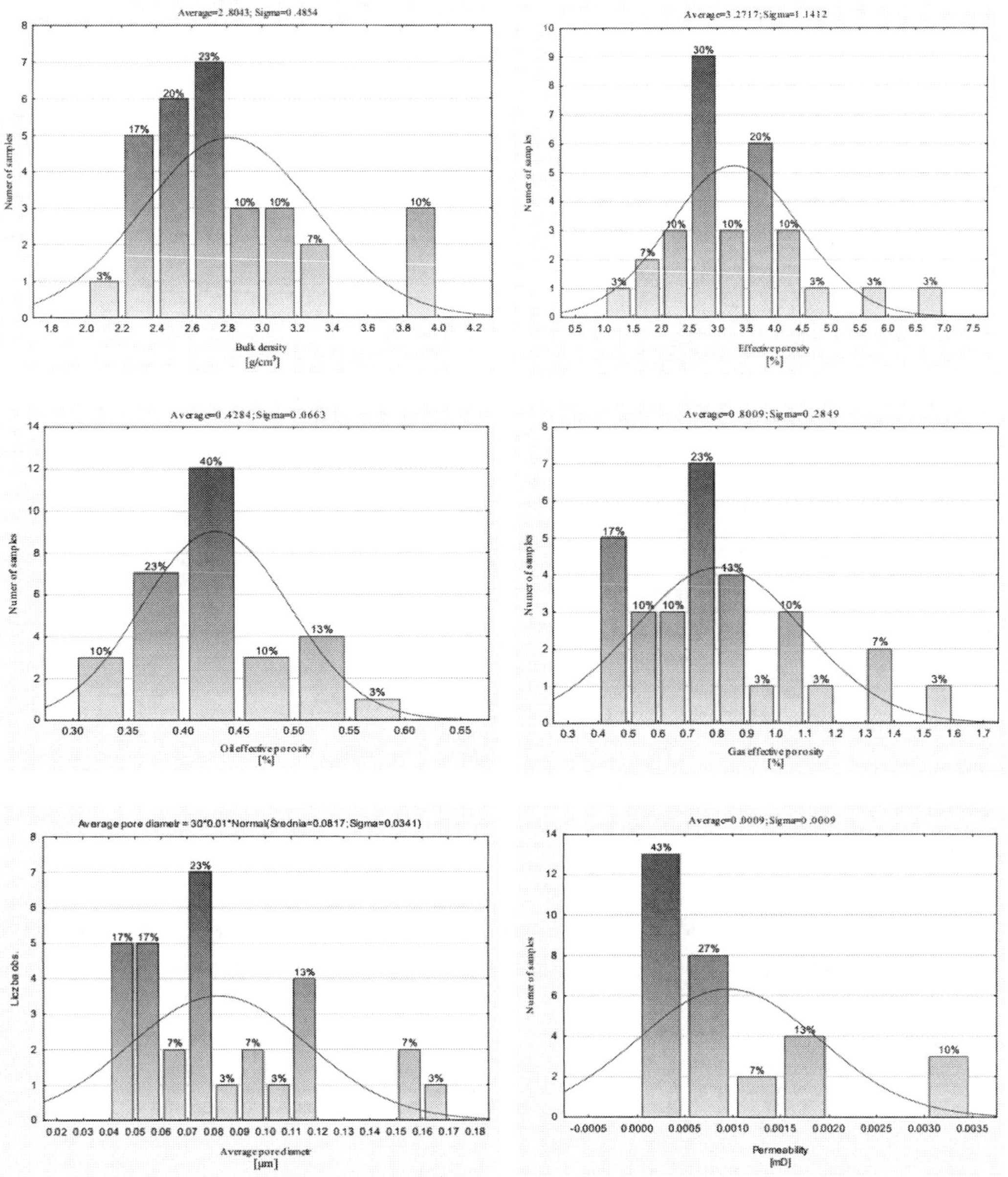

Fig. 5. Distribution of effective porosity.

Table 1. Statistical description of the mercury porosimetry measured data.

	Bulk density	Skeletal density	Total intrusion mercury volume	Compactness	Effective porosity	Oil effective porosity	Gas effective porosity	Average pore diameter	Total pore area	Permeability
	[g/cm3]	[g/cm3]	mL/g	[%]	[%]	[%]	[%]	[µm]	[m2/g]	[mD]
Min	2.10040	2.146254	0.005300	82.20105	1.36180	0.341	0.440	0.041	0.1840	0.0000110
Max	3.94010	4.219400	0.017800	99.32500	6.61930	0.600	1.560	0.162	1.1980	0.0030700
Avg	2.80427	2.927337	0.010836	92.11118	3.27166	0.428	0.801	0.082	0.6383	0.0009413

4 Conclusions

The aim of the petrophysical studies using the porosimetric method was to identify the pore space structure of the sandstones of the selected area. This means that not only the porosity qualifies the rocks to potentially reservoir in a particular class, but also the other physical parameters of the space, as well as the size and distribution of the pore diameter, the fracture aperture, the surface area and the pore distribution geometry as a function of their diameter, in the range of overcapillary, capillary and subcapillary areas. Summarize on the basis of the preliminary analysis:

1. Inoceramian Sandstones of the study area are the reservoir rock of a pore-fracture type reservoir space of a mono-modal character.
2. Pore space consist of micropores, which is associated with the "tight" type rock.
3. The analysis showed low values of permeability and a variation in effective porosity (1,3-6,6%)
4. Presented petrophysical characteristic shows that the Inoceramian sandstones show features of unconventional reservoir rocks.

The presented results were considered preliminary, due to limited data. To confirm the results of the study, the authors plan to repeat the study using helium porosimetry in the next phase of the work.

References

1. P. Such, Prace naukowe INiG (in Polish), **183**, 153 (2012)
2. F. Javadpour, D. Fisher, M. Unsworth, Journal of Canadian Petroleum Technology, **46**, 10 (2007)
3. S.M. Aboujafar, EAGE Reservoir Characterization and Simulation Conference, 1-8 (2009)
4. P. Karnkowski, „Geos" AGH Kraków, **2** (1993)
5. J. Waraszyńska , *Karpaty Polskie*. Wydawnictwa Uniwersytetu Jagiellońskiego (1995)
6. E. Stupnicka, *Geologia regionalna Polski*, Wydawnictwa Uniwersytetu Warszawskiego (1997)
7. http://galaxy.agh.edu.pl/~k-se/x/pl/team/laboratorium-porozymetryczne/ access 18.06.2017
8. K. Słupczyński, R. Semyrka, E. Bobula, Polish Journal of Mineral Resources, Geosynoptics GEOS, Kraków, **3**, 191–195 (2001)
9. P. Such, Prace naukowe INiG, **113**, 1–84 (2002)

Comparison of recent S-wave indicating methods

Katarzyna Hubicka[1,*] and *Jakub* Sokolowski[2,**]

[1] *Faculty of Pure and Applied Mathematics, Janiszewskiego 14a, Wroclaw University of Science and Technology, Poland*
[2] *Machinery Systems Division, Na Grobli 15, Wroclaw University of Science and Technology, Poland*

Abstract. Seismic event consists of surface waves and body waves. Due to the fact that the body waves are faster (P-waves) and more energetic (S-waves) in literature the problem of their analysis is taken more often. The most universal information that is received from the recorded wave is its moment of arrival. When this information is obtained from at least four seismometers in different locations, the epicentre of the particular event can be estimated [1]. Since the recorded body waves may overlap in signal, the problem of wave onset moment is considered more often for faster P-wave than S-wave. This however does not mean that the issue of S-wave arrival time is not taken at all. As the process of manual picking is time-consuming, methods of automatic detection are recommended (these however may be less accurate). In this paper four recently developed methods estimating S-wave arrival are compared: the method operating on empirical mode decomposition and Teager-Kaiser operator [2], the modification of STA/LTA algorithm [3], the method using a nearest neighbour-based approach [4] and the algorithm operating on characteristic of signals' second moments. The methods will be also compared to well-known algorithm based on the autoregressive model [5]. The algorithms will be tested in terms of their S-wave arrival identification accuracy on real data originating from International Research Institutions for Seismology (IRIS) database.

1 Introduction

The indication of seismic event moment of arrival is vital in the processing of seismic signals, e.g. when the problem of event's energy evaluation is considered (which is mostly done by computing the integral of the registered signal from it's moment of arrival till the end). Moreover, the result has to be rescaled in accordance to the distance between the sensor and the event's hypocenter (due to inverse-square law). Naturally when this approach is conducted one has to estimate the moment of wave arrival and the distance (which is commonly solved with the aid of onset moment indicated by at least four different single-axial sensors).

Single excitation indicate various waves which differ in terms of propagation patterns. They can be divided into [6]: body waves, which travel through the Earth interior, and surface waves which propagate along the interface between differing media. As their name suggest the main difference between the body and surface waves is the place of their occurrence. Another difference is their

*e-mail: 221667@student.pwr.edu.pl
**e-mail: jakub.sokolowski@pwr.edu.pl

carrier frequency which for body waves is much higher thus their identification on seismometer data is relatively easy.

Thebody waves can be qualified as Primary waves (P-waves, pressure waves) and Secondary waves (S-waves, shear waves). As the P-waves are the fastest one (especially faster than the S-wave), sensors register their moment of arrival as first. When the sensors are relatively close to the event epicentre the delay between event's onset moments are small thus the indication of arrivals other than P-wave is difficult. The indication of waves arrival may be considered as a case of finding a structural break detection [7–10]. Recently couple new methods to indicate the S-wave arrival were created. However, the results of each are obtained from different set of data. Thus this is quite hard to tell which of them is the most accurate. It was decided to compare these new methods altogether with one well-known method. All of the methods employs the same real data set.

2 Methodology

2.1 Kirbas-Peaker Algorithm

Kirbas-Peaker algorithm [2] was proposed in 2016. The method is based on Empirical Mode Decomposition (EMD) and Teager-Kaiser energy operator (TKEO).

First part of the algorithm, EMD, is useful for processing natural signals as it doesn't need assumption about its linearity and stationarity [11]. The EMD divides the time series into Intrinsic Mode Functions (IMF) which form complete and nearly orthogonal basis for the signal. Each mode function fulfils two constraints: having only one extreme between zero crossing, and the mean value of their upper and lower envelope must equal zero.

TKEO is a nonlinear operator which is used to estimate the energy of signal. For n-th sample of discrete signal $x[n]$ it is defined as [12]

$$\Psi(x[n]) = x[n-1]^2 - x[n]\,x[n-2]\,. \tag{1}$$

In this method the mean TKEO is used. It is described as:

$$\Psi(x[n]) = \log\left(\frac{1}{N}\sum_{m=m-N+3}^{n} x[n-1]^2 - x[n]\,x[n-2]\right), \tag{2}$$

where N is the window length. Kirbas-Peaker algorithm has the following steps:

1. The signal is filtered by 10-point moving average and IRR Butterworh band-pass filter [13].
2. The resulting signal is divided into windows of N samples.
3. The EMD algorithm is applied to each window. The IMF with the most visible S-wave is taken.
4. In each window the mean TKEO is calculated.
5. First window with mean TKEO exceeding predefined threshold th_s implies the period with S-wave onset.
6. The sample with maximum TKEO value (in the corresponding window) is chosen as actual S-wave arrival moment.

2.2 Rawles-Thurber Algorithm

The method [4] was presented in 2015 and uses a nearest neighbours-based approach of Nikolov [14]. A method does not need to use parameters estimated from data, but uses data itself to build a model. The main issue is to aggregate the set of reference data with P-waves and S-waves picks which is compared to the analysed seismic signals. The method may somehow be considered as a clusterisation problem [15, 16]. The Rawles-Thurber algorithm uses 3 traces of a signal. To indicate the S-wave, firstly the P-wave onset is pointed. The method consists of following steps:

1. In the first step, two reference data sets are built: for P-waves from vertical trace and for S-waves from horizontal traces. Each reference set consists of positive $\mathcal{R}_+$ and negative examples $\mathcal{R}_-$, where positive examples are N-length windows surrounding the P-wave/S-wave onsets and negative are windows with length of N containing for example noise or P-wave/S-wave coda.

2. The next move is windowing and data preprocessing by applying the bandpass filter. All of signal's traces are divided into moving-windows of the same length as reference windows. Each window is scaled as follow: the absolute amplitude of window is divided by the mean of its first half.

3. For each window of N-length, the similarity metric $d(r, s)$ to reference examples is computed, where $d(r, s)$ is the sum of squares of the Euclidean distance between the sample of reference signal r_i and observed signal s_i:

$$d(r, s) = \sum_{i=1}^{N} (r_i - s_i)^2 . \tag{3}$$

4. Subsequently the score function $R(s)$ is determined for each window as the ratio of the negative to positive distances of the particular trace:

$$R(s) = \frac{\sum_{r \in \mathcal{R}_-} d(r, s)}{\sum_{r \in \mathcal{R}_+} d(r, s)} . \tag{4}$$

   As for S-wave identification two horizontal traces are used, to appoint the S-wave score function $R_S(s)$, two horizontal scores $R_{H1}(s)$ and $R_{H2}(s)$ are multiplied:

$$R_S(s) = R_{H1}(s) \cdot R_{H2}(s) . \tag{5}$$

5. The centre of the window with the biggest value of score $R_P(s)$ for a vertical trace indicates the moment of P-wave onsets, and the biggest $R_S(s)$ – the moment of S-wave onset. In addition, the $R_P(s)$ and $R_S(s)$ scores must exceed the thresholds and the difference between P-wave and S-wave picks has to be located in user-supplied range.

2.3 Tasic-Runovc Algorithm

Tasic-Runovc approach [3] is a modification of STA/LTA algorithm and needs three-component seismic station monitors with velocities in east-west (E), north-south (N) and vertical (V) directions. The STA/LTA algorithm is based on the energy analysis by the ratio of the short-term average (STA) energy to long-term average (LTA) energy level from the same seismogram [17]. In T-R method the modification of STA/LTA is applied through the transformation of signal $f(x_i)$. The nominator still gives the information about the sensitivity for local changes in the amplitude as the STA segment, however the denominator informs about the amplitude variation from the sample i to the end of the signal. The algorithm is as following:

1. Each trace of signal is saved as vectors $\mathbf{Y_E}$, $\mathbf{Y_N}$, $\mathbf{Y_V}$:

$$\mathbf{Y_E} = (e_1, e_2, \dots, e_k) \quad (6)$$
$$\mathbf{Y_N} = (n_1, n_2, \dots, n_k) \quad (7)$$
$$\mathbf{Y_V} = (v_1, v_2, \dots, v_k). \quad (8)$$

   The length of the signal is k and an i-th sample is in the range $1 < i_p < i_s < k$, where i_p indicates the moment of P-wave and i_s – moment of S-wave arrival.

2. To estimate the relative total energy, the vector $\mathbf{O}$ consisting of the sum of squares of the components of 3 traces is used:

$$o_i = e_i^2 + n_i^2 + v_i^2 \, . \quad (9)$$

3. To appoint the S-wave pick, the $f(x_i)$ function is made:

$$f(x_i) = \frac{\frac{1}{l}\sum_{j=i}^{i+l-1} |x_j|}{\frac{1}{k-i+1}\sum_{j=i}^{k-l} |x_j|} \quad \text{for } i = 1, \dots k-l \, . \quad (10)$$

   where l is the window's length.

4. The arguments of the function above are components of vectors $\mathbf{Y_E}$, $\mathbf{Y_N}$ and $\mathbf{O}$. The next step is to make a characteristic vector $\mathbf{C}$ with components:

$$c_i = f(e_i) f(n_i) f(o_i) \ \text{ for } i = 1, \dots n-k \, . \quad (11)$$

5. If the sample of the characteristic function exceeds the user-supplied threshold as the first one, sample i_s is taken as the moment of S-wave arrival.

2.4 Sokolowski-Wylomanska-Zimroz algorithm

The algorithm is expansion of P-wave indicating algorithm [18]. The procedure is based on the characteristic of signals' empirical second moment. It is known that when X_t is consisted of i.d.d. random variables with mean μ and variance σ^2 they fulfill [19]:

$$E\sum_{i=1}^{n} X_i^2 = \sum_{i=1}^{n} EX_i^2 = \sum_{i=1}^{n} (E(X_i - EX_i)^2 + (EX_i)^2) = n(\sigma^2 + \mu^2). \quad (12)$$

This clearly induces that when the cummulative sum of squared signal $\sum_{i=1}^{n} x_i^2$ is considered, one has to be aware that if x_i is i.i.d. the regression line can be fitted. It is common to assume that the seismic noise fulfils the assumption of being i.d.d. However, when signal's part after the P- or S-wave onset moment is considered the signal is clearly not i.i.d.

It was noticed that the statistics after the P-wave onset moment (and before the S-wave moment) becomes concave. It was also noted that when applying the logarithm to the statistics the resulting function is in general concave from the beginning till the onset moment and from the onset moment till the end (if on whole its course it is concave it means that no impulse was registered).

In the paper it was also observed that the logarithmic statistic when considered for the part of signal which includes the seismic wave is well approximated with exponential function. This led the authors to the following algorithm:

1. Determine the moment of P-wave arrival (p) for input signal s_t of length n. Take $x_t = s_{t+p-1}$ for $t \in \{1, 2, ..., n - p + 1\}$.

2. Compute $L_t = \log(\sum_{i=1}^{t} x_i^2)$ for $t \in \{1, 2, ..., n - p + 1\}$.

3. For $i = \{3, 4, ..., n - p + 2\}$ let $L_t^1 = L_t$ for $t = 1, 2, ..., i$, $L_t^2 = L_{t+i}$ for $t = 1, 2, ..., n + 1 - p - i$. Fit $e^{a_1 x + b_1} + c_1$ to L_T^1 and $e^{a_2 x + b_2} + c_2$ to L_T^2 sum their squared errors (SSE). Find i which minimises SSE.

4. S-wave arrival is indicated for $i + p$.

2.5 Takanami-Kitagawa Algorithm

Algorithm of Takanami-Kitagawa [5] was proposed in 1988 and uses the AR model to analyse seismic signal. Despite the fact of seismic signals being non-stationary, they can be divided into two subseries and each of them can be approximated by the AR model.
The method to obtain the moment of S-wave arrival is as follows:

1. The signal is divided into two subseries before and after the unknown moment of S-wave arrival:

$$x_n = \sum_{m=1}^{M(1)} a_m^1 x_{n-m} + \varepsilon_n^1, \quad (1 \leq n < s_1) \tag{13}$$

$$x_n = \sum_{m=1}^{M(2)} a_m^2 x_{n-m} + \varepsilon_n^2, \quad (s_1 \leq n \leq N) \tag{14}$$

   where ε_n^i is a Gaussian noise with mean zero and variance σ_i^2, a_m^i is the autoregressive coefficient, $M(i)$ is the order of the i-th model, s_1 is the unknown arrival time and N is the length of series. The [13] is considered as the "noise" part of signal and [14] as the "earthquake" part of signal, where S-wave appears. The length of "noise" window increases together with the s_1, while the "earthquake" window's length decreases.

2. To measure the quality of estimated model, the Akaike Information Criterion (AIC) is used:

$$AIC = -2(\log L_{max}) + 2(M_P)\,, \tag{15}$$

   where L_{max} is the maximised value of the likelihood function for the model and M_P is the number of independently estimated parameters.
   For each potential time of S-wave arrival s_1 AIC is calculated for the first part of signal AIC_i^1 and second part AIC_i^2.

3. Then the sum of two AIC is taken:

$$AIC_i = AIC_i^1 + AIC_i^2\,. \tag{16}$$

4. The sample i_{min} with minimum value of AIC_i is searched.

5. The sample $i_{min} + 1$ responds the S-wave pick.

3 Applications

The above-mentioned algorithms are tested on two signal sets from International Research Institutions for Seismology (IRIS) database with reference picks. Each of them contains the events from West Coast of South America.

The first set consists of 40 signals registered in Brazilian seismic station SAML from different earthquakes in 2003-2017, while the second one contains 100 series from LVC station events registered in 2002-2017. Both data sets are sampled at 20 Hz, the magnitudes of events are above 5.5 and the distances between stations and earthquakes are between 7°and 20°.

If the method does not provide otherwise, the input seismic signals are not denoised as such step can make onset moment indication easier [20, 21]

In K-P, R-T and T-R algorithms, the setting of few parameters is essential and most of them is determined by the minimum mean squared or absolute errors between reference and auto S-wave picks on the mesh with different values of parameters.

In Kirbas-Peaker approach the optimal length of window N, the number of IMF and threshold th_s are chosen from the mesh of mean absolute errors between references and auto picks for particular parameters.

The least error is reached for north-south trace and parameters which are showed in table 1.

Table 1. Parameters used in Kirbas-Peaker algorithm

	1st data set	2nd data set
length of window N (in samples)	72	50
the number of IMF	1	5
S-wave threshold th_s	0.7207	0.05

In Rawles-Thurber method the thresholds th_P and th_S are appointed from the mesh of mean absolute errors between picks and the length of window was taken as in Rawles-Thurber paper [4]. The range of distance between P-wave and S-wave picks is determined on the base of reference picks in data sets. A different reference set from other observed signals (39 for 1st set and 99 for 2nd set) is created for each signal. The parameters used in this method are presented in table 2.

Table 2. Parameters used in Rawles-Thurber algorithm

	1st data set	2nd data set
length of window N (in samples)	67	67
P-wave threshold th_P	0.0001	0.03
S-wave threshold th_S	0.0001	0.0001
minimum distance between P and S-wave SP_{min} (in samples)	200	900
maximum distance between P and S-wave SP_{max} (in samples)	3100	3100

To chose the best parameters for Tasic-Runovc algorithm the mesh with mean squared errors is applied. Optimal threshold th_S and the length of window l are presented in Table 3.

Table 3. Parameters used in Tasic-Runovc algorithm

	1st data set	2nd data set
length of window l (in samples)	50	100
S-wave threshold th_S	2	2

In Takanami-Kitagawa algorithm, the AR models are matched to data starting from the sample $i_P + 200$, where i_P is the moment of P-wave onset and ending on the end of the signal. In the first step of fitting autoregressive models the "noise" part consists of the samples i_n:

$$i_n \in [i_1, \ldots, i_{199}],$$

where $i_1 = i_P + 200$ and the "earthquake" part from the samples i_e:

$$i_e \in [i_{200}, \ldots, i_N],$$

where N – the length of signal. The first and second part was moving on the right in the each consecutive step (the last element in "noise" part and the first one in "earthquake" part are changing) and the best models of AR are fitted to each part. Maximum order of AR model to first part of signal is 10 and to second part – 15 in both data sets.

3.1 Application to the particular signal

In the figures below the application of all five methods to the one signal from the first analysed set of data is shown. The upper panels of all figures present the signal with indicated reference S-wave onset.

Figure 1 presents the implementation of Kirbas-Pekaer to the exemplary signal. The lower panel illustrates the mean TKEO value computed for each window of signal. The horizontal solid line represents the S-wave threshold th_S and the distinguished box indicates the window exceeding the value of threshold as the first one. The sample with maximum value of mean TKEO in that window responds the S-wave pick.

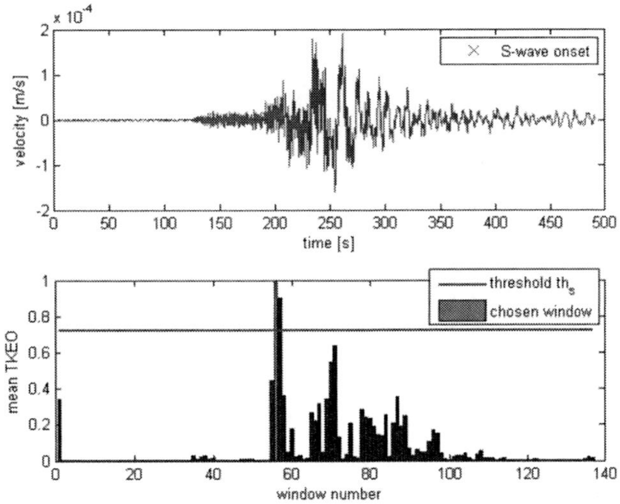

Figure 1. Application of Kaiser-Peaker method to the particular signal.

The implementation of Thurber-Rawles algorithm was presented in the figure 2. The lower panel illustrates the score function R_S, which was computed for each window of the signal. The dot represents the minimum value of R_S and thereby the moment of S-wave onset.

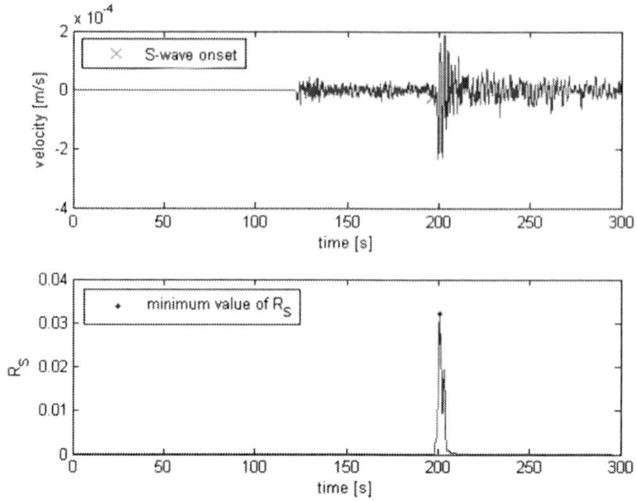

Figure 2. Application of Rawles-Thurber method to the particular signal.

In Figure 3 the application of Tasic-Runovc method can be seen altogether with the statistic which determines the results. It can be seen that with the S-wave onset moment the statistics considerably rises which confirms the algorithm idea.

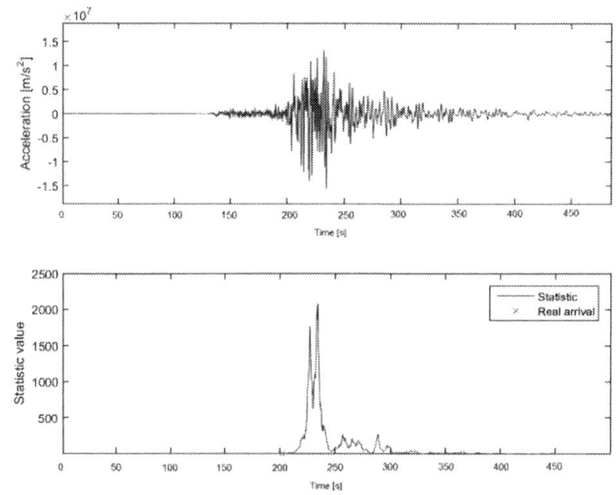

Figure 3. Application of Tasic-Runovc method to the particular signal.

In Figure 4 the exemplary signal is presented in upper figure and its L statistic. It can be seen that this statistic consists of 3 consequent concave functions which have their limits in moments of P- and S-wave arrivals. Thus when the algorithm finds these limits it finds the moment of wave arrival.

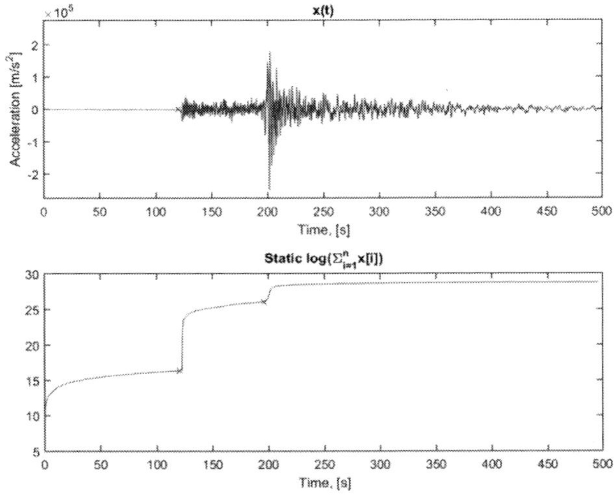

Figure 4. Application of Sokolowski-Wylomanska-Zimroz method to the particular signal.

Figure 5 presents the application of Takanami-Kitagawa algorithm. The lower box pictures the sums of AIC values from "noise" and "earthquake" auto-regressive models. The cross indicates the sample i_{min} with minimum sum of AIC and the sample i_{min} + 1 represents the S-wave onset.

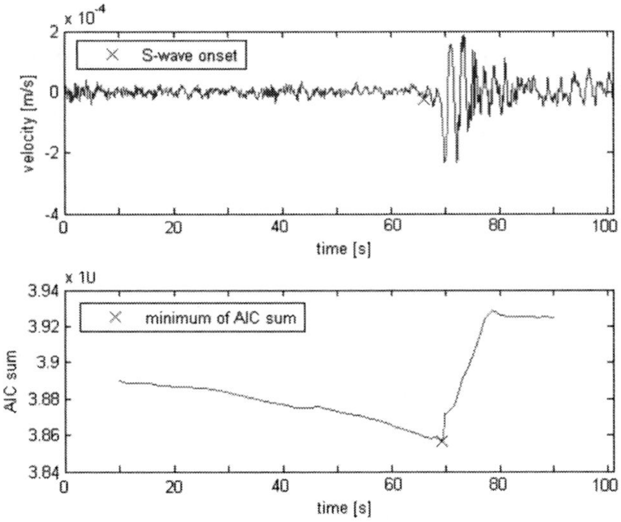

Figure 5. Application of Takanami-Kitagawa method to the particular signal.

3.2 Algorithms' results

In this section all methods are compared. Their results are summarised by such statistics as number of events whose evaluated S-wave arrival are closer than 10 seconds, mean of absolute differences between the estimation and real moment, its standard deviation, and standard deviation of differences. In Table 4 the results of algorithms tested on the first set (which consists of 40 events) are presented. The bold numbers symbolise the algorithm with best performance in specific statistics.

Table 4. Algorithms' results for the first set

	K-P	R-T	T-R	S-W-Z	T-K
$\#i : \lvert s_i - \hat{s}_i \rvert \leq 10$ s	7	7	**15**	14	**15**
$\frac{1}{n}\sum_{i=1}^{n} \lvert s_i - \hat{s}_i \rvert$ [s]	46.31	42.59	36.30	**30.37**	34.00
$\mathrm{std}(\lvert s_i - \hat{s}_i \rvert)$ [s]	68.90	44.37	51.20	**32.90**	33.23
$\mathrm{std}(s_i - \hat{s}_i)$ [s]	50.65	32.97	36.25	**29.71**	29.76

In Table 4 it can be seen that the Takanami-Kitagawa and Tasic-Runovc algorithms has the most number of indications closer than 10 seconds (15 out of 40 which is 37.5%). However other statistics (mean value of errors, and their standard deviations) clearly show the superiority of Sokolowski-Wylomanska-Zimroz method. The Kirbas-Peaker algorithm seems to be the least accurate.

Table 5. Algorithms' results for the second set

	K-P	R-T	T-R	S-W-Z	T-K
$\#i : \lvert s_i - \hat{s}_i \rvert \leq 10$ s	14	10	1	26	**31**
$\frac{1}{n}\sum_{i=1}^{n} \lvert s_i - \hat{s}_i \rvert$ [s]	41.97	58.51	31.18	**17.71**	27.57
$\mathrm{std}(\lvert s_i - \hat{s}_i \rvert)$ [s]	36.57	41.58	20.70	**16.07**	28.31
$\mathrm{std}(s_i - \hat{s}_i)$ [s]	55.05	67.49	31.04	**22.93**	36.46

In Table 5 the results of methods applied to the second set (100 signals) are presented. This time the results are almost the same as in Table 4. The biggest change is the small number of accurate picks of Tasic-Runovc algorithm (Previously 37.5% of indications and now 1%). Also the results of Sokolowski-Wylomanska-Zimroz algorithm seem to be more accurate than for the previous set.

4 Conclusions

In this paper the problem of S-wave arrival has been presented. Within this four relatively new algorithms indicating moment of S-wave arrival were compared altogether with one earlier and more commonly used method. The algorithms were tested on real data which was gained from the International Research Institutions for Seismology database.

The results for the first data-set provides that the Takanami-Kitagawa and Tasic-Runovc algorithms have the most number of accurate picks for, however they are not the most accurate algorithms, as the Sokolowski-Wylomanska-Zimroz indicated picks that were on average most accurate with the results having the lowest variance (and in this terms the Takanami-Kitagawa is the second best). The results also show that the Kirbas-Peaker is the least one and the Rawles-Thurber is last but one in terms of results accuracy.

For the second data-set only Takanami-Kitagawa algorithm has the biggest number of accurate picks (Sokolowski-Wylomanska-Zimroz has the second best number and this time Tasic-Runovc method has only one accurate pick). In terms of accuracy once again S-W-Z algorithm is the most accurate one, Takanami-Kitagawa has the second best average picks, however greater standard deviation than Tasic-Runovc. This time the Kirbas-Peaker algorithm has the fourth best results and Rawles-Thurber is the least accurate one.

The aggregated results provide that the Sokolowski-Wylomanska-Zimroz algorithm despite being second the best in terms of number of accurate picks has the most accurate results on average and the least average variance of results thus it is recommended to use.

In the paper no attention was paid on the computational complexity of algorithms which may be a proposition for future work.

This work was financed by the Grant no. 0401/0128/17.

References

[1] S. J. Gibowicz, A. Kijko, *An Introduction to Mining Seismology* (ACADEMIC PRESS, INC., San Diego, 1994)
[2] I. Kirbas, M. Peker, Neural Computing and Applications, 1-11 (2016)
[3] I. Tasic, F. Runovc, Acta Geotechnica Slovenica, **6**, 46-55 (2009)
[4] C.Rawles, C. Thurber, Geophysical Journal International, **202.2**, 1164-1179 (2015)
[5] T. Takanami, G. Kitagawa, Journal of Physics of the Earth, **36.6**, 267-290 (1988)
[6] P.M. Shearer, *Introduction to seismology* (Cambridge University Press, Cambridge, 2009)
[7] D. Kucharczyk, A. Wylomanska, R. Zimroz, Physica A: Statistical Mechanics and its Applications, **471**, 499-511 (2017)
[8] R. Hossa, R. Makowski, R. Zimroz, International Journal of Rock Mechanics and Mining Sciences, **91**, 29-39 (2017)
[9] D. Kucharczyk, A. Wylomanska, J. Obuchowski, R. Zimroz, M. Madziarz, Shock and Vibration, **2016**, (2016)
[10] R. Zimroz, M. Madziarz, G. Zak, A. Wylomanska, J. Obuchowski, Journal of Vibroengineering, **17.6**, 3111-3121 (2015)
[11] A. Erturk, M.K. Gullu, S. Erturk, IEEE Trans Geosci Remote Sens, **51**, 2787-2798 (2013)
[12] J.F. Kaiser, ICASSP-90, **1**, 381-384 (1990)
[13] S. Butterworth, Wireless Engineer, **7.6**, 536-541 (1930)
[14] S. Nikolov, M.Sc. thesis (2012)
[15] J. Sokolowski, J. Obuchowski, M. Madziarz, A. Wylomanska, R. Zimroz, Journal of Vibroengineering, **18.3**, 1654-1667 (2016)
[16] M.Polak, J. Obuchowski, M. Madziarz, A. Wyłomańska, R. Zimroz, Journal of Vibroengineering, **18.1**, 267-275 (2016)
[17] C.Y. Bai, B.L.N. Kennett, Journal of Seismology, **5**, 217-231 (2001)
[18] J. Sokolowski, J. Obuchowski, R. Zimroz, A. Wylomanska, E. Koziarz, Shock and Vibration, **2016**, (2016)
[19] J. Gajda, G. Sikora, A. Wylomanska, Acta Physica Polonica B, **44.5**, 1015-1035 (2013)
[20] F. Chaari, J. Leskow, A. Napolitane, R. Zimroz, A. Wylomanska, *Cyclostationarity: Theory and Methods III*, (Springer International Publishing, Cham, 2017), 51-68
[21] A Wyłomańska, R Zimroz, J Janczura, J Obuchowski, IEEE Transactions on Industrial Electronics, **63.9**, 5612-5621 (2016)

Seismic signals hard clipping overcoming

Paula Olszowa[1,⋆] and *Jakub* Sokolowski[2,⋆⋆]

[1]*Faculty of Pure and Applied Mathematics, Janiszewskiego 14a, Wroclaw University of Science and Technology, Poland*
[2]*Machinery Systems Division, Na Grobli 15, Wroclaw University of Science and Technology, Poland*

Abstract. In signal processing the clipping is understand as the phenomenon of limiting the signal beyond certain threshold. It is often related to overloading of a sensor. Two particular types of clipping are being recognized: soft and hard. Beyond the limiting value soft clipping reduces the signal real gain while the hard clipping stiffly sets the signal values at the limit. In both cases certain amount of signal information is lost. Obviously if one possess the model which describes the considered signal and the threshold value (which might be slightly more difficult to obtain in the soft clipping case), the attempt of restoring the signal can be made. Commonly it is assumed that the seismic signals take form of an impulse response of some specific system. This may lead to belief that the sine wave may be the most appropriate to fit in the clipping period. However, this should be tested. In this paper the possibility of overcoming the hard clipping in seismic signals originating from a geoseismic station belonging to an underground mine is considered. A set of raw signals will be hard-clipped manually and then couple different functions will be fitted and compared in terms of least squares. The results will be then analysed.

1 Introduction

In the data acquisition system the problem of good quality data collection is vital. The problem of data validation in the real world measurement system is renowned [1]. The quality of the data is poor mostly due to harsh environmental conditions (e.g. in monitoring of such machines as wind turbines [2] or mining machines [3–5]) or complexity of analysed process [6–8]. In the seismic signal processing one has to deal with plenty of difficulties. They include such problems as: small signal-to-noise ratio, missing data, signal distortions due to nonlinear effect of the data acquisition or transmission systems.

The signal clipping is one of the most common problems met in the signal processing field. The problem arises due to the limited range of values which the common sensors can register. The example of hard clipping can be seen in Figure 1

The negative impact of the clipping is obvious. The information of the signal is lost thus such values as the total energy which is travelling through the sensor is impossible to acquire [9].

One of the solutions implemented in order to overcome the clipping problem is the artificial division of signal values during the period of high values, thus producing the soft clipping. Such solution

⋆e-mail: 221708@student.pwr.edu.pl
⋆⋆e-mail: jakub.sokolowski@pwr.edu.pl

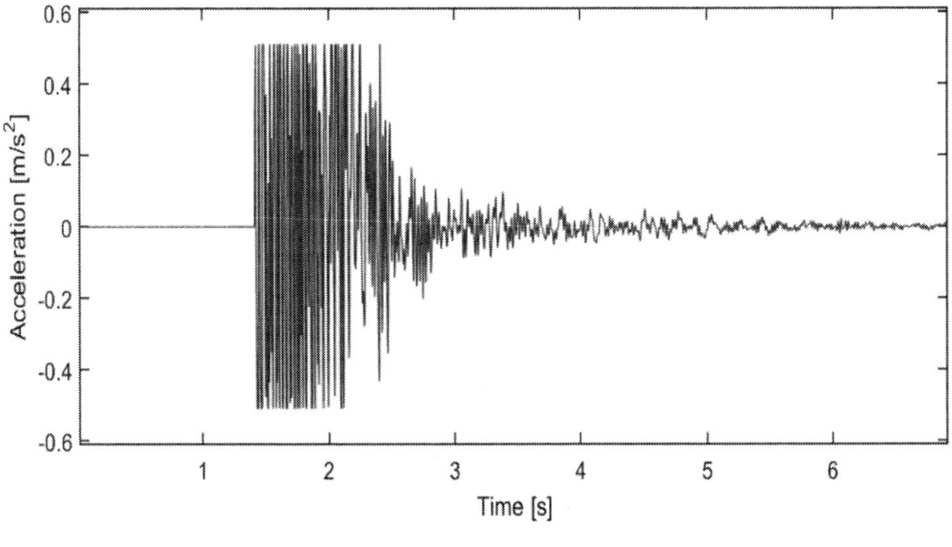

Figure 1. The example of hard clipped signal

may be problematic as it requires to gather the information of the clipping periods. However, some sensors lack this kind of solution, thus some alternative is recommended.

One might suppose that the solution for real value evaluation during the hard clipping can be solved with use of the signal which is registered on parallel sensor (where the signal is not clipped). Unfortunately, due to different propagation paths these signals might be significantly diverse, which rejects such approach.

Due to the above mentioned fact, to solve the problem of choosing the proper function one has to work on the non-clipped signals, artificially clipping them afterwards and compare evaluated values to vanilla signals. It is commonly assumed that the seismic signals take the form of an impulse response of some specific system [10]. This may lead to belief that the sine wave may be the most appropriate to fit. However, it is recommended to be tested.

The main idea of this paper is to bring the signal artificially under hard clipping. Then basing on the signal samples which occur just before and after the clipping, the signal is estimated. With comparison of different functions fitted to artificially clipped periods, one can acquire information which function may estimate the clipping moments in the most accurate way.

Such results may provide the solution, which may be used for sensors which lack the advanced options.

2 Methodology

During the analysis of signals the artificial limit of clipping is estimated. It is done for non clipped signals and then the artificial limit is $\alpha = p \cdot max(x)$ for the upper clipping or $\alpha = p \cdot min(x)$ for the lower clipping, where $p \in (0, 1)$ and x is the signal. The main points needed to assign the estimation function are selected according to the following algorithm:

1. Find the first index of a sample, whose value is above the limit of clipping (the top clipping) or below the limit of clipping (the bottom clipping) and take this point as x_1 by changing its value to α.

2. Find the following point after x_1, which is below (the top clipping) or above (the bottom clipping) the straight line of clipping, then take the previous point and call it x_2, change its value to α.

3. Do points 1 and 2 for the whole signal.

The interval for which the clipping is estimated contains in $[x_1 - k, x_2 + k]$, where k is the number of points before or after the clipping.

To estimate the real signal, the basic functions are needed. They are described with the formulas:

- Two linear functions: $f(x) = a_1 \cdot x + b_1 \mathbb{1}_{[p1,c]} + a_2 \cdot x + b_2 \mathbb{1}_{(c,p2]}$,
- Quadratic function: $f(x) = (a \cdot x^2 + b \cdot x + c)\mathbb{1}_{[p1,p2]}$,
- Sine function: $f(x) = a \cdot sin(b \cdot x + c)\mathbb{1}_{[p1,p2]}$,
- Power function: $f(x) = (a \cdot |x - \bar{x}|^b + c)\mathbb{1}_{[p1,p2]}$,
- Exponential function: $f(x) = (e^{a \cdot x^2} + b)\mathbb{1}_{[p1,p2]}$,

where p_1, p_2 stands for left and right limit of single clipping. The main reason behind selection of these functions was their simplicity (before testing more complicated functions it is justified to check the simpler).

To make the function estimation more precise, it is necessary to take the proper number of points before and after clipping into account during the approximation. Depending on the taken p, the number k may be different. Additionally the considered observations are divided by length of clipping interval (lengths from 2 to 5 and lengths greater 5).

Sum of squared errors is defined as:

$$SSE = \sum_{i=1}^{n} (x_i - \widehat{x_i})^2, \tag{1}$$

Where n is length of the clipping period, x_i is the signal value before clipping, $\widehat{x_i}$ is estimated value. SSE is used to estimate the error with which the solution fits to the real signal.

3 Applications

In order to efficiently evaluate the individual function potential, one has to test it on the real data. In this section the details of analysed data and results of tested functions are presented.

The data-set on which the functions are tested consists of 38 real seismic signals acquired in an underground mine which were basically not clipped. In Figures 2-6 the results of proposed functions are contained. In order to compare their effectiveness the tested function is the same in each figure. The exemplary function clipping level is 0.25.

In Figure 2 one can see the example fitting of the linear function. It can be seen that the function has the tendency to fit values more extreme (for the absolute value of signal, the estimated values greater than the actual ones).

In Figure 3 one can see the second order polynomial fitted to the artificially clipped signal. The estimated values have the tendency to be closer to the mean value of the signal than the actual values.

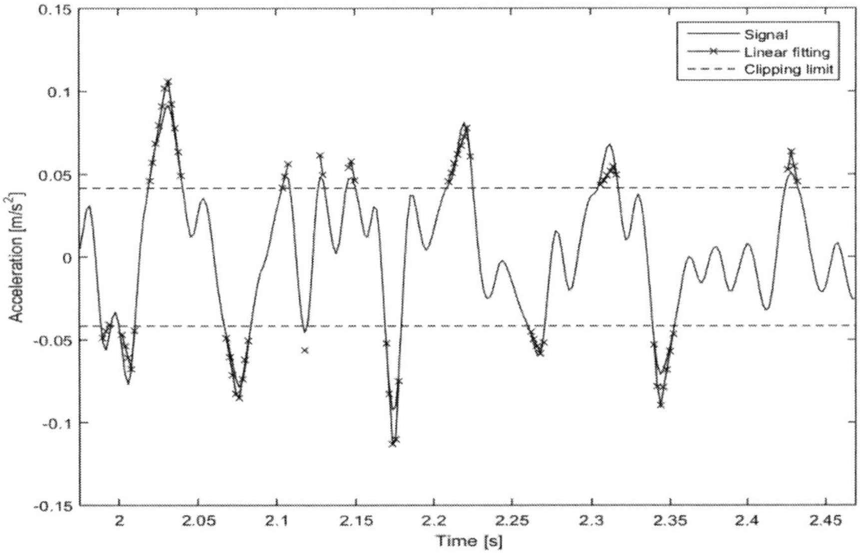

Figure 2. The example of linear function fitting used on clipped signal fragment

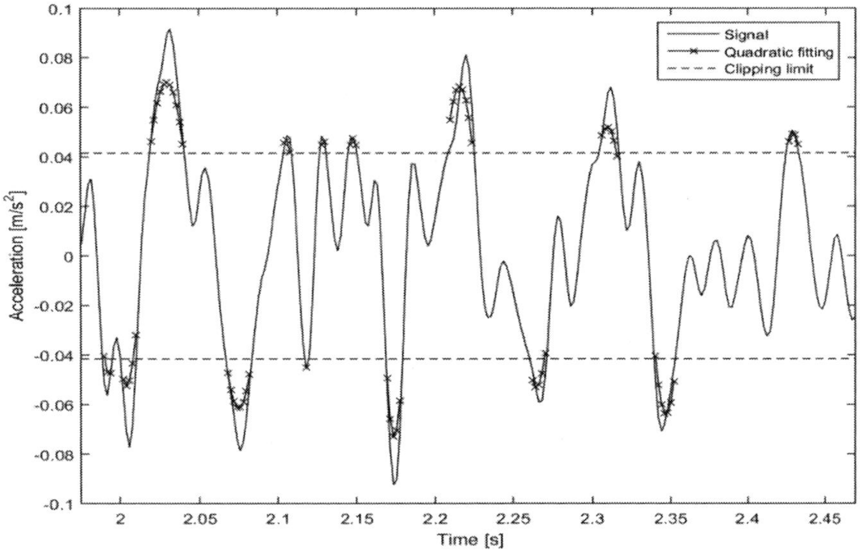

Figure 3. The example of second order polynomial function fitting used on clipped signal fragment

The fitting of the sine function can be seen in Figure 4. Clearly the values of the estimated function for the example signal are closer to actual values than for that in Figure 3.

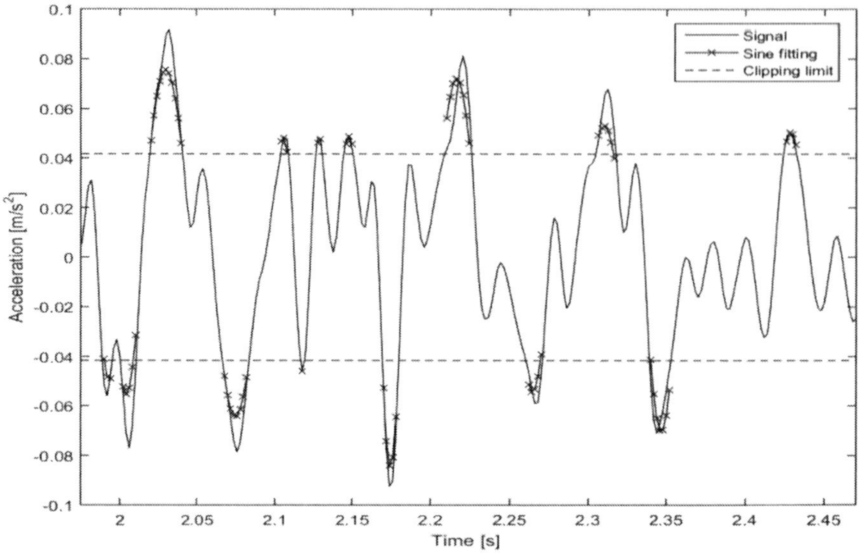

Figure 4. The example of sine function fitting used on clipped signal fragment

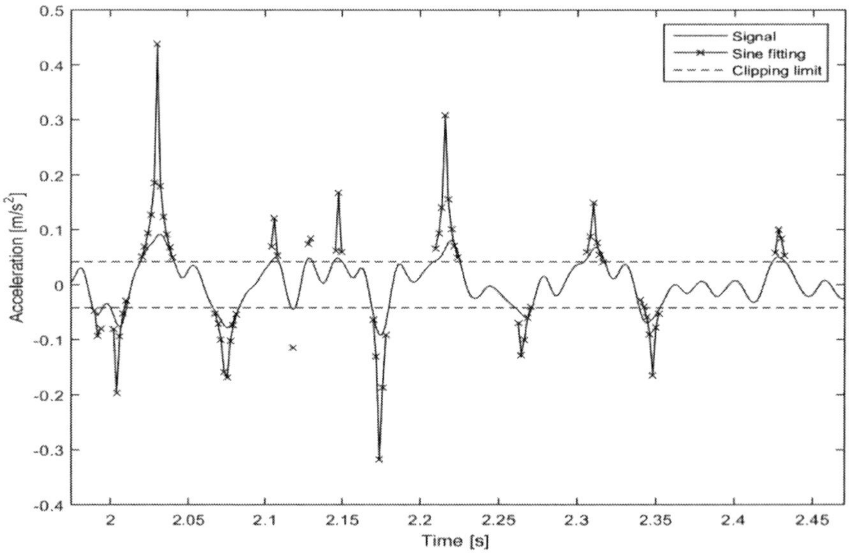

Figure 5. The example of power function fitting used on clipped signal fragment

In Figure 5 the power function fitted to the clipping is included. It can be seen, that the proposed function has the tendency to estimate values more extreme than the actual values.

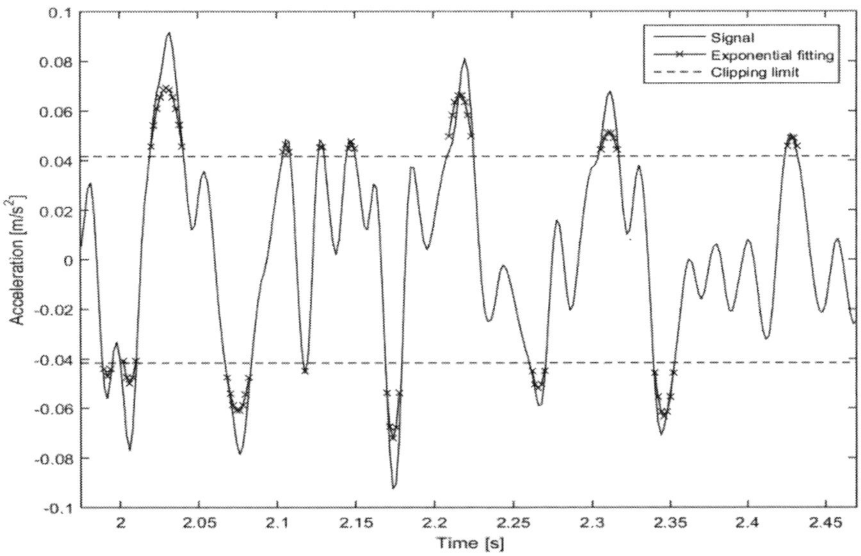

Figure 6. The example of exponential function fitting used on clipped signal fragment

The clipping with fitted exponential function can be seen in Figure 6. The results of this function seem to be similiar to those of the quadratic function (fitted function too smooth).

In Figure 7 the collective results for all different functions can be seen for one upper clipping (a), and one lower (b). Both examples show that the power and linear functions have the values steeper than the actual and the rest of the functions have the results considerably smoother (with considerably similar values). Moreover, the lack of the real signal symmetry can be seen (however, this can be due to the sampling frequency).

The summary of the functions fitting potential is presented in Table 1 for short clippings and in Table 2 for long ones.

Table 1. Clipping ≤ 5 samples

Clipping level	Linear	Quadratic	Sine	Power	Exponential
0.25	0.0157	0.0090	**0.0079**	0.0222	0.0101
0.40	**0.0156**	0.0221	0.0197	0.1980	0.0248
0.55	**0.0241**	0.0393	0.0364	0.0276	0.0421

In Table 1 the sum of squared errors between fitted functions and real signal for short clippings are included. It can be seen that for the clipping level 0.25 the sine function is the best fitted and the quadratic function is the second best. For clipping levels 0.40 and 0.55 the linear function is the best fitted function and the sine function is the second best for the 0.40 level, and power function for 0.55. The power function has the worst results for the 0.25 and 0.40 levels and the exponential function for 0.55. The good results of power function for the exemplary signal seem not to double for the rest of the signals.

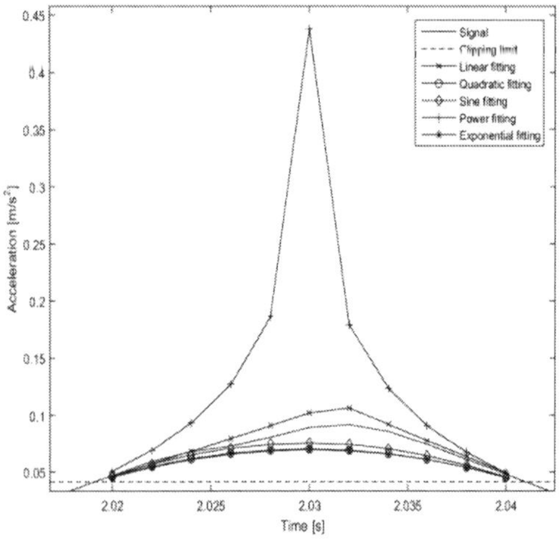

a)

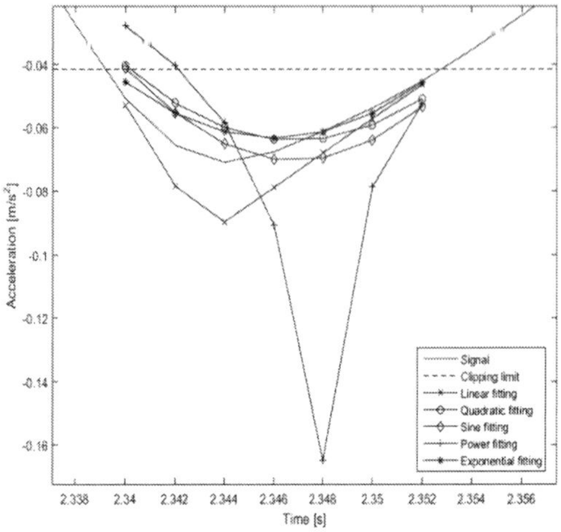

b)

Figure 7. Comparison of all tested functions on a) upper and b) lower clipping

Table 2. Clipping > 5 samples

Clipping level	Linear	Quadratic	Sine	Power	Exponential
0.25	0.0091	0.0014	0.0015	**0.0011**	0.0013
0.40	0.0096	0.0051	**0.0042**	0.0054	0.0075
0.55	0.0490	0.0631	0.0458	0.0938	**0.0252**

In Table 2 the results for the longer clippings are included. In this case the power, sine and exponential functions have the best fitting for the respectively 0.25, 0.40, 0.55 clipping levels and the exponential, quadratic and sine functions are the second best. In case of longer clippings the power function is considerably more accurate for clipping level 0.25 and 0.40.

Unfortunately none of the considered functions proved to be undoubtedly better than the other. This may be due to the considerably small testing dataset.

4 Conclusions

In this paper a problem of seismic signals clipping overcoming has been shown. In order to find the best function which could be fitted to hard-clipped data, five different functions were considered. Four of them were symmetric, and one (linear) not.

The function were tested on a real non-clipped seismic signals originating from an underground mine. To exhaust the subject, the signals were clipped on different level and each of these levels were considered separately. Also the clippings were divided into short (≤ 5 samples) and long.

The results proved to be ambiguous, as for different clipping level and length of clipping, different functions proved to be the best fitted. It is suspected that this may due to relatively small testing data-set. In future work it is suggested to enlarge testing set in order to make the results clear. Also it is recommended to check fitting potential of other non-symmetric function.

This work was financed by the Grant no. 0401/0128/17.

References

[1] A. Ray, IEEE Control Systems **11/2**, 44-49 (1991)

[2] A. Jablosnki, T. Barszcz, M. Bielecka, Measurement **44/10**, 1954-1967 (2011)

[3] A. Jablosnki, T. Barszcz, Mechanical Systems and Signal Processing **38/1**, 248-263 (2013)

[4] P. Kepski, T. Barszcz, Diagnostyka **4/64**, 25-30 (2012)

[5] P. Kepski, T. Barszcz, in: T. Fakhfakh, W. Bartelmus, F. Chaari, R. Zimroz, M. Haddar, *Condition Monitoring of Machinery in Non-Stationary Operations, Part 4* (Springer, Berlin, 2012) 461-469

[6] R. Zimroz, J. Wodecki, R. Krol, M. Andrzejewski, P. Sliwinski, P. Stefaniak, in. C. Drebenstedt, R. Singhal, *Mine Planning and Equipment Selection* (Springer, Cham, 2014) 1285-1294

[7] P. Stefaniak, R. Zimroz, J. Obuchowski, P. Sliwinski, M. Andrzejewski, Procedia Earth and Planetary Science **15**, 797-805 (2015)

[8] P. Stefaniak, R. Zimroz, W. Bartelmus, M. Hardygora, Advances in Condition Monitoring of Machinery in Non-Stationary Operations **4**, 250-263 (2016)

[9] S.Q. Li, IEEE Transactions on Communications, **37/11**, 1192-1202 (1989)

[10] R. Hossa, R. Makowski, R. Zimroz, International Journal of Rock Mechanics and Mining Sciences, **91**, 29-39 (2017)

Indicators of recent geodynamic activity in the Książ Castle area (Świebodzice Unit, Sudetes) in the light of structural analysis and geodetic measurements

Damian Kasza[1,*], *Aleksander* Kowalski[2], *Jurand* Wojewoda[2], and *Marek* Kaczorowski[3]

[1]Wroclaw University of Science and Technology, Wyb. Wyspianskiego 27, 50-370 Wroclaw, Poland
[2]University of Wroclaw, Pl. Uniwersytecki 1, 50-137 Wroclaw, Poland
[3]Space Research Centre of Polish Academy of Sciences, Bartycka 18A, 00-716 Warsaw, Poland

Abstract. Indicators of recent geodynamic activity in the Książ Castle area are registered by the measuring instruments of the SRC PAS (Space Research Centre of Polish Academy of Sciences) Geodynamic Laboratory at Książ. Over 40 years of continuous observations from quartz horizontal pendulums (since 1974) and over 10 years of observations from water-tube tiltmeters (since 2002) have documented irregularly repeatable strong signals related to the relative displacement of blocks in the rock substrate, on which Książ Castle is located. These signals have dip (rotational) and vertical strike-slip components. Also, the presence of a horizontal strike-slip component is evidenced by geometric anomalies (deformations) of the shape of the Pełcznica river valley, which directly correspond to the orientation of the main faults in the area. Recent geodynamic activity is documented by destruction of the construction elements in the castle complex. Instrumental indicators of movement, geodetic measurements and structural analysis of the rock massif have allowed for constructing a model showing the main unconformity surfaces in the analysed rock massif. Sinistral, NE–SW and ENE–WSW-oriented strike-slip faults prevail in the laboratory corridors, along with perpendicular WNW–ESE and NW–SSE-oriented dextral and normal faults. Most dislocations are accompanied by zones of intense cataclasys, secondary silification, and Fe and Mn mineralization. Generally, the faults were formed due to reactivation of joint fractures cutting the steeply N- and S-dipping (at 75–90°) deposits of the Książ Conglomerate Formation.

1 Introduction

The Świebodzice Unit (Sudetes, SW Poland) belongs to parts of Poland with recent geodynamic activity. This activity was confirmed by multiannual measurements with application of instruments working in the underground Space Research Centre of Polish Academy of Sciences (SRC PAS) Geodynamic Laboratory (GL) at Książ Castle, i.e. quartz

* Corresponding author: damian.kasza@pwr.edu.pl

horizontal pendulums (HP) and water-tube tiltmeters (WT). The registered phenomena are incidental in character, reflecting in form of relative dislocation of rock blocks and substrate tilting [1-5]. The dislocations take place in the Książ massif mainly along of the discontinuoity surfaces, i.e. faults within the massif. These include strike-slip and oblique-slip faults, but also low-angle thrusts, documented in the underground excavations of the GL and in natural exposures in the vicinity of Książ Castle.

One of the most probable indicators of geodynamic activity in the vicinity of Książ Castle are well-documented shape anomalies of the Pełcznica river valley [5]. These anomalies are most probably caused by the kinematics of the rock substrate, along fault surfaces, which constrained the modification of the river valley shape [6]. The idea of tectonic foundation of the Pełcznica valley and other river valleys incising the Świebodzice Unit (e.g. Czyżynek, Szczawnik and Lubiechowski Potok valleys) was confirmed by morphometric analyses conducted by Wojewoda [7]. These analyses, including determination and quantification of indicators of river channels torsion, have indicated the presence of numerous structural anomalies in the valley shapes, which correspond both in range and orientation to faults determined by mapping surveys in the Świebodzice Unit by other authors.

This paper presents the results of geodetic measurements and structural analyses, which indicate that the concept of the tectonically-induced development of the Pełcznica river valley and the Książ Castle area is correct. The obtained results indicate that the confirmed faults and fault zones were formed due to reactivation of pre-existing joint sets cutting the sedimentary rocks of the Książ and Chwaliszów formations. These faults indicate recent activity.

2 Characteristics of the study area

The Świebodzice Unit (ŚU), in older literature referred to as depression, trough, synclinorium or basin [7-10], is a small, rhomb-shaped, regional geological structure with an area of about 100 km^2, located in the Central Sudetes between Wałbrzych to the south, Strzegom to the north and Świdnica to the east (Fig. 1). The Świebodzice Unit is composed mainly of Carboniferous sedimentary rocks, traditionally assigned to the Pogorzała, Pełcznica, Chwaliszów and Książ formations. Blocks of Upper Devonian rocks, forming isolated and likely allochthonous rock bodies of various sizes, subordinately occur in the ŚU structure. From the south, the ŚU bounds with the Sowie Mts. Gneiss Massif along the Szczawienko Fault; from the east and north it is separated from the Fore-Sudetic Block by the Sudetic Marginal Fault; and from the west it adjoins with the Intra-Sudetic Synclinorium along the Strumyk (Struga) Fault. From the north, the boundary of the ŚU with the metamorphic rocks of the Kaczawa Complex (Cieszów Unit) is not defined precisely [7, 11, 12] and is most probably a system of smaller, generally sinistral strike-slip faults [12]. According to Teisseyre [13], the sedimentary rocks forming the ŚU are locally strongly folded, and the particular mesofolds have approximately W-E-oriented axes. The ŚU was for the first time distinguished as a separate structural element of the Sudetes in the 1950-ties [14, 15]. Authors of older geological maps concluded that the ŚU is an integral part of the Intra-Sudetic Synclinorium [16, 17].

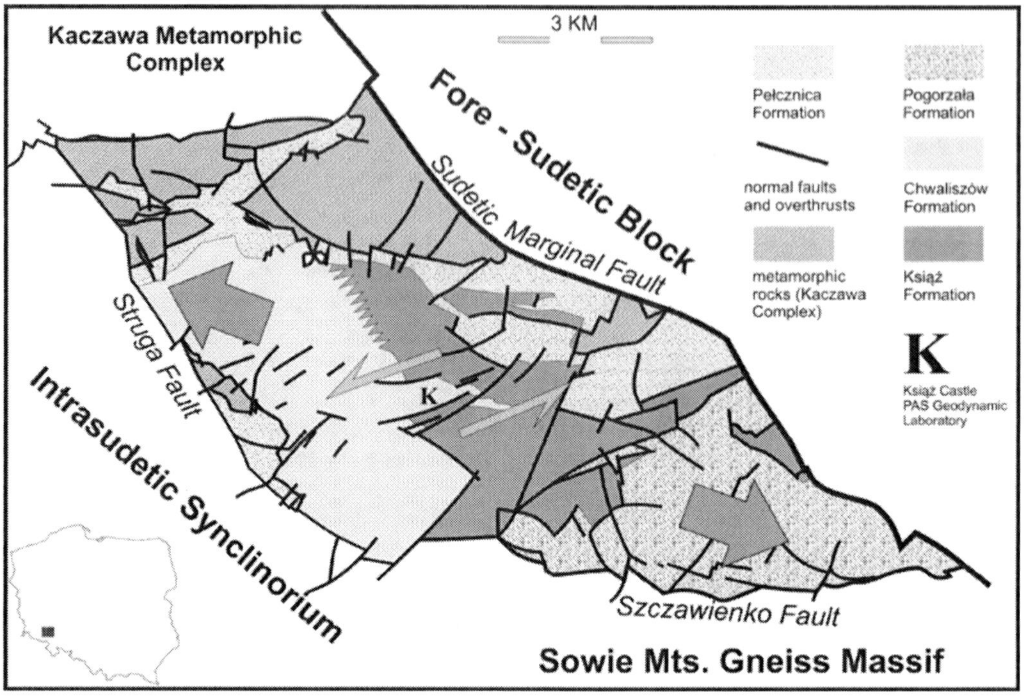

Fig. 1. Simplified geological map of the Świebodzice Unit [5].

The SRC PAS Geodynamic Laboratory at Książ is located within the Książ massif (Książ morphological elevation), situated in the central part of the ŚU.

2.1 Underground Geodynamic Laboratory at Książ

The Geodynamic Laboratory at Książ, belonging to the Space Research Centre of the Polish Academy of Sciences has been established in the underground excavations under the Książ Castle courtyard (Fig. 2). The corridors were excavated for military purposes at the end of World War II by the Nazi Todt Organization [18-20]. After World War II, the object was managed by the Silesian Military District, and in the 1970-ties was made available to the Polish Academy of Sciences for scientific research.

The first research laboratory in the underground of the Książ massif, the Geophysical Observatory of the Institute of Geophysics, Polish Academy of Sciences, was established following the initiative of Prof. Roman Teisseyre, who installed seismographs at Książ. In 1974, owing to Prof. Tadeusz Chojnicki, the research activity was extended to studies of tidal signals with application of quartz horizontal pendulums. In the early 2000s, thanks to Prof. Marek Kaczorowski, the underground SRC PAS Geodynamic Laboratory was established. New instruments were constructed, allowing for expanding the range of the research onto non-tidal processes, including e.g. kinematic effects in the rock massif [5, 21, 22].

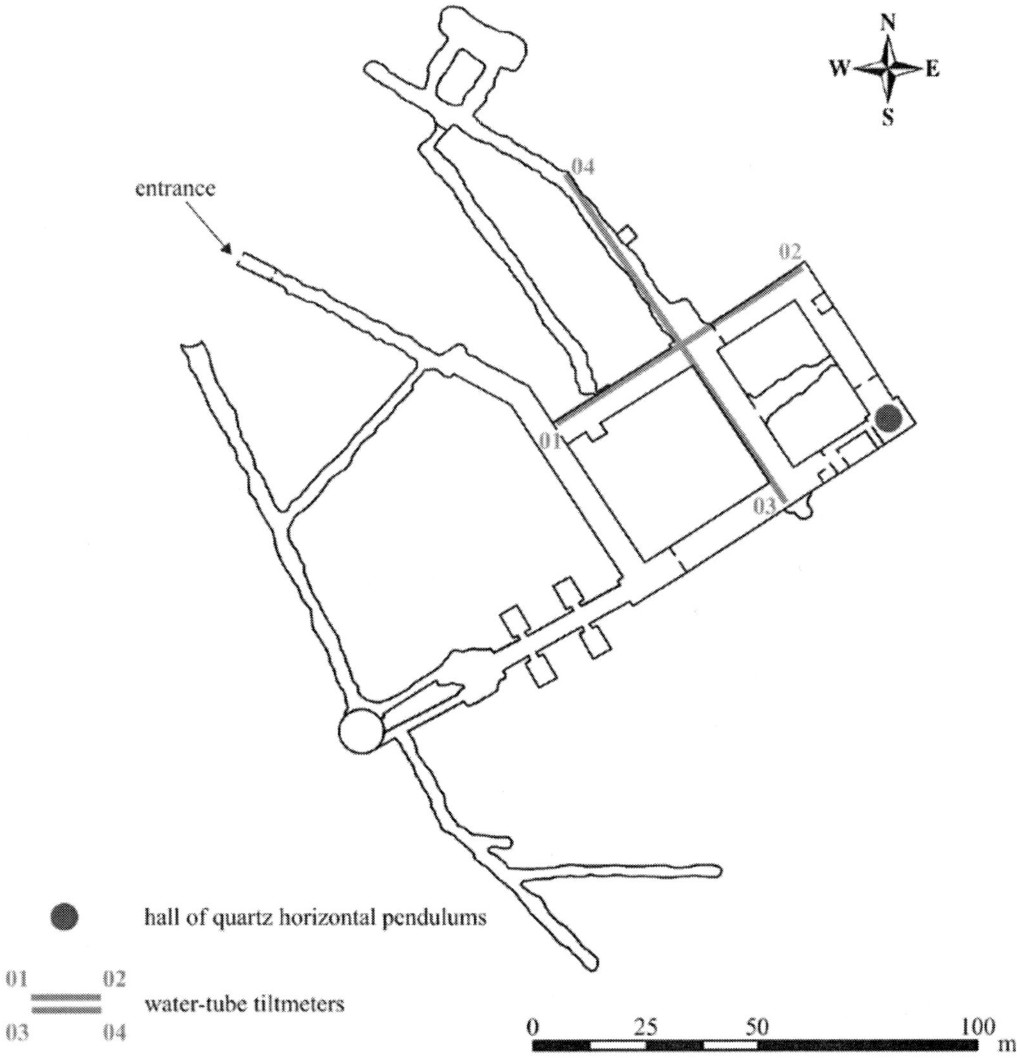

Fig. 2. Sketch-map of the underground excavations under the Książ Castle courtyard.

3 Indicators of recent geodynamic activity

Data on the effects of geodynamic activity in the Książ area are supplied by the GL measuring instruments. At first, they were quartz horizontal pendulums. The over 40-year continuous record of data indicates the presence of relatively short-term periods, when the pendulums relatively rapidly changed their measuring azimuths. This phenomenon was interpreted as being resulted from tilting of the rock massif (lithosphere fragment), on which the horizontal pendulums were installed, caused by relative displacement of tectonic nature [1]. Effects of geodynamic processes were confirmed in the Książ massif after the installation of subsequent instruments – water-tube tiltmeters. At present these instruments supply the most precise information in terms of quality and quantity on the short-term kinematic phenomena in the Książ massif.

3.1 Water-tube tiltmeter – composition and measuring principle

The water-tube tiltmeter is composed of two main elements: a hydrodynamic system (pipe), half-filled with water, and two interference gauges measuring the change of fluid level at its ends [2]. The instrument works on a principle that uses the hydrodynamic property of water attempting to attain hydrodynamic equilibrium [23]. When deformation effects occur, caused e.g. by Earth tides, symmetric changes of water level take place at the ends of the hydrodynamic system as a result of change of vertical. Using algorithms [3], the values of these changes are calculated to vertical movement. In the case of the GL at Książ, the hydrodynamic system (WT) is cut by several, recently active unconformity surfaces [5, 24].

3.2 Characteristics of the registered tectonic phenomena

The effects of kinematic phenomena during periods of increased activity in the Central Sudetes [25] are recorded as asymmetric changes of the signal representing water level oscillations at both ends of the water-tube tiltmeters. This asymmetry is caused by dislocation of blocks along unconformity surfaces, i.e. faults of the Książ massif [5]. Since their installation, the WT at Książ have registered several tens of phenomena with amplitudes several times exceeding tidal amplitudes (from 0.05 to 1.5 mm; Fig. 3). Moreover, these effects occurred during different intervals in the year and had different durations.

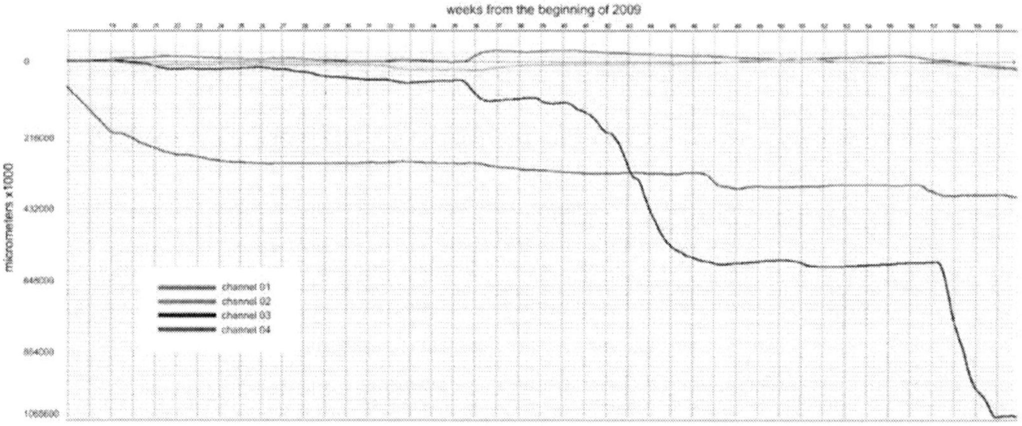

Fig. 3. Records from four measuring channels of water-tube tiltmeters from 2009.

4 Methodology of geological research

Mapping surveys in the Pełcznica river valley and the vicinity of Książ Castle were conducted in 2010–2017. They comprised the documentation of natural exposures and observations in the GL underground excavations. Structural measurements included bedding planes, fractures and faults. Particular attention was drawn on expressions of hydrothermal mineralization related with fault zones. Field measurements on the surface were made with application of Nomad Trimble and Columbus V-900 GPS receivers. Measurements of structural elements were marked on shaded relief maps based on the LiDAR digital terrain model (DTM), being the effect of aerial laser scanning conducted in Poland in 2011–2014. MicroDEM, Global Mapper 15.0 and Surfer (Golden Software) v. 9.0 software were used. The mapping materials and geological data were worked out and processed in the PUWG 1992 projection.

Measurements of dislocation zones, whose orientation is visible in the GL underground corridors, were made using geodetic techniques. An oriented total station with stabilized points of the horizontal control network was used for fault measurement. Spot heights locating each fault in the GL corridors (their coordinates) were also processed in the PUWG 1992 projection. The minimum number of spot heights necessary for determining the strike and dip of a fault zone is 3. Depending on the measurement conditions (complexity of the zone visible on the sidewalls and ceiling, presence of screens), the number of measured points varied from 3 to 9. As a result, a cloud of points representing spot heights along particular dislocation zones was obtained for each zone. During determination of directional parameters of each fault, an assumption was made that within the corridor cross-section the fault surface may roughly correspond to a plane. Accordingly, determination of the fault zone parameters was reduced to determining the plane equation in R^3. The plane equation was determined from the levelling of the set of spot heights at a condition of the minimum sum of squares of the distance of measured spot heights in each fault to the calculated plane [24]. The effect of these actions is presented in Fig. 4.

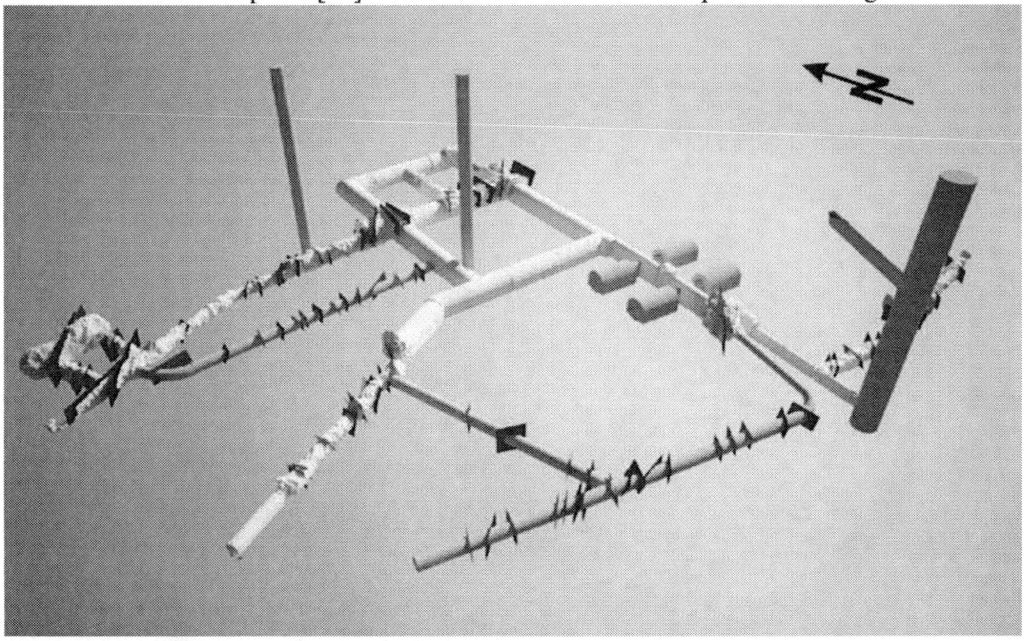

Fig. 4. Visualisation of the determined fault planes (blue) in the underground corridors of the Geodynamic Laboratory (grey and yellow).

5 Mapping survey in the Książ massif – results of structural analysis

Carboniferous deposits assigned to the Książ and Chwaliszów formations are well-exposed in the vicinity of Książ Castle. The Książ Formation mostly consists of coarse-grained, poorly-sorted conglomerates with single clasts that reach maximally up to 2 m in diameter. Most of the framework grains constitute gneisses, migmatites and granites with an admixture of sub-angular limestone clasts. In close vicinity of Książ Castle, conglomerates of the Książ Formation pass laterally to the north into deposits of the Chwaliszów Formation [9, 10], which consists of poorly-sorted, polymictic conglomerates interbedded with lithic sandstones. Deposits of the Książ and Chwaliszów formations that crop out in the Książ massif are strongly deformed (Figs. 5, 6). The bedding planes of the conglomerates and sandstones are W–E-trending and with steep dips (65–90°) to the N–NNW and S–SSE (Fig. 7A). This produces a visual effect of the presence of fold structures in the study area [9]. The investigated rocks are cut by joints of three main sets (j1, j2 and j3 respectively, cf. Fig. 7C, 7D), NW–SE and W–E-oriented dextral faults, and NE–SW-oriented sinistral faults (Fig. 7B). Numerous minor (secondary) fault-related structures were documented. They included mesostructural kinematic indicators [26] associated with brittle and strike-slip tectonics. Slickensides, grooves, hackles, striated ridges, low-angle R (Riedel) shears, *en echelon* cracks and high-angle antithetic R' shears were measured and documented (Fig. 6). Numerous extensional, NE–SW-trending extensional fractures associated with faults are filled with clay gouge, sometimes calcitified and impregnated with hematite mineralisation. This mineralisation is probably linked with the youngest stage of faulting. The measurements were plotted using a stereographic projection on the lower hemisphere.

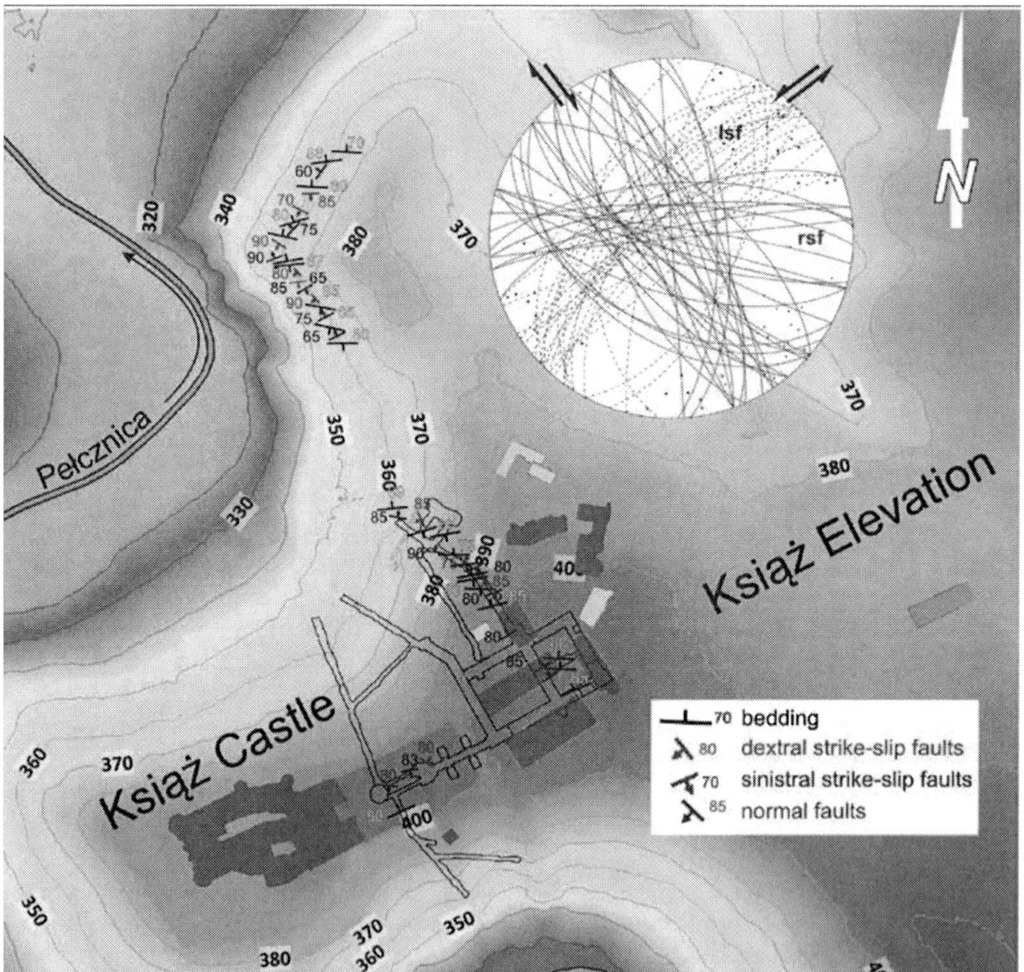

Fig. 5. Map showing the orientation of bedding planes in the deposits of the Książ Formation, and of the main fault structures in the Książ massif. Upper right diagram (Schmidt net, lower hemisphere) presents the orientation of the planes of dextral strike-slip faults (rsf, red) and sinistral strike-slip faults (lsf, blue), determined in the underground excavations of Książ Castle and in exposures on the surface.

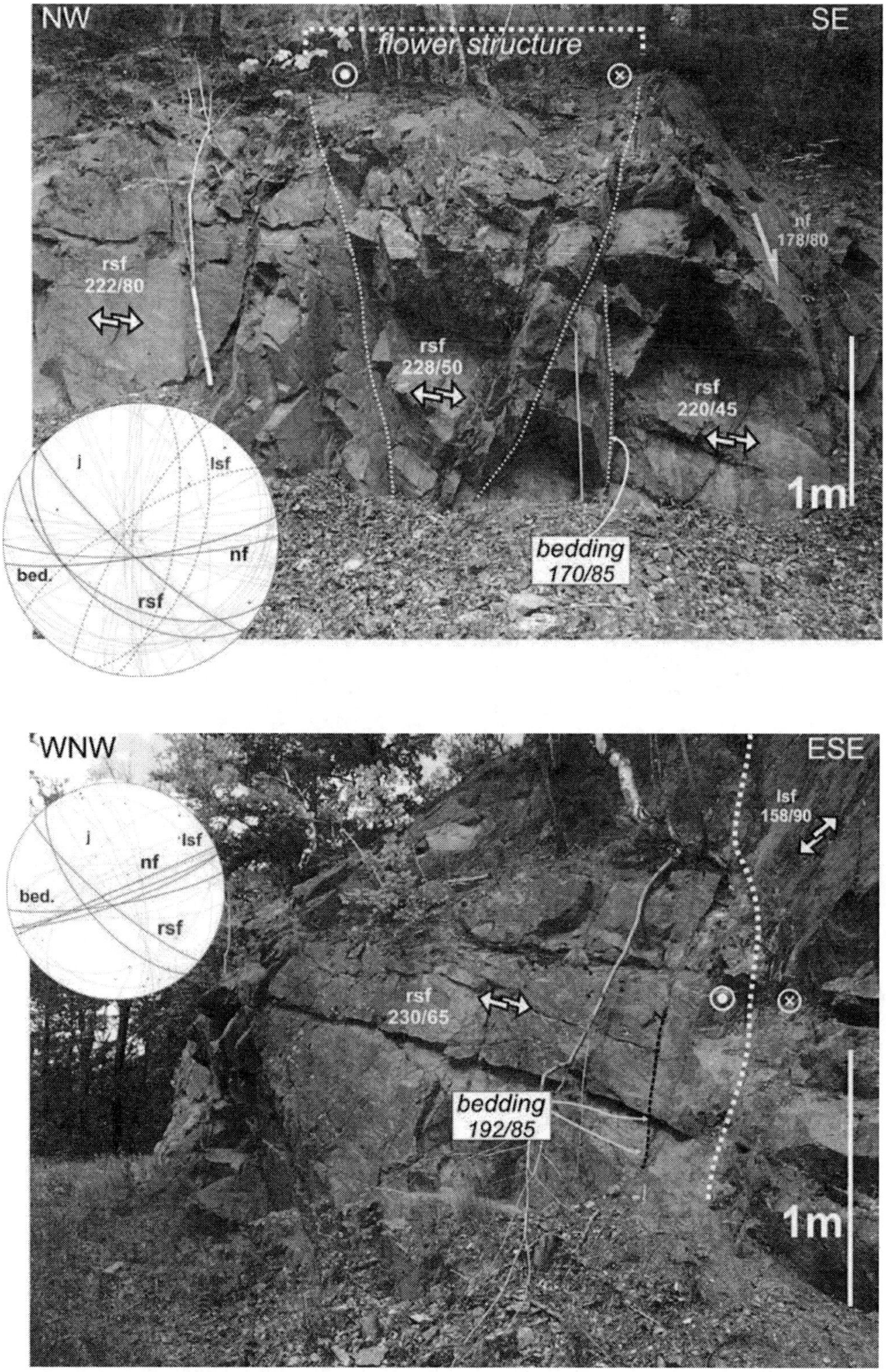

Fig. 6. Exposures of the Książ Formation to the north of Książ Castle. Explanations: bed. - bedding, j – joints, nf – normal faults. For other explanations see Fig. 5.

 https://doi.org/10.1051/e3sconf/20182900021

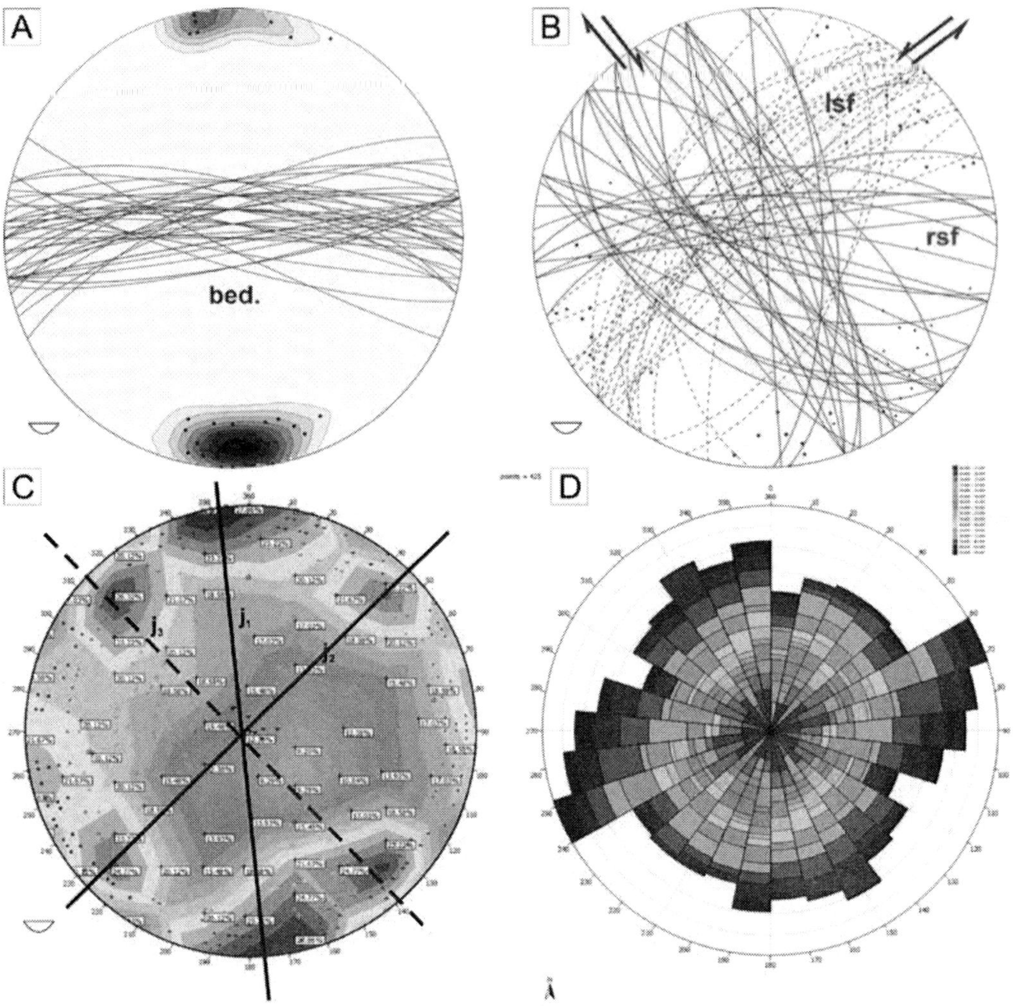

Fig. 7. Diagram showing the orientation of bedding planes (A; Schmidt net, lower hemisphere), strike-slip faults (B), joint fractures j1, j2, j3 (C), and composite rosette diagram of fractures in the Książ massif (520 measurements). Colour scale corresponds to the percentage contribution of particular classes of the fracture plane dip angles. For other explanations see Figs. 5, 6.

5 Discussion of the results

The presented investigations indicate the prevalence of oblique-slip and dip-slip faults in the formation of the central part of the Świebodzice Unit, at an insignificant contribution of thrusts and reverse faults. This conclusion is supported by morphometric analyses and sedimentological studies conducted in the ŚU by Wojewoda [7, 10]. Although our studies confirm the general kinematic scheme presented for the ŚU by Porębski [27] (simple, dextral shear along the bounding W–E and NW–SE-oriented dislocations), there is no evidence for the change of tectonic regime at the Frasnian to Famennian/early Tournaisian boundary from transtension to transpression, especially in relation to the new age-determinations for the deposits filling the Świebodzice basin [10]. Porębski [27] pointed to the significant role of the Intra-Sudetic Fault, interpreted as the major strike-slip zone controlling the Świebodzice basin, which did not find confirmation in mapping surveys.

The prevalence of strike-slip and normal zones points to the passive evolution of the central part of the ŚU (Książ elevation), mainly linked with transtension, and in consequence with the formation of an intra-basinal high in the area of the present day vicinity of Książ Castle. The process of formation of morphological and structural intra-basinal elevations within pull-apart basins was documented in experiments [28] and postulated for the ŚU by Wojewoda [7]. The main dislocations of the Książ massif are sinistral, NE–SW-oriented strike-slip faults, interpreted here as high-angle R' faults, complementary to the bounding dislocations of the ŚU. Sinistral strike-slip faults are reflected in the structural anomalies of the Pełcznica river valley determined during mapping surveys and DTM LiDAR morphometric analyses [10]. Although the determination of the exact time of the Pełcznica valley formation and of the largest activity of strike-slip and normal faults is difficult based on structural analyses, this process was most likely related to the stage of the last tectonic inversion (Neogene) in the Sudetes. Furthermore, the results of geodetic measurements indicate the recent tectonic activity of this area [1-4]. This activity was induced by sharp movement of the fault walls of both strike-slip and normal faults [25]. At the same time, the influence of rock massif collapse was also eliminated [29], as well as the effects of surface deformation of areas linked with former mining activity [30, 31] in the neighbouring areas (Wałbrzych Coal Basin).

6 Conclusions

Geodetic and geological investigations conducted in the Geodynamic Laboratory at Książ Castle supply new data on the displacement mechanism of blocks of the rock massif, on which the castle complex is built. Results of structural analyses and kinematic indicators recognized within the fault structures confirm the general kinematic trend proposed for the central part of the Świebodzice Unit by Wojewoda [7, 10]. Furthermore, field investigations conducted on the surface and in the underground excavations of Książ Castle confirm that the currently registered tectonic phenomena result from displacement along the dislocation zones that were active also in the past. These dislocations contributed to the formation of the present-day form of the Pełcznica river gorge, modified the channel course of this river, and influenced the present-day seismic activity of the area.

Noteworthy, indicators of the neotectonic activity are dangerous for the stability of the construction elements of the Książ Castle complex, recently deweloping its damages. Analysis of the above presented 3D model of the Książ massif structure confirms that part of the main, currently active dislocations zones run directly below the castle complex. Further geodetic surveys and structural investigations will aim at recognizing dislocation zones that may pose potential hazard to the Książ Castle.

This work was financed by the Polish Statutory Research Grant no. 0402/0137/16 and Grant no. 0401/0128/17. LiDAR data used for the construction of the topographic map (Fig. 5) were used under the licence DIO.DFT.DSI.7211.1619.2015_PL_N, issued by the General Surveyor of Poland to the Faculty of Earth Sciences and Environmental Management, University of Wrocław.

References

1. T. Chojnicki, P.A. Blum, Artificial Satellites, vol. **31**, no. 3, 123-129 (1996)
2. M. Kaczorowski, Artificial Satellites, vol. **39**, No. 2, 147-154 (2004)
3. M. Kaczorowski, *Earthquake Source Asymmetry, Structural Media and Rotation Effects*, Springer-Verlag Berlin Heidelberg, 493-520 (2006)

4. M. Kaczorowski, Acta Geodynamica et Geomaterialia, Vol. **6**, No. 3(155), 369-381 (2009)
5. M. Kaczorowski, J. Wojewoda, Acta Geodynamica et Geomaterialia, Vol. **8**, No. 3(163), 249-261 (2011).
6. S.A. Schumm, J.F. Dumont, J.M. Holbrook, *Active Tectonics and Alluvial Rivers* (Cambridge University Press, 2002)
7. J. Wojewoda, VI Polska Konferencja Sedymentologiczna POKOS 6 - "Granice Sedymentologii", 28.06.2016-1.07.2016, Chęciny - Rzepka, 145-148 (2016)
8. H. Teisseyre, Biul. Inst. Geol. **106**, 1-60 (1956)
9. S. Porębski, Geologia Sudetica **16,** 99-190 (1981)
10. J. Wojewoda, Przegląd Geologiczny **64** (9), 597-603 (2016)
11. S. Porębski, *Obszary źródłowe: Zapis w Osadach*, J. Wojewoda (ed.), T. **1**, 35-52 (1997)
12. Z. Cymerman, Prace Państwowego Instytutu Geologicznego **200**, 1-64 (2014)
13. H. Teisseyre, Biuletyn Instytutu Geologicznego **222**, 77-106 (1968)
14. H. Teisseyre, Rocznik Polskiego Towarzystwa Geologicznego **21**, 296-310 (1952)
15. H. Teisseyre, Biuletyn Instytutu Geologicznego **106**, 5-48 (1956)
16. G. Berg, E. Dathe, E. Zimmermann, Geologische Karte von Preussen und benachbarten Bundesstaaten, 1:25 000, Blatt Freiburg (1910)
17. R. Cramer, L. Finckh, E. Zimmermann, Geologische Karte von Preussen und benachbarten Bundesstaaten, 1:25 000, Blatt Schweidnitz (1921)
18. M. Dudziak, *Tajemnica Gór Sowich – przewodnik*, Wydawnictwo JMK (1996)
19. P. Kruszyński, *Podziemia w Górach Sowich i zamku Książ*, Wydawnictwo Muzeum Gross-Rosen (2004)
20. P. Zagożdżon, K. Zagożdżon, Hereditas Minariorum **3**, 267-279 (2016)
21. M. Kaczorowski, Centrum Badań Kosmicznych Polskiej Akademii Nauk, unpublished report (2010)
22. D. Kasza, Mining Science, Vol. **21**, 23-31 (2014)
23. F.R. Moulton, Astrophysical Journal **50**, 346-355 (1919)
24. D. Kasza, M. Kaczorowski, R. Zdunek, R. Wronowski, Acta Geodynamica et Geomaterialia, vol. **11**, No. 3, 225-232 (2014)
25. J. Stemberk, B. Kostak, S. Cacoń, Tectonophysics **487**, 1-12 (2010)
26. J.P. Petit, Journal of Structural Geology 9, 597-608 (1987)
27. S. Porębski, Obszary źródłowe: Zapis w Osadach, T. 1, J. Wojewoda (ed.), WIND, 35-52 (1997)
28. J.E. Wu, K. McClay, P. Whitehouse, T. Dooley, Marineand Petroleum Geology **26**, 1608-1623 (2009)
29. D. Kasza, Współczesne ruchy tektoniczne w aspekcie stabilności kompleksu zamkowego Książ, PhD dissertation, Wrocław University of Science and Technology, unpublished (2016)
30. J. Blachowski, W. Milczarek, P. Grzempowski, Acta Geodynamica et Geomaterialia, vol. **12**, No. 3(179), 227-236 (2015)
31. W. Milczarek, J. Blachowski, P. Grzempowski, Acta Geodynamica et Geomaterialia, Vol. **14**, No. 1(185), 41-52 (2017)

Reduction of atmospheric disturbances in PSInSAR measure technique based on ENVISAT ASAR data for Erta Ale Ridge

Anna Kopeć[1,*]

[1]Wroclaw University of Science and Technology, Faculty of Geoengineering, Mining and Geology, 27 Wyb. Wyspianskiego St., 50-370 Wroclaw, Poland

Abstract. The interferometric synthetic aperture radar (InSAR) is becoming more and more popular to investigate surface deformation, associated with volcanism, earthquakes, landslides, and post-mining surface subsidence. The measuring accuracy depends on many factors: surface, time and geometric decorrelation, orbit errors, however the largest challenges are the tropospheric delays. The spatial and temporal variations in temperature, pressure, and relative humidity are responsible for tropospheric delays. So far, many methods have been developed, but researchers are still searching for the one, that will allow to correct interferograms consistently in different regions and times. The article focuses on examining the methods based on empirical phase-based methods, spectrometer measurements and weather model. These methods were applied to the ENVISAT ASAR data for the Erta Ale Ridge in the Afar Depression, East Africa

1 Introduction

The Interferometric Synthetic Aperture Radar (InSAR), is a modern technique for measuring the deformation of the Earth's surface, using electromagnetic waves emitted by instruments installed on satellites. The electromagnetic wave, covering twice the distance between the satellite, and the measuring object, is exposed to the effects of various external factors. One of them is the Earth's atmosphere. Currently, the reduction of impact of the atmosphere, is one of the biggest challenges for measurements with the InSAR techniques.

2 The research object (Erta Ale Ridge)

The volcanic ridge Erta Ale, is situated in the area of the Afar Depression, in Eastern Africa, at the junction of the three tectonic plates: Nubian (African), Arabic, and Somali. This is an

* Corresponding author: anna.kopec@pwr.edu.pl

area of active continental rift, the northern fragment of East African Rift (Fig. 1). In the area of Afar, the Arab and Somali Plates move away from the stable Nubian Plate [1]. The Erta Ale region is of shield volcano nature, which is characteristic of the continental rift, still there lie also stratovolcanoes. The volcanic chambers occur very shallowly beneath the surface of the Earth (< 2 km). In the very crater of Erta Ale, there is an exposed lava lake [2]. Eruptions, closely related to the seismic shocks, can cause considerable dislocations, both in the vertical and in the horizontal plane [3].

The important object in the region of the Erta Ale Massif is the volcanic group Alu Dallafila, situated 30 km south of the main crater of Erta Ale. Alu - Dallafila is made up of basalt volcanic cone - Alu, stratovolcano- Dalafilla and basalt volcanic area Alu South lying between them (Fig. 1). Until 2008, only fumarolic activity was observed. On November 3rd, 2008 the eruption began, which started with the earthquake. The eruption was associated with significant displacement of the terrain, reaching the value of up to 1.9 m for the Alu cone and about 1 m in the Alu South region [3]. Today, it is the only one observed eruption of this area in the Erta Ale region.

The Afar Depression Area, due to the lack of vegetation and the prevalence of desert areas, volcanic covers and saltwater lakes, allows for very large numbers of Pernament Scatter (PS) points.

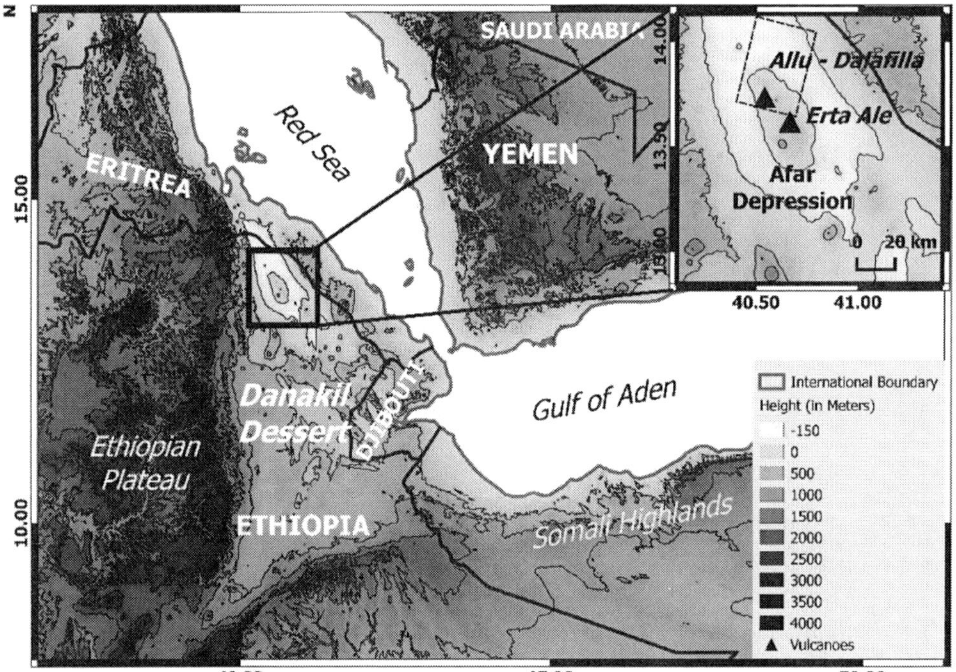

Fig. 1. The research object Erta Ale Ridge – dashed polygon indicate the ground area of InSAR data.

3 PSInSAR measurements

The calculations used 45 ENVISAT-ASAR satellite imaginery, from path 049, taken in the period since November 28th, 2003 to September 17th, 2010. The imagery area includes the northern part of the Erta Ale Massif - including Alu - Dallafila volcanoes, measuring approximately 29 x 38 km and the area of approximately 1 100 km^2. The data cover the period before and after the earthquake in November 2008. The interferograms were created using the Delft Object-oriented Radar Interferometric Software (DORIS) [4]. PSInSAR processing was performed in Stanford Method for Persistent Scatterers (StaMPS) software [5]. A total of 205,793 PS points were identified, characterized by a large and stable wave reflection time. All interferograms refer to the master image taken on November 21st, 2008, so the significant land displacements, due to earthquakes and eruptions of November 3rd, 2008, would be visible to images taken prior to the master image. The results obtained, confirm the previous research carried out by Pagli [3]. In the area of the Alu cone, the terrain was reduced by more than 1 m (Fig. 2).

It should be noted, that the calculation results, for both the displacements and the delays are set in radians, formula (1) should be applied to change the unit to centimeters.

$$\Delta\phi_{cm} = \frac{\lambda \cdot \Delta\phi}{4\pi} \qquad (1)$$

where:
ϕ_{cm} – phase change in centimeters,
λ – the radar wavelenth,
$\Delta\phi$ – phase change in radians.

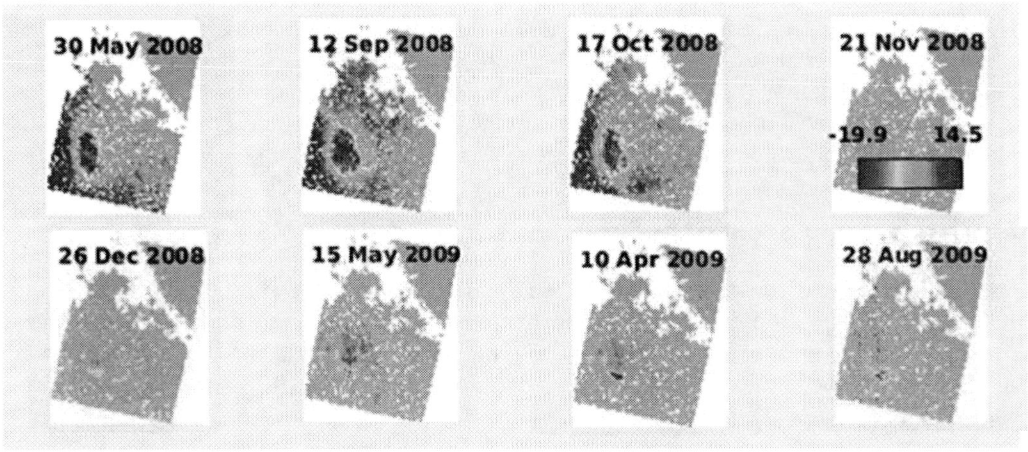

Fig. 2. Unwraped phase reduced by DEM and orbits errors for selected images, values are set in radians.

4 Atmospheric delay in the InSAR measurements

The inhomogeneity of the atmosphere (the ionosphere and the troposphere) and its variability in time and space, cause change in the signal speed on the geometrical path between the antenna and the surface of the terrain. It has been observed in the 1990's that this may create problems in the interpretation of interferograms [6]. The atmospheric delay is also referred to as an atmospheric signal, or an atmospheric phase screen (APS). Measurement errors, resulting in atmospheric delay, can range up to ± 15 - 20 cm [7], which can have a significant impact on the identification and interpretation of deformation. Conditions in the troposphere and the ionosphere have the greatest impact on the signal course. Due to the differential nature of the InSAR technique measurements, the difference between two interferogram images, creating the given interferogram, is more significant than the value of the atmospheric delay.

In the ionosphere, the course of an electromagnetic wave is disturbed by changes in the Total Electron Content (TEC) along with its propagation path. Because of the greater width and dispersion nature of the phenomenon, the ionospheric refraction has a much greater effect on the propagation of longer waves ranging from P and L. Signal disturbances for C waves are minimal and do not affect the InSAR measurements [8].

In the case of C waves, atmospheric delay is often referred to as a tropospheric delay because it is caused by tropospheric refraction. In the troposphere there is a high spatial variability of such factors as temperature, pressure, and relative air humidity. The tropospheric refraction can be divided into two components: the dry hydrostatic (N_{hydr}) and the wet hydrostatic (N_{wet}). The tropospheric delay for the signal phase is determined by the change of the two refraction components along the geometrical path between the antenna and the terrain surface (2).

$$\phi_{tropo} = \frac{-4\pi}{\lambda}\frac{10^{-6}}{cos\theta}\int_{h}^{h_{top}}(N_{hydr} + N_{wet})dx \qquad (2)$$

where:
θ - the incidence angle.

4.1 The wet component of the tropospheric refraction

The wet component of the tropospheric delay depends on the distribution of relative air humidity, which changes dynamically, both in the vertical and horizontal plane. These changes are turbulent and result in many factors, including: thermal convection, differences in direction and speed of wind at different altitudes, friction, and extensive weather systems. The delay resulting from these disturbances is very difficult to model [7, 9].

4.2 The dry component of the tropospheric refraction

The dry component of the tropospheric delay is represented by the vertical temperature distribution and dry air pressure. The troposphere is divided into an infinite number of layers characterized by various refraction values, and their vertical span is variable over time. This is important for the terrains with large areas of height differences [9].

5 Determination and reduction of delay

Different approaches are used to determine the tropospheric delay in InSAR techniques and additional data from: spectrometers [7, 10], weather models [7, 11], GNSS measurements [12] are used. However, data from different sources are characterized by different temporal and spatial resolution in relation to the interferometric data. The approach based on determining delay in an empirical way, directly from the interferograms [7, 13] is also applied. In this paper, delay has been determined by means of three methods: the spectrometric, empirical (linear), and based on weather data models.

5.1 Empirical method - linear

Tropospheric delay, empirically designated directly from interferograms, is specified by the linear relationship between the phase and topography (3) [13].

$$\Delta\phi_{tropo} = K_{\Delta\phi}h + \Delta\phi_0 \qquad (3)$$

where:
$K_{\Delta\phi}$ is the estimate of the top-phases relation,
h - ground level,
$\Delta\phi_0$ – constant coefficient of deviation from the entire inerferogram.

Determining the delay is not differentiated into wet and dry refraction component, this is the total relative delay between the *master* and the *slave* image.

The phase-based and topography-based methods are not effective in the areas where there is no large height difference and errors are mainly due to seasonal and local atmospheric variability.

5.2 The spectrometric method

The spectrometric data allow to measure the total amount of Precipitable Water Vapor (PWV). The PWV is determined by the ratio of radiation absorbed on the channel sensitive and insensitive to the presence of water vapor in the air [14]. Based on the PWV, a Zenith Wet Delay (ZWD) can be determined - formula 4, followed by wet tropospheric delay formula 5 [14],

$$ZWD = \Pi \cdot \mathrm{PWV} \tag{4}$$

where Π is the PWV transformation coefficient for zenithal delay. Π usually takes on the value of about 6.2.

$$\phi_{tropo}^{wet} = \frac{-4\pi}{\lambda} \frac{ZWD}{\cos\theta} \tag{5}$$

The spectrometric data allow to determine only the wet delay component that is associated with turbulent tropospheric disorders. Spectrometers are passive sensors that perform measurements exclusively in daytime and under clear conditions. The calculations used data from the MERIS spectrometer, which operated on the ENVISAT platform. This avoids errors associated with temporal and spatial interpolation.

5.3 The ERA-I weather model

Weather models are generated based on the surface and satellite data, providing information on pressure, temperature and relative air humidity. They allow both, wet and dry refraction to be determined. ZWD is determined in the same way as for spectrometric measurements - formulas 4 & 5. The Zenith Hydrostatic Delay (ZHD) is calculated on the basis of the Saastamoinen model [15] - formula 6, where the hydrostatic delay is determined by the relation between the latitude (φ), the height (h) and the atmospheric pressure (P).

$$ZHD = 0.0022768 \frac{P}{1 - 0.00266\cos 2\varphi - 0.00028h} \tag{6}$$

The ERA-I model of the resolution of about 80 km was used at the calculations. The temperature and relative humidity values are determined for 37 levels with a certain atmospheric pressure. The data acquisition interval is 6 hours. Because of the low temporal and spatial resolution, the data require interpolation.

6 Results

The reduction of delay was based on the three above described methods. It has been proven which of these methods provides the best results for the data, which illustrate a significant, rapid land displacement. The methods were compared in qualitative and quantitative terms. Qualitatively, by the data accessibility, comparing the spatial distribution of delays with terrain topography. Quantitatively, by comparing the numerical values of delays and their Root Mean Square Error (RMSE), calculated with respect to wet-field spectrometry and weather model for dry components.

The results of the empirical delay calculation by the linear method show a strong dependence on the displacement values. The delay takes on the values even up to -22 rad (about 100 cm), only for the images taken before the earthquake. The spatial location of the delay coincides also with the location of the land displacement. For other interferograms, the delay is not varied (Fig. 3a).

The spectrometric method allowed for the determination of delay for 30 out of 44 interferograms. This was due to the lack of data availability for some images - the measurement was taken under unclear conditions, or at night. It has been assumed that the state of the sky should not exceed 20%. The calculated wet component of the delay reaches values up to approx. ± 20 rad (approximately ± 90 cm). However, the most of the values oscillate around a few radians. The delay is not explicitly correlated with topography. For several images, such as of May 30th and September 12th, 2008, and April 10th, June 19th and October 2nd, 2009, the clear limit of variation of delay in the Allu-Dallafila Massif can be seen (Fig. 3b).

The results of calculations based on data of the ERA-I weather model show the smallest delay values. For the hydrostatic component, the maximum is approx. ± 0.1 rad (approx. ± 0.4 cm) while for the wet component is approx. ± 3 rad (approx. ± 13 cm). It can be seen for the most interferograms, that the values are interpolated (Fig. 3c, d). This is due to the fact, that the resolution of the model is only 80 km, which is not enough for such a small area (29 x 38 km).

It has been verified to what extent the delays determined by the linear method and based on the weather model, for the both components, depart from the delay results of the spectrometric method combined with the dry component of the weather model. The results are presented in Table 1 and, testify to the considerable variation of linear method with respect to the spectrometric method and the weather model. RMSE for the linear method is 0.63 cm larger, compared to the weather model.

Table 1. RMSE comparison of investigation methods.

Reference - spectrometric method combined with the dry component of the weather model	Empirical linear method	Weather model (both components)
RMS	RMSE	
2.02 *cm*	2.65 *cm*	2.02 *cm*

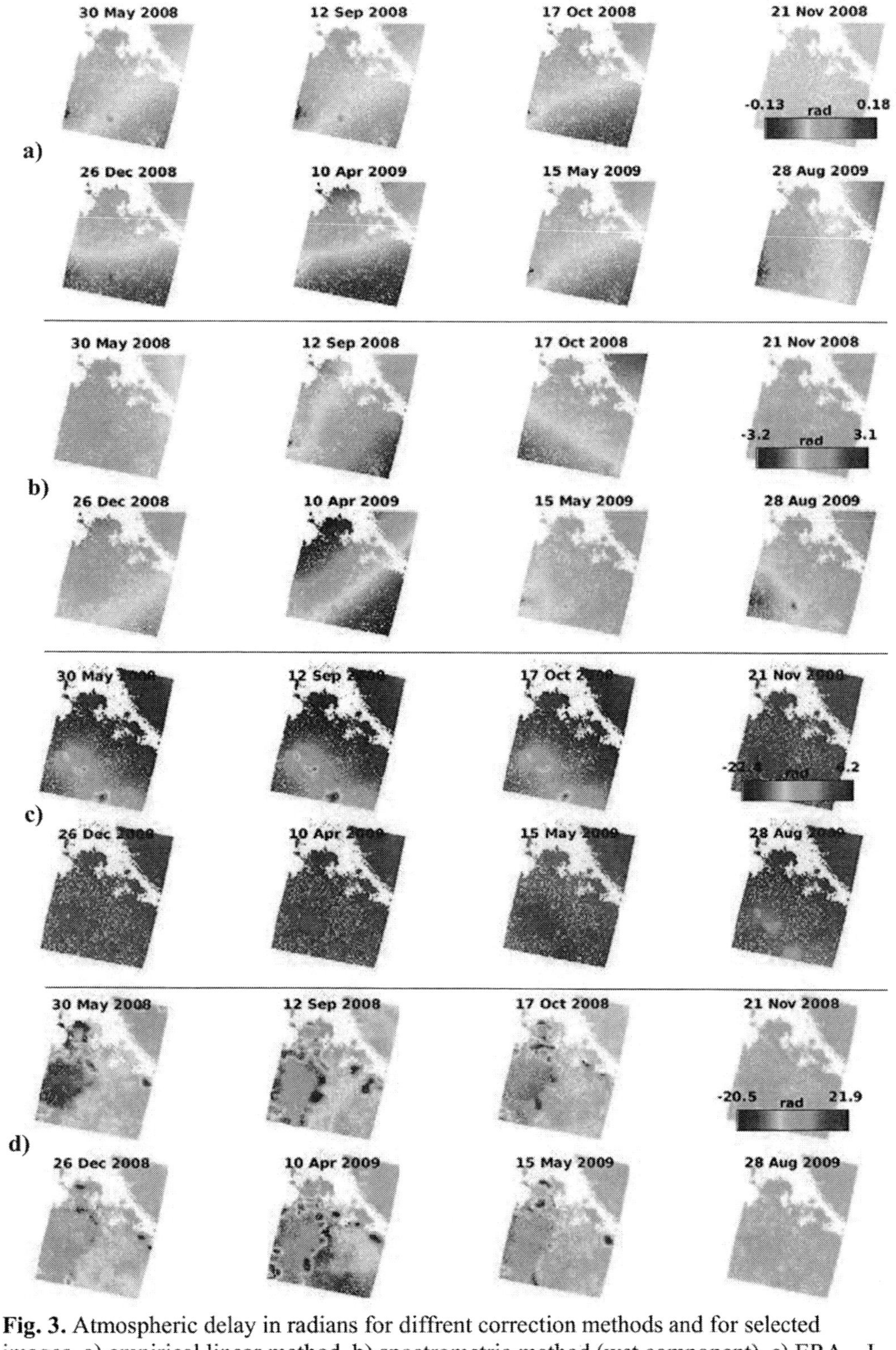

Fig. 3. Atmospheric delay in radians for diffrent correction methods and for selected images, a) empirical linear method, b) spectrometric method (wet component), c) ERA – I model (dry component), d) ERA – I model (wet component).

6 Summary

The atmospheric delay in InSAR measurements can take large values that may have a significant impact on the calculation results. Choosing the right reduction method is not easy and depends on many factors. In this case, the linear method, working well for areas with different topography, should not be used due to the impact of high displacement on the delay value. The spectrometric method seems to be the most uninfluenced, but in some cases the data are not available. The weather model, due to its low resolution, does not provide sufficiently accurate results.

The atmospheric constituent of the SAR signal is determined by various factors, that affect its value and spatial distribution, in different ways. The choice of the reduction method, should be individually selected for the particular research area and data set.

Partially processed SAR images were used in the paper in the form of single look complex (SLC) images. which have been obtained under the grant number 18310: *Identification of rock mass surface deformations on abandoned mining areas* (project manager, Wojciech Milczarek, Ph.D., Eng.), granted by the European Space Agency. The DORIS software [4] and StaMPS software [5] were used for the PSInSAR calculations. The tropospheric correction was performed using the TRAIN software [7]. The data from the European Space Agency - the spectrometric MERIS data, and the data from the European Centre for Medium-Range Weather Forecasts (ECMWF) - the ERA-I weather model, were used to determine the tropospheric delay. This work was financed by the Grant no. 0401/0128/17.

References

1. S. Dumont, Y. Klinger, A. Socquet, C. Doubre, E. Jacques, J. Struct. Geol. **95**, 48 (2017)
2. F. Barberi, J. Varet, Bull. Volcanol. **34**, 848 (1970)
3. C. Pagli, T.J. Wright, C.J. Ebinger, S-H. Yun, J.R. Cann, T. Barnie, A. Ayele, Nat. Geosci. **5**, 284 (2012)
4. B. Kampes, R, Hanssen, Z. Perski, *FRINGE* (2003)
5. A. Hooper, D. Bekaert, K. Spaans, M. Arikan, Tectonophysics **514**, 1-13 (2012)
6. R. Goldstein, Geophys. Res. Lett. **22**, 2517 (1995)
7. D.P.S. Bekaert, R.J. Walters, T.J. Wright, A.J. Hooper, D.J. Parker, Remote Sens. Environ. **170**, 40 (2015)
8. A.L. Gray, K.E. Mattar, Geophys. Res. Lett. **27**, 1451 (2000)
9. R.F. Hannsen, *Radar interferometry: Data interpretation and error analysis*, Kluwer Acad. Publ., Dordrecht (2001)
10. Z. Li, E.J. Fielding, P. Cross, R. Preusker, Int. J. Remote Sens. **30**, 3343 (2009)
11. M.-P. Doin, C. Lasserre, G. Peltzer, O. Cavalié, C. Doubre, J. Appl. Geophys. **69**, 35 (2009)
12. F. Onn, H.A. Zebker, J. Geophys. Res. **111**, 2517 (2006)
13. Y.N. Lin, M. Simons, E.A. Hetland, P. Muse, C. DiCaprio, Geochem. Geophys. Geosyst. **11** (2010)
14. Z. Li, J.-P, Muller, P. Cross, P. Albert, T. Hewison, R. Watson, J. Fisher, R. Bennartz, *MERIS User Workshop* (2003)
15. J. Saastamoinen, *Atmospheric Correction for the Troposphere and Stratosphere in Radio Ranging Satellites, in The Use of Artificial Satellites for Geodesy*, AGU, Washington, D.C. (1972)

First experience with Remote Sensing methods and selected sensors in the monitoring of mining areas – a case study of the Belchatow open cast mine

Jaroslaw Wajs[1,*]

[1]Wroclaw University of Science and Technology, Faculty of Geoengineering, Mining and Geology, 27 Wyb. Wyspianskiego St., 50-370 Wroclaw, Poland

Abstract. The paper presents satellite imagery from active SENTINEL-1A and passive SENTINEL-2A/2B sensors for their application in the monitoring of mining areas focused on detecting land changes. Multispectral scenes of SENTINEL-2A/2B have allowed for detecting changes in land-cover near the region of interest (ROI), i.e. the Szczercow dumping site in the Belchatow open cast lignite mine, central Poland, Europe. Scenes from SENTINEL-1A/1B satellite have also been used in the research. Processing of the SLC signal enabled creating a return intensity map in VV polarization. The obtained SAR scene was reclassified and shows a strong return signal from the dumping site and the open pit. This fact may be used in detection and monitoring of changes occurring within the analysed engineering objects.

1 Introduction

Monitoring of mining areas with application of geodetic measurements is the basic task of mining surveys. Classical geodetic measurements include linear probing of the terrain by observation lines, and surface probing such as measurement with use of a regular grid within the surveyed mining area [1]. Classical observation systems in open cast mines use surface, photogrammetric and satellite GNSS geodetic measurements as a source of input data for GIS 3D databases [2]. The paper presents selected applications of satellite remote sensing in the monitoring of mining areas and their surroundings with regard to assessing the influence of open cast mining on the environment [3, 4]. Issues of remote sensing monitoring are presented in macro scale [5]. The processed remote sensed data have allowed for large-surface analysis as a tool for identifying local anomalies and identification of potential hazards [6].

* Corresponding author: jaroslaw.wajs@pwr.edu.pl

2 Motivation

Being the member of the European Space Agency since 2012, Poland has become a beneficiary of the Copernicus programme, which continues the European GMESS programme. Products of satellite images in the Copernicus programme allow for acquisition of high-resolution radiometric, spatial and temporal remote sensing data. The paper analyses scenes from passive sensors SENTINEL-2A and 2B. A subsequent part of the research were scenes from active sensors SENTINEL-1A and 1B. The present state of sensor development for 2016 was presented by Toth et al. [7], and the chances and challenges for future years were discussed by Ban [8].

3 Study area

The analyses were performed in the PGE GiEK KWB Belchatow open pit lignite mine located in central Poland, Europe. The KWB Belchatow object comprises two excavation areas. The first exploited area was the Belchatow exposure, marked with an A in Fig. 1. The outline of the open pit and of the internal dumping site is marked with a black rectangle. Letter B marks the Szczercow excavation. Letters D and E depict areas representing the exploited lignite deposit. The ROI included the Szczercow external dumping site and is marked with a white square and the letter E.

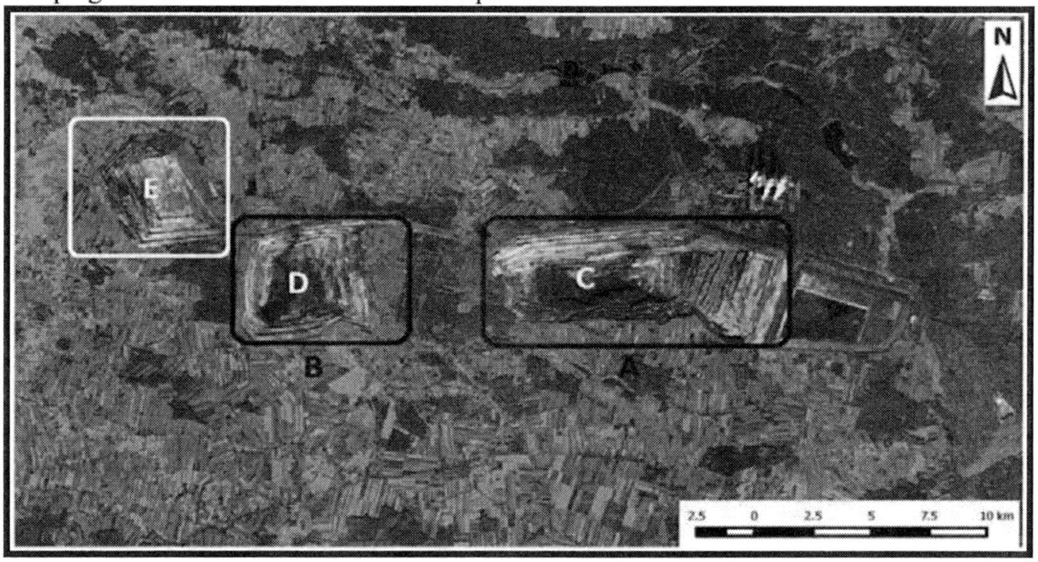

Fig. 1. Portion of Landsat 8 scene of the Belchatow open pit lignite mine. The combination is in a false colour 6-5-4 band acquired on 24th August 2014.

4 Methodology

The methodology of data analysis was based on the study of available remote sensing sensors for ROI analyses. A subsequent stage was the elaboration of the processing path of SENTINEL-2 and SENTINEL-1 data. The research included selecting algorithms, tools, data processing and spectral classification focused on detection of changes within the analyzed Szczerców dumping area.

4.1 Analysed Remote Sensing sensors

The first part of the research focused on the overview of remote sensing sensors used for the analysis. Data presented in Table 1 were classified with regard to type of sensor, spatial and temporal resolution, return time, and application for monitoring of mining areas. The data indicate that RapidEye observation system have the best field resolution, reaching 5 m on the equator. Rapideye imagery may thus become a competition for SENTINEL-2 MSI scenes. SENTINEL-2 is a European wide-swath, high-resolution, multispectral operational imaging mission within the GMES (Global Monitoring for Environment and Security) programme, jointly implemented by the EC (European Commission) and ESA (European Space Agency) for global land observation. It was launched on 23rd June 2015 and the first imagery of ROI was acquired in August 2016. The full mission specification of the twin satellites flying in the same orbit but phased at 180°, is designed to give a high revisit frequency of 5 days at the equator. The twin satellites of SENTINEL-2 provide a continuity of USGS services: SPOT and LANDSAT imagery datasets science 1972. SENTINEL-2 carries out a Multi-Spectral Optical Instrument payload that samples 13 spectral bands: four bands at 10 m, six bands at 20 m and three bands at 60 m spatial resolution. The orbital swath width is 290 km. The SENTINEL-2 mission orbit is sun-synchronous, with 143 orbits per cycle for a single satellite. Sun-synchronous orbits are used for consistent maintenance of the angle of sunlight on Earth's surface. SENTINEL-2A and SENTINEL-2B occupy the same orbit, but are separated by 180°. The mean orbital altitude is 786 km. The orbit inclination is 98.62° and the Mean Local Solar Time (MLST) at the descending node is 10:30 (am). This value of MLST was chosen as a compromise between the suitable level of solar illumination and minimization of the potential cloud cover.

The second analyzed sensor is SENTINEL-1, which continues the C-band SAR Earth Observation heritage of ESA's ERS-1, ERS-2 and ENVISAT, and Canada's RADARSAT missions. The satellite is equipped with a C-band Synthetic Aperture Radar (SAR), which acquires the Interferometric Wide Swath mode (IWS): 240 km swath, 5 m x 20 m resolution, and a single-look complex. SENTINEL-1 has a near-polar, sun-synchronous orbit with a 12-day repeat cycle and 175 orbits per cycle for a single satellite. Both SENTINEL-1A and SENTINEL-1B share the same orbit plane with a 180° orbital phasing difference. With both satellites operating, the mission images the entire Earth every six days.

Table 1. Overview of remote sensing sensors used for the analysis.

Name	Funding	Launch	Country	Constell ation	Sensor type	GSD range (m)	Swath width (km)	Revisit time (day)	Access
RapidEye	Commercial	2008	Germany	5 satellites	5 MSI	5 x 5	77	1- 5,5	Paid
Sentinel-1	ESA	2014	European Union	Dual satellites	C-band SAR	5 x 5 5 x 20 25 x 40	80, 250, 400	6	Free
Sentinel-2	ESA	2015	European Union	Dual satellites	13 MSI	10 x 10 20 x 20 60 x 60	290	5	Free

4.2 Sentinel 2 MSI data pre-processing

The first stage of data processing from the SENTINEL-2A/2B remote sensing sensors was acquisition of products from the Copernicus Open Access Hub. SENTINEL-2 data are acquired on 13 spectral bands in the visible and near-infrared (VNIR), and short-wavelength infrared (SWIR) spectra, as shown in Table 2. The SENTINEL-2A satellite delivered its first imagery from space on 29th June 2015. The first images of ROI were with a cloud cover. Monitoring of the Belchatow open pit mine by SENTINEL-2A began on 25th November 2016. L2C ESA products are post-proceed since 2017. Level-2A products are organized in ortho-rectified tiles of 100 km x 100 km in UTM WGS84 projections. Data processing includes Scene Classification and Atmospheric Correction applied to Top-Of-Atmosphere (TOA) Level-1C orthoimage products. The main output of Level-2A is an orthoimage Bottom-Of-Atmosphere (BOA) product with corrected reflectance. The first scene of the analyzed ROI in the L2C standard was acquired on 29th March 2017.

Multispectral imagery acquired by Sentinel 2A (MSI - Multi Spectral Instrument) were used to elaborate colour compositions. Fig. 2 present test colour compositions for the analysed images from 4th November 2017. They were used as input data for further analyses and were the source of data for classifying land use and land cover, and land change monitoring in the analysed Szczercow dumping site. The methodology of land change monitoring was based on the Natural Colour RGB, Colour Infrared and NDVI index (Normalized difference vegetation index) colour compositions.

Table 2. List of SENTINEL-2 spectral bands.

Band	Resolution (m)	Central wavelength (nm)	Band width (nm)	Purpose
B01	60	443	20	Aerosol detection
B02	10	490	65	Blue
B03	10	560	35	Green
B04	10	665	30	Red
B05	20	705	15	Vegetation classification
B06	20	740	15	Vegetation classification
B07	20	783	20	Vegetation classification
B08	10	842	115	Near infrared
B08A	20	865	20	Vegetation classification
B09	60	945	20	Water vapour
B10	60	1375	30	Cirrus
B11	20	1610	90	Snow / ice / cloud discrimination
B12	20	2190	180	Snow / ice / cloud discrimination

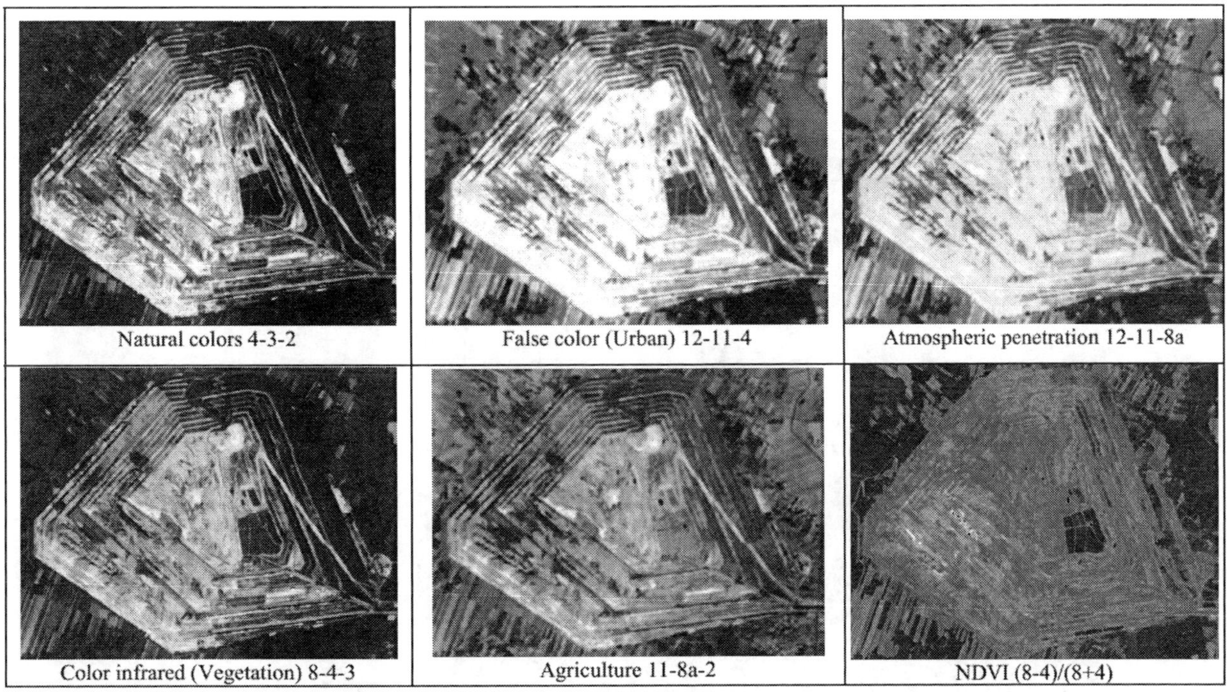

Fig. 2. Example of a selected colour composition of SENTINEL-2A/2B on 4th November 2017.

4.3 Sentinel 2 MSI change detection

The procedure of detecting changes based on spectral classification of images was described by Yoshi et al. [9]. The first stage of spectral classification is sub-division of the image into sub-sets; this process is known as segmentation. The main aim of the classification is assignment of the selected segments with similar statistical characteristics and their aggregation into subsets. The classification procedure of remotely sensed imagery leads to select areas with a specific land use and land cover. Detection of changes within the monitored objects requires difference analyses. Based on two temporal states T1 and T2, change detection is possible for time delta T. In the analyzed ROI, analyses of changes were made using image differencing. The stage of image classification, when potential errors may be summed up in further analyses, was omitted.

Results of the analyses are presented in Fig. 3a-c. The analyses show that application of SENTINEL-2 images allows for obtaining information on land use and land change in the analyzed ROI – Szczercow dumping area.

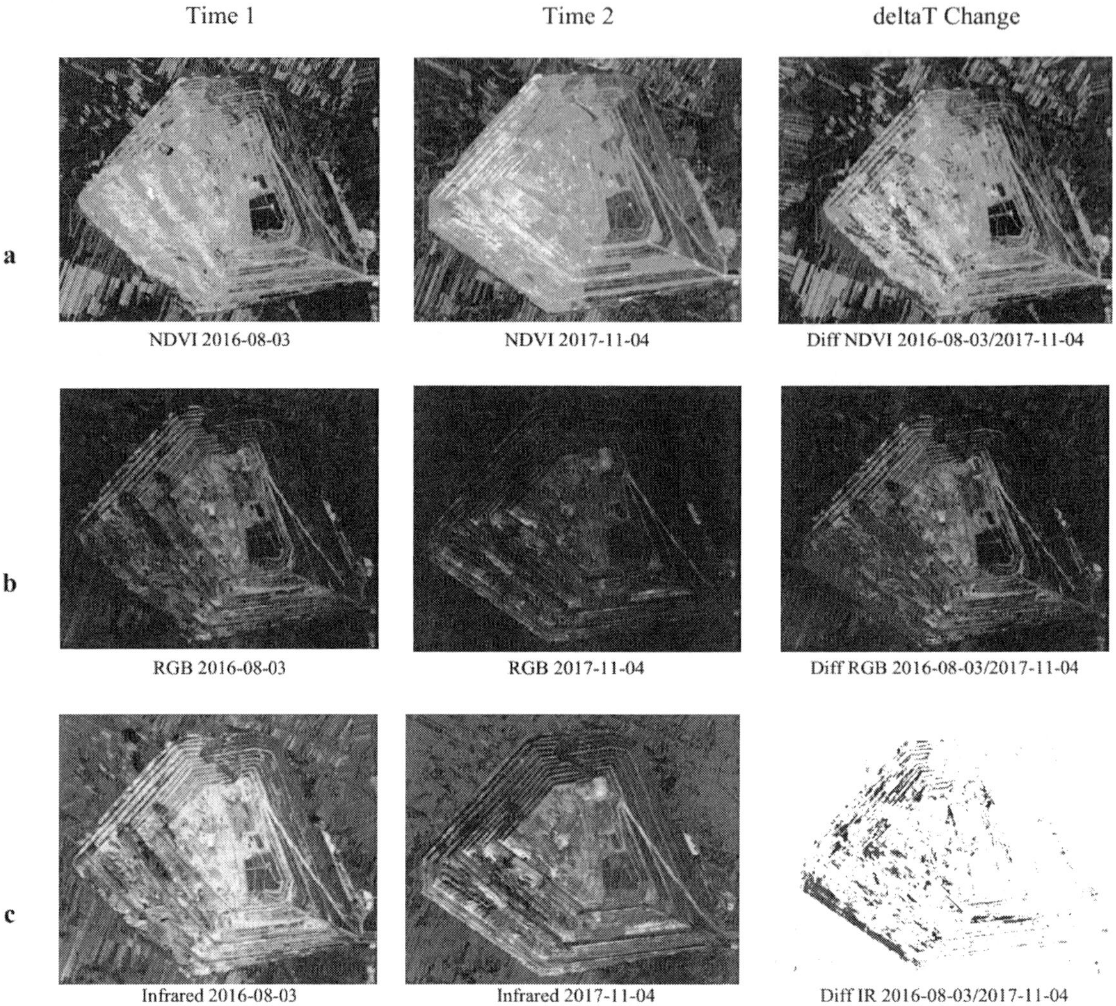

Fig. 3. Results of Sentinel 2A change analysis. (a) NDVI spectral band combination, (b) RGB colour combination, (c) Infrared band combination.

4.4 Synthetic-aperture radar data processing for land classification

The third part of this report presents analyses of Synthetic Aperture Radar SAR SENTINEL-1A/1B radar images [10]. The analysis focused on a Single Look Complex from SENTINEL-1A in VV polarization from August 2016. Numerous SAR signal processing procedures focus on detection of pseudo vertical changes using pairs of images and difference algorithms (InSAR) [11, 12]. SAR signal processing was made with the use of the SENTINEL-1 Toolbox (S1TBX) and SENTINEL-2 Toolbox (S2TBX). SNAP is the software platform for the three Sentinel Toolboxes, developed by the European Space Agency (ESA) for exploitation of SENTINEL-1, SENTINEL-2 and SENTINEL-3 missions under GNU GPL open source license. In this report, the SENTINEL-1A scene was cropped to the ROI area, which significantly reduced the processing time. Next, radar image filtration was made with use of Speckle Filtering in the SNAP toolbox. The final image was georeferenced in the WGS 84 system with application of Range Doppler Terrain Correction applications. SLC VV intensity is visualized in Fig. 4. The next stage of SAR imaging was its classification controlled by the K-means clustering methodology. Results of this classification are presented in Fig. 5. Blue and yellow colors depict regions characterized by strong return of the radar signal. The image shows that the regions represent industrial areas, buildings and mining areas that are not covered by vegetation. Marked with blue and yellow in Fig. 5, the processed SAR image indicates strong signal return from the dumping site and from the excavation pit of the mine.

Fig. 4. SENTINEL-1A SLC geometric referenced intensity VV image.

Fig. 5. Sentinel 1A SLC classified intensity VV image.

5 Results

Analysed remote sensing imagery allowed to perform large-surface analysis as a tool for detecting local anomalies and identification of potential hazards. Classification of passive scenes allowed for monitoring land use and land change, and determining the changes as XY components. The data supply information on the area, in which other independent geodetic measurement techniques may be used. The detected changes may induce high-resolution techniques for 3D modelling for obtaining 3D data. The analysed SAR active remote sensing data allowed for monitoring of land use in the Belchatow open cast mine area as an active sensor. These analyses indicate that monitoring of land change and the active state of the open pit mine in macro scale with application of remote sensed imagery is possible. The analyses show that processing of Copernicus SENTINEL scenes with application of numerous revisits above the ROI allows for obtaining reliable information on the actual state of the ROI.

6 Conclusions

Results of analysis of SENTINEL-2A/2B passive scenes have allowed for obtaining continuous information in form of raster images with field resolution of 10 m. Spectral analysis of SENTINEL-2A/2B colour compositions enables detection of land change and modelling on XY change in the surrounding of the Belchatow open cast lignite mine. The final product in form of GRID surfaces is a perfect derivative for the GIS database of the mine. Application of data from the SENTINEL-2A/2B passive sensor for the analyzed ROI is a perfect product for terrain analysis in the case, when the area of the open-cast mine is not covered with snow. Another restriction maybe the influence of Earth's atmosphere on the acquisition of MSI images. The presence of clouds significantly decreases the quality and volume of information on multispectral images. Further investigations will use data from HR and VHR imageries from nanosatellites. Contrary to passive remote sensors. SENTINEL-1A/1B active SAR remote sensing is a source of data that are independent from weather conditions, clouds or angle of sunlight. Further investigations on SENTINEL-1A/1B data processing will focus on determining line of sight (LOS) changes in the analyzed area for monitoring of potential pseudo-vertical movements.

This work was financed by the Polish Statutory Research Grant no. 0402/0146/16 and Grant no. 0401/0128/17. Gratitude is expressed to Wojciech Milczarek, PhD Eng., for technical and substantial support during analysis of data from the active SENTINEL-1A/1B sensor.

References

1. Polish Journal of Laws, pos. 1941 (2015)
2. G. Dawidziak, Z. Skalski, Materiały Sympozjum Warsztaty "Zagrożenia naturalne w górnictwie", Wyd. IGSMiE PAN, Kraków (2004)
3. Z. Kasztelewicz, M. Zajączkowski, Polityka Energetyczna **13**, 227-243 (2010)
4. J. Zawadzki, K. Przeździecki, Z. Miatkowski, Journal of environmental management, **166**, 605-614 (2016)
5. S. Mularz, W. Drzewiecki, T. Pirowski, Roczniki Geomatyki, T. **V**, Z. 8 (2007)
6. Z. Miatkowski, S. Lewiński, W. Kowalik, A. Sołtysik, J. Turbiak, Woda Środowisko Obszary Wiejskie **16** (2006)
7. C. Toth, G. Jóźków, ISPRS Journal of Photogrammetry and Remote Sensing, **115**, 22-36 (2016)
8. Y. Ban, M. Salvatore, E. Lars, International Journal of Remote Sensing, Vol. **38**, No. 7, 1759-1764 (2017)
9. N. Joshi, M. Baumann, A. Ehammer, R. Fensholt, K. Grogan, P. Hostert, M. Jepsen, T. Kuemmerle,
P. Meyfroidt, E. Mitchard, J. Reiche, C. Ryan, B. Waske, Remote Sensing, **8**(1) (2016)

10. A. Ferretti, C. Monti-Guarnieri, F. Prati, *InSAR Principles: Guidelines for SAR Interferometry Processing and Interpretation* (ESA Publications, 2007)
11. T.R. Lauknes, J. Dehls, Y. Larsen, K.A. Høgda, D.J. Weydahl, Fringe 2005 Workshop, Proceedings of the Conference held 28 November - 2 December, 2005 in Frascati, Italy, ESA SP-610 (2006)
12. W. Milczarek, J. Blachowski, P. Grzempowski, Acta Geodyn. Geomater., Vol. **14**, No. 1(185), 41-52 (2017)

3D modeling of underground objects with the use of SLAM technology on the example of historical mine in Ciechanowice (Ołowiane Range, The Sudetes)

Jarosław Wajs[1,*], *Damian* Kasza[1], *Paweł P.* Zagożdżon[1], and *Katarzyna D.* Zagożdżon[1]
[1]Wrocław University of Science and Technology, Faculty of Geoengineering, Mining and Geology, 27 Wyb. Wyspiańskiego St., 50-370 Wrocław, Poland

Abstract. Terrestrial Laser Scanning is a currently one of the most popular methods for producing representations of 3D objects. This paper presents the potential of applying the mobile laser scanning method to inventory underground objects. The examined location was a historic crystalline limestone mine situated in the vicinity of Ciechanowice village (Kaczawa Mts., SW Poland). The authors present a methodology for performing measurements and for processing the obtained results, whose accuracy is additionally verified.

1 Introduction

In the case of classic measurement methods applied for provide an inventory of underground objects, time required to perform survey work is relatively long. Such methods produce XYZ coordinates of specific points. Custom designed measurement networks can be applied for generate derivative products, such as horizontal and vertical cross sections, or volumetric measurements. Over the past two decades, rapid development of laser technologies has allowed producing high resolution 3D models.

The specific topology of such engineering objects as underground excavations frequently causes the measurements to require a significant amount of work [1, 2]. Therefore, the dedicated measurement systems must be mobile and available for use in extreme conditions which exist in objects of this type [3]. For example, measurements of control points poses a real challenge in excavations of limited volume (limited vertical cross-section area).

This paper presents the results of research on applying *SLAM* (Simultaneous Localization And Mapping) technology in the acquisition of measurement and navigation data. The test object was a historic mine with a chamber in Ciechanowice (Ołowiane Mountains, the Sudetes range), in which crystalline limestone was mined [4, 5].

[*] Corresponding author: jaroslaw.wajs@pwr.edu.pl

2 Characteristic of examined object

The analyzed mine is located in the southern part of the Kaczawa Mountains, in the area of a small morphological unit called the Ołowiane (Lead) Mountains and – in administrative terms – in the peripheral, western part of Ciechanowice village (Fig. 1). The inlet of the adit is situated at a steep slope, about 50 meters above the Bóbr River (about 460 m a.s.l.). The mine was probably operating in the 19th or 18th century and the rock exploited (crystalline limestone) was used for firing lime in the lime kiln situated nearby (400 m to S).

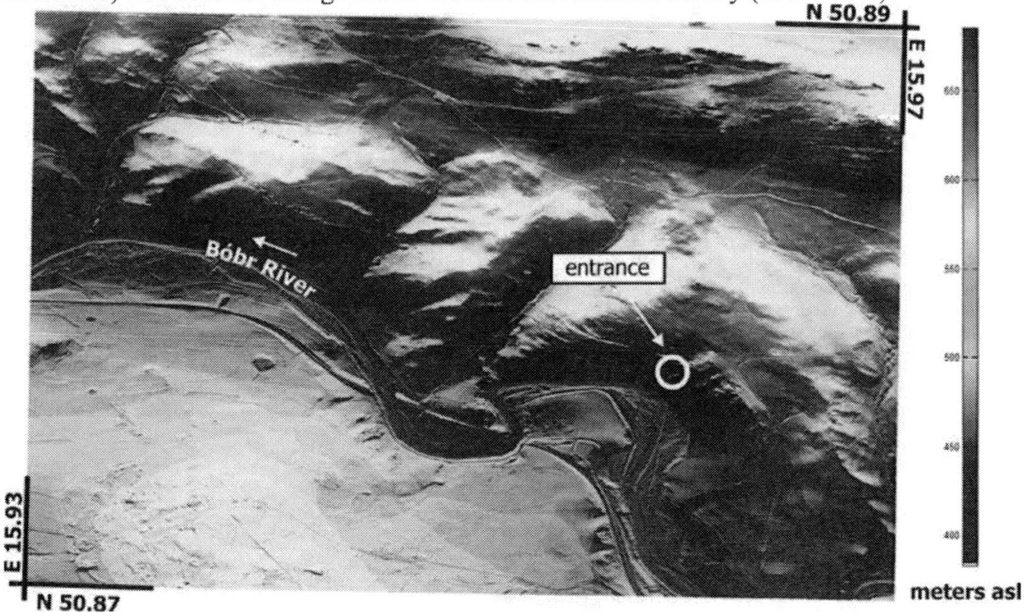

Fig. 1. Location of historic mine in Ciechanowice.

2.1 State of preservation and geology

The mine consists of an about 90-meter long adit and a mining chamber. The adit runs straight towards NE. It is about 1.8 m wide and its height varies between 1.5 m and 2.3 m.

The excavation is well preserved – small rocky screes are found only at the inlet and about 65 m from the entrance. A large exploitation chamber extends towards NW–SE (Fig. 2). Its length is about 30 m, the width is up to 8 m, and the height (in some places) exceeds 5 m. The surface of the floor in the chamber is very irregular. Original level of the floor is preserved on the extension of the tunnel, but in the northern part of the chamber it is mostly covered with loose material (probably barren rock from other parts of the excavation), while the floor of southern part of chamber rises rapidly and large rock fragments separated from the roof cover to the floor. Behind the chamber, there is the final section of the adit, about 16 m in length.

The adit runs within rocks known as greenstones or greenstone schists [6-9]. These rocks are structurally diverse, and show partial folding. Fine lenses of pegmatite can be seen in several places. The chamber has been drivaged in crystalline limestone, although now (due to progressive falling off rock blocks from the roof) the walls of its southern part are also formed by greenstone. Two lithological varieties are distinguished among the limestones: light-colored massive and laminated. The contacts of all mentioned rocks are clearly visible [4].

In the excavations, a specific ecosystem has been evolved, inhabited by species of insecticidal fungi [10]. Additionally, the exploitation chamber comprises many clearly visible hand-drilled blastholes which remain after breaking rock with black powder. For these reasons, the mine is systematically researched as an object interesting from the viewpoint of geology, biology, and history of mining, and is being prepared for protection.

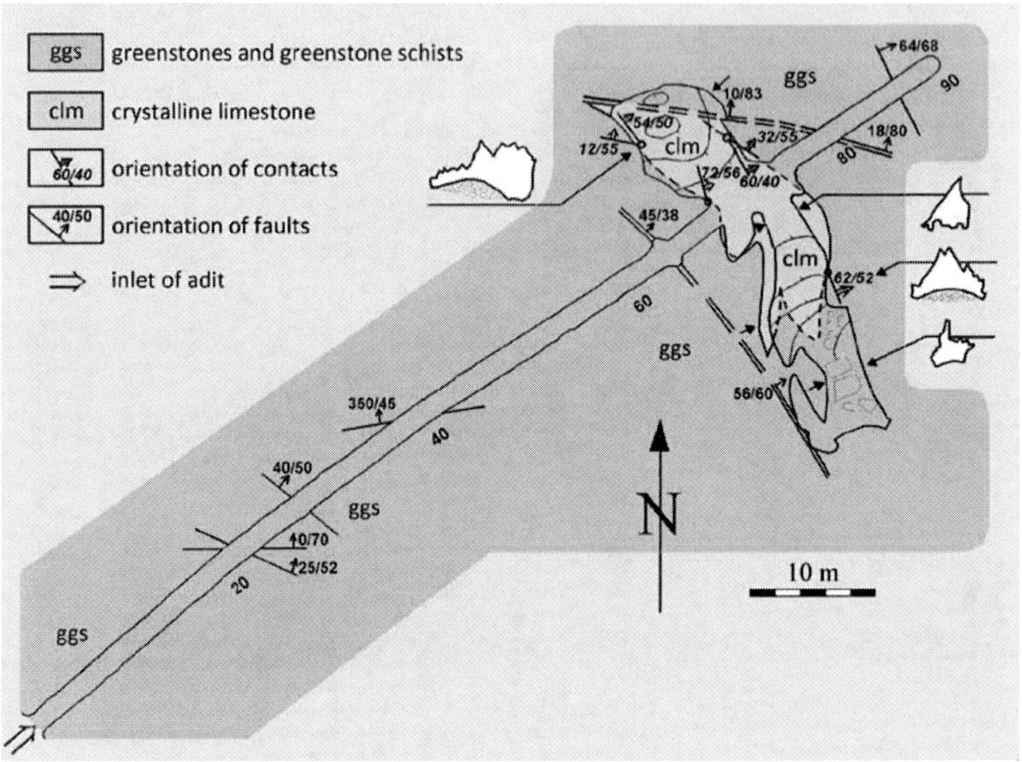

Fig. 2. Geological map of the historic mine in Ciechanowice.

3 Survey work

As is typically the case in single hole excavations, the geodetic control network in the Ciechanowice adit is an open polygon, which offers limited accuracy. A more advantageous solution consists in constructing a closed polygon, in which the first measurement point coincides with the last point, in this case the point being located at the entrance to the adit.

In many cases, performing measurements with the use of *TLS* (Terrestrial Laser Scanning) technology is very time consuming due to a large number of measuring stations [11]. This is because the extensive range of terrestrial lasers cannot be efficiently used in objects of this type. Errors, i.e. incomplete 3D models, frequently result from inappropriate choice of stations and the consequent discontinuity of the mesh surface. In order to minimize the influence of the above negative factors, the present research has been performed using SLAM technology.

3.1 SLAM measurement technology

Implementation of the SLAM technology in positioning allows collecting the measurement data in parallel with the navigation data [12]. Kalman filtering [13] allows integrating

LiDAR (Light Detection and Ranging) observations with the *IMU* (Inertial Measurement Units) data. The analyzed object was scanned with the use of the GeoSLAM ZEB-REVO handheld revolving laser-scanner.

The scanner is equipped with a revolving 360° camera capable of producing high-resolution point clouds. Alongside with the LiDAR measurements, as the scanner passes through the survey environment, the sensor's 3D trajectory is recorded on the basis of the built-in IMU.

3.2 Measurement sessions

The analyzed object was measured in two sessions, which provided data for *IQC* (Internal Quality Control). The first measurement session lasted 15 minutes and covered the total *AOI* (Area of Interest). The subsequent quality control session was independent and lasted 35 minutes.

4 Data processing

The obtained raw measurement data were processed using dedicated GeoSLAM Desktop application. The generated metric products were based on a primary result, i.e. on the adjusted 3D trajectory along with its conditioning. The processed point clouds were recorded in the binary standard ISPRS.LAS 1.3.

An additional product generated by the GeoSLAM Desktop software are comprised *pct_shaded* point clouds as well as point clouds whose color corresponded to the distance value in relation to the trajectory in PLY format, representing every 11th point from the original set. During the first measurement session, 29,606,712 points were recorded (0.8 GB capacity). This measurement provided reference data for further analyses. The resulting data set comprised 78,469,866 points (2.15 GB).

The resulting point cloud may be analyzed using various methods what was discussed by Samieinejad et al. [14]. The obtained 3D models may be conditioned by visual inspection of the model's continuity, volumetric analysis [15], or in a point-to-point approach, as is the case in this work.

The point-to-point methodology of point cloud processing has been described by Holst et al. [16]. At present, the most frequently used cloud point analysis algorithms include Cloud-to-Cloud (C2C) analysis, Multiscale-Model-to-Model-Cloud (M3C2) analysis, Cloud-to-Mesh (C2M) and Mesh-to-Mesh (M2M) analysis. The main criterion in the point cloud analyses was Internal Quality Control. The analyses were performed in the Cloud Compare application, using the C2C function and the M3C2 distance analysis [17].

The analyses were divided into a total test and a local test. The total test provided information on the integrity of the measurement range and on potential gross errors (Figs. 4, 6), while the local test consisted in inspecting the point cloud after the outlier values have been discarded (Figs. 3, 5).

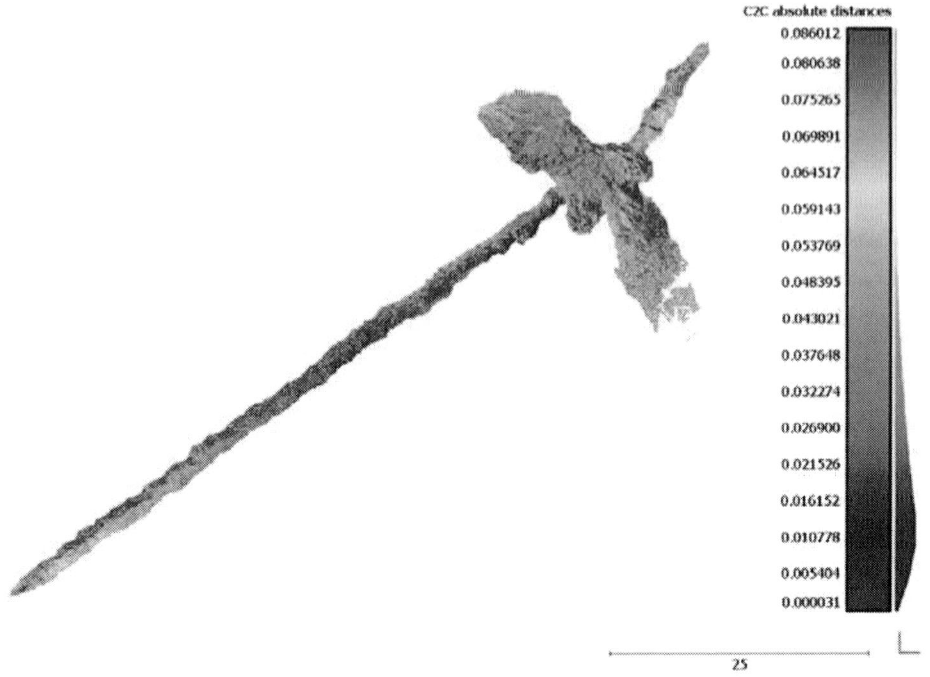

Fig. 3. Local test results analyzed using the C2C distances algorithm.

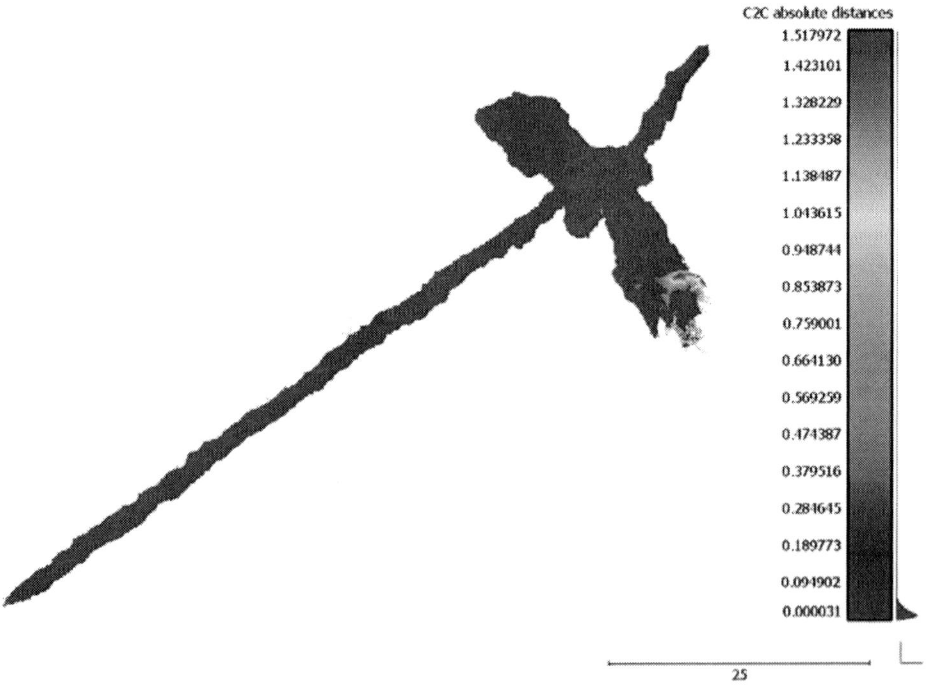

Fig. 4. Global test results analyzed using the C2C distances algorithm.

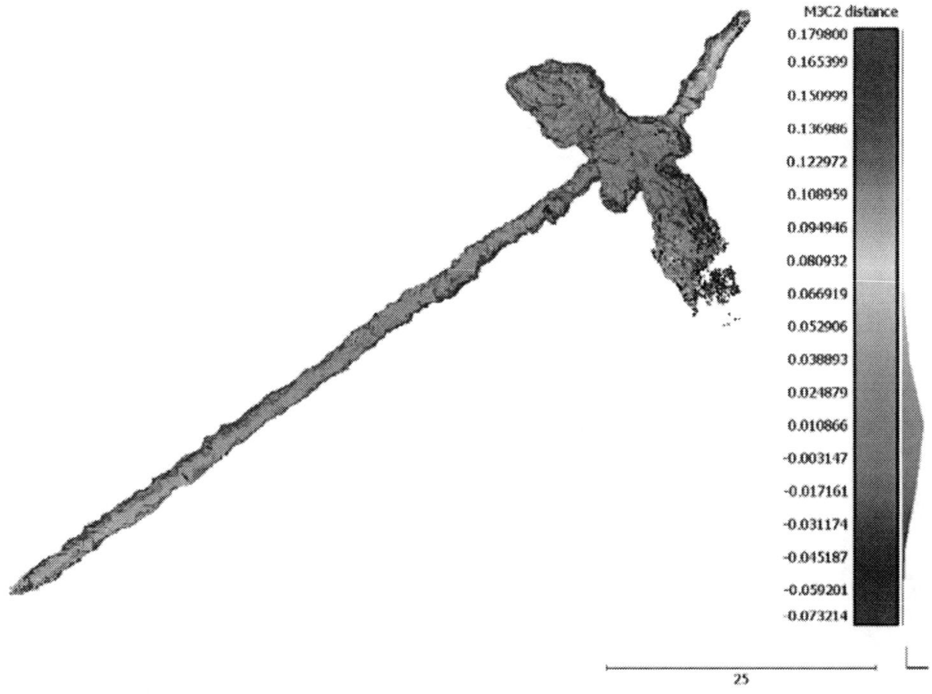

Fig. 5. Local test results analyzed using the M3C2 distances algorithm.

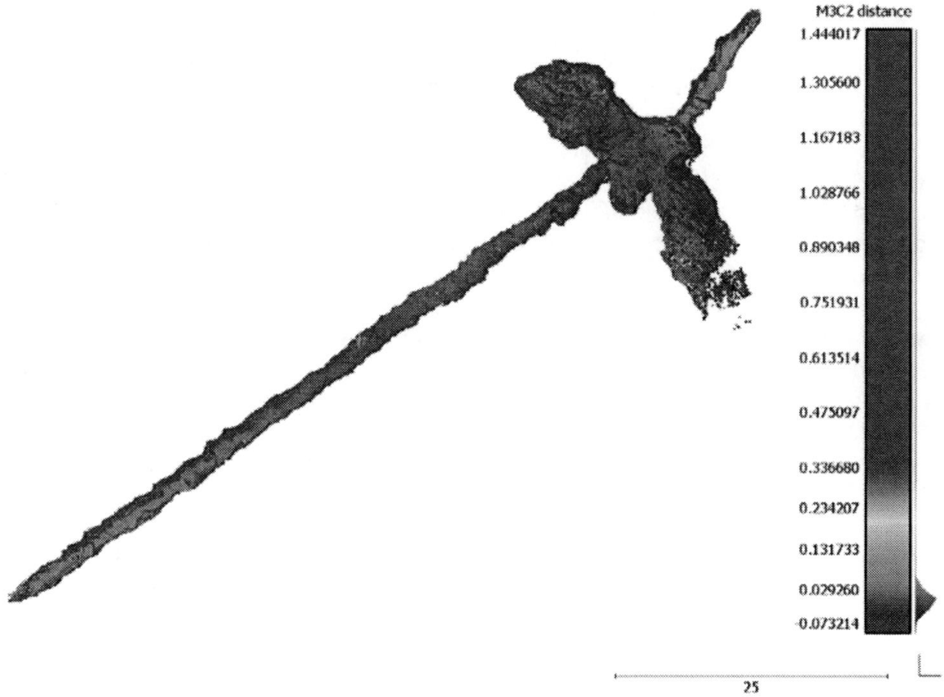

Fig. 6. Results of the global test analyzed using the M3C2 distances algorithm

4.2 Construction of the 3D model

Prior to developing a mesh model in the MeshLab application, the packet of the combined measurement data required further processin. The points located outside the excavation (fragments of the surface area and vegetation) were removed manually, as was the so-called noise, i.e. points located in the immediate vicinity of the scanned area, but not included in this area (location error of laser beam deflection point). In addition, a *Poisson-disk sampling* filter [18] was applied to the set in order to unify the point cloud (minimum distance between the adjacent points was set at 50 mm). This operation reduced point density in the locations where scans overlap, and hence improved the balance of point distribution in the scanned area.

After the above operations have been carried out, the developed mesh models were tested for their accuracy depending on the four defined parameters of the *Poisson surface reconstruction* algorithm [19, 20]. *Octree depth* – the most important of these parameters – defined the recursion degree of the tree data structure, and consequently offered greater accuracy to the developed mesh model. For the defined density of the scanned surface points and for the value of this parameter being >12, no increase in the quality of the obtained models was observed. The remaining factors, which influence the use of the computing power, the minimum number of points visible in the tree branch and the isosurface compensation method were left at default levels (8; 1; 1).

5 Discussion

The Internal Quality Control of the data collected from two independent measurements performed with the GeoSLAM ZEB-REVO scanner showed that the obtained sets of points have various density, which is directly related to the survey data acquisition time. Both for the C2C and for the M3C2 method, global tests showed that maximum distance values between the analyzed clouds were at an average of 0.02 m and at a maximum of 1 m, thus enabling the identification of those locations which were not scanned twice. The C2C analysis in the local test showed the compared point clouds to be consistent. The average value of shift between the clouds was 2 cm. Maximum distance values for the analyzed clouds were 7.5 cm. The local test based on the M3C2 method indicated that the average distances between the analyzed data sets were at 1.5 cm. Importantly, the results based on the M3C2 method represent an actual value of distances between the clouds, with allowance for negative values. Histograms from the analyses based on the C2C algorithm represent distance as an absolute value.

The mesh-type structure based on the point cloud was used to construct a local geological model. The metrical model of the excavation allowed the representation to be supplemented by prior measurements of the tectonic structures [4]. Due to the high resolution of the mesh surface, it was possible to identify and plot the fault zones and the contacts of petrographic variations (Fig. 7).

Fig. 7. 3D geological model of the surveyed area: adit (grey), faults (yellow) and body of crystalline limestone (green).

6 Conclusions

The results of the analyses show that the final product, i.e. the adjusted point cloud, represents a metric model of actual world-scale objects. If all the construction requirements are observed when using the SLAM technology to adjust the trajectory, it is possible to obtain a cloud of points unbiased with such errors as drift. Moreover, the analyses indicate that the two compared data sets comprise cohesive and repeatable data. The global test showed that the analyzed data also comprise areas with "insufficient data" – the so-called gaps, which result from different measurement rate and from different trajectories in the two measurement sessions. Internal quality control was positive. The obtained mean distance values between the analyzed point clouds were 4-5 cm, which is in accordance with the specification provided by the manufacturer. The results are shown in Fig. 8. On the Fig 8a the global integrity of the analyzed point cloud is 7 cm. The local test at Fig 8b shows that the mean biases is 2 cm. At the Fig 8c and 8d the presented M3C2 offset is about 5 cm.

Terrestrial Laser Scanning is currently one of the fastest and most effective methods for acquiring information about an object's topography It allows fast and precise reconstruction of scanned objects in virtual reality, and in addition it enables them to be referenced against the already existing numerical information, such as LiDAR data.

The presented method of data processing and analysis proves useful in a number of cases. Primarily, it offers a significant aid in works related to geological surveying. The unique possibility to observe the rock mass from the inside (by penetrating excavations) and to perform the above described works allowed the construction of a digital, spatial geological model of the surveyed area.

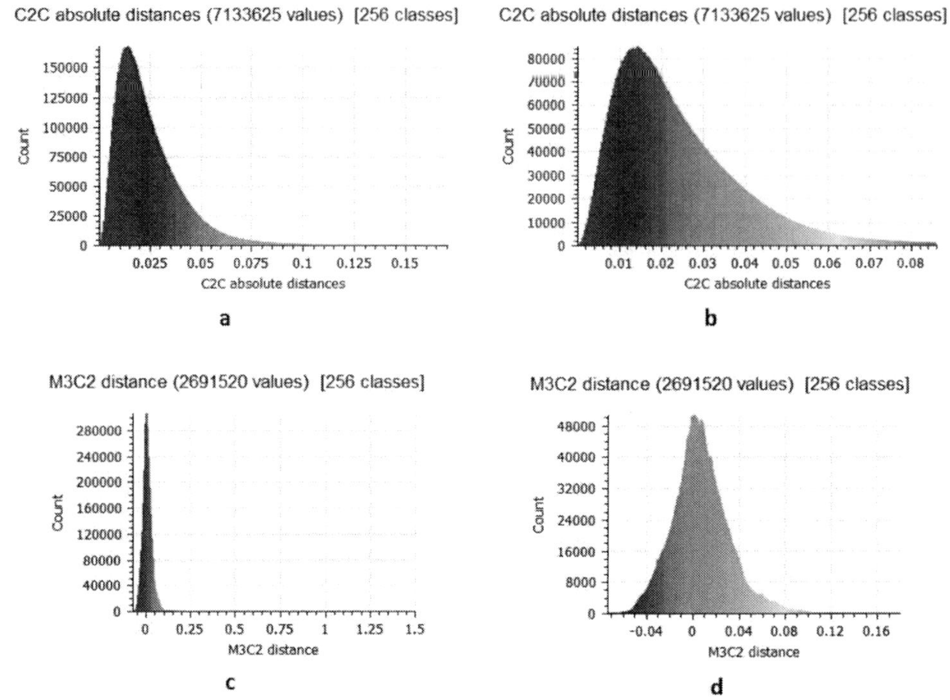

Fig. 8. Histograms of biases in (a) C2C global tests, (b) C2C local tests, (c) M3C2 global tests, and (d) M3C2 local tests.

This work was financed by the Polish Statutory Research Grant no. 0402/0137/16, 0402/0146/16 and Grant no. 0401/0128/17.

References

1. R. Kershaw, Helictite **41**, 87-94 (2012)
2. C. Baselgia, M. Bosse, R. Zlot, C. Holenstein, *Field and Service Robotics, Results of the 8th International Conference*, K. Yoshida and S. Tadokoro (eds.), Springer Tracts in Advanced Robotics **92**, 541-554 (Berlin, 2014)
3. C. Holenstein, R. Zlot, M. Bosse, IEEE/RSJ International Conference on Intelligent Robots and Systems, 3830-3837 (San Francisco, CA, 2011)
4. P. Zagożdżon and K. Zagożdżon, Miesięcznik WUG **4** (152) (2007)
5. J. Wajs, D. Kasza, P. Zagożdżon, K. Zagożdżon, XVIIth Conference of PhD Students and Young Scientists, May 23-26, Szklarska Poręba, Poland (2017)
6. G. Berg, Geologische Karte von Preussen und benachbatren deutschen Ländern. Erläuterungen zu Blatt Kupferberg, II Aufl. Preuss. Geol. Landesanst (1938)
7. G. Berg, *Geologische Karte von Preussen und benachbatren deutschen Ländern. Blatt Kupferberg*, II Aufl. Preuss. Geol. Landesanst. (1940)
8. J. Szałamacha, Detailed geological map of the Sudetes 1:25000. Sheet Janowice Wielkie, Wyd. Geol. (1956)
9. J. Szałamacha, Explanations to detailed geological map of the Sudetes 1:25000. Sheet Janowice Wielkie, Wyd. Geol. (1969)

10. W. Pusz, J. Grzeszczuk, P. P. Zagożdżon, W. Kita, Pol. J. Environ. Stud. (to be published)
11. M. Gallay, J. Kaňuk, Z. Hochmuth, J. D. Meneely, J. Hofierka, V. Sedlák, International Journal of Speleology, **44** (3), 277-291 (2015)
12. B. Mu, S. Y. Liu, L. Paull, J. Leonard, J. P. How, IEEE/RSJ International Conference on Intelligent Robots and Systems, 4602-4609 (2016)
13. R.E. Kalman, J. Basic Eng **82**(1), 35-45 (1960)
14. M. M. Samieinejad, N. Hosseini, K. Ahangari, Journal of Mining & Environment, Vol. **8**, No.3, 455-465 (2017)
15. J. Wajs, Mining Science, vol. 22, 75-83 (2015)
16. C. Holst, B. Schmitz, H. Kuhlmann, *Terrestrisches Laserscanning 2016 (TLS 2016)*, C. Holst (Ed.), 39-58 (2016)
17. D. Lague, N. Brodu, J. Leroux, ISPRS Journal of Photogrammetry and Remote Sensing, **82**, 10-26 (2013)
18. M. Corsini, P. Cignoni, R. Scopigno, IEEE Transaction on Visualization and Computer Graphics, Volume **18**, Number 6, 914-924 (2012)
19. M. Kazhdan, M. Bolitho, H. Hoppe, *Eurographics Symposium on Geometry Processing*, Alla Sheffer and Konrad Polthier (eds.), The Eurographics Association, 61-70 (2006)
20. M. Bolitho, M. Kazhdan, R. Burns, H. Hoppe, Advances in Visual Computing, Lecture Notes in Computer Science, Springer, Volume **5875**, 678-689 (2009)

Participatory GIS in design of the Wroclaw University of Science and Technology campus web map and spatial analysis of campus area quality

Jan Blachowski[1,*], *Jakub* Łuczak[1], and *Paulina* Zagrodnik[1]

[1]Wroclaw University of Science and Technology, Faculty of Geoengineering, Mining and Geology, 27 Wyb. Wyspianskiego St., 50-370 Wroclaw, Poland

Abstract. Public participation geographic information system (GIS) and participatory mapping data collection methods are means that enhance capacity in generating, managing, and communicating spatial information in various fields ranging from local planning to environmental management. In this study these methods have been used in two ways. The first one, to gather information on the additional functionality of campus web map expected by its potential users, i.e. students, staff and visitors, through web based survey. The second, to collect geographically referenced information on campus areas that are liked and disliked in a geo-survey carried out with ArcGIS Online GeoForm Application. The results of the first survey were used to map facilities such as: bicycle infrastructure, building entrances, wheelchair accessible infrastructure and benches. The results of the second one, to analyse the most and the least attractive parts of the campus with heat and hot spot analyses in GIS. In addition, the answers have been studied with regard to the visual and functional aspects of campus area raised in the survey. The thematic layers developed in the results of field mapping and geoprocessing of geosurvey data were included in the campus web map project. The paper describes the applied methodology of data collection, processing, analysis, interpretation and geovisualisation.

1 Introduction

According to [1] Public Participation Geographic Information Systems (PPGIS) pertains to the use of geographic information systems (GIS) to broaden public involvement in policymaking, as well as to the value of GIS to promote the goals of non-governmental organizations, community-based organizations etc. The concept may be traced back to Carver [2]. The article [1] provides a comprehensive view of PPGIS: its origins, potential, criticisms and evolution to date. Another term, Participatory Geographic Information Systems (PGIS) refers to participatory approach to spatial planning, spatial information and communication management [3]. The main difference as distinguished by [4] between PGIS and PPGIS is that the first one is used to collect and display community collected information, whereas the

* Corresponding author: jan.blachowski@pwr.edu.pl

other one (PPGIS) works with existing public data and visualising them according the needs of the community. The study [1] also provides a discussion of the difference between PPGIS and PGIS and the current campaign to merge the two perspectives and practices under a single acronym. Schlossberg and Shuford [5] note inconsistencies in how practitioners and scholars utilize the term PPGIS across uses and applications. They provide background on the two terms and present a two-plane matrix to help users and researchers provide a more methodical and deliberate context for PPGIS. Recent examples of PPGIS applications include: use of public participation GIS for park visitor management with focus on mountain biking [6], assessment of the value of public lands and social landscape metrics [7]. Studies also focus on quality of PPGIS crowd-sourced spatial data as in the example of PPGIS for conservation planning [8].

Furthermore, other terms found in literature that are associated with Participatory GIS include: Volunteered Geographic Information (VGI), citizen science, crowdsourcing, collaborative or web mapping and mashups. VGI are the tools to create, assemble, and disseminate geographic data provided voluntarily by individuals [9]. Examples of VGI include OpenStreetMap or WikiMapia - online editable maps. Citizen science or crowd-sourced science can be thought as public participation in scientific research [10]. In this scope, collaborative mapping is the aggregation of web mapping and user-generated content. Applications, use mainly two types of collaboration, i.e.: edition of the map itself or overlay layers to the basemap. Exampe uses include support for humanitarian objectives such as disease mitigation [11] or response to effects of natural hazards [12]. Whereas, crowdsourcing is the idea of using the power of a crowd to collect data that is too vast, heterogeneous, or expensive to be collected by other types of sensors. Web mapping is the process of using internet maps delivered by geographic information systems (GIS). A web map on the Internet is both served and consumed, thus web mapping is more than just web cartography, it is a service by which consumers may choose what the map will show [13]. Web GIS facilitates public participation systems using web GIS tools to provide public participants with data, analysis tools, and various communication channels. Mashups are defined by [14] as 'web applications that combine data from multiple sources to form a new integrated resource'. The study by [15] analysing the impact of using a concept map-based 'mashup' to provide user contributed (volunteered) ease of access information to travellers with limited mobility is an example. Peng [16] has proposed a taxonomy of a web-based public participation systems that includes: web browsing, static map images (clickable maps), communication channels (discussing and voting), interactive map-based search query and analysis, scenario building and on-line editing. Authors of the project Linkage – linking systems, perspectives and disciplines for active bioversity governance that utilised workshop and internet based participatory mapping methods list the following main advantages and disadvantages of these two approaches (Table 1) [17].

Table 1. Advantages and disadvantages of participatory mapping methods [17].

Workshop	Internet based survey
High control of mapping process	Low control of mapping process
Likely higher involvement of participants	Risk of low interest of potential respondents
Limited number of participants	Possible participation of greater and representative number of people
Limited time for collection of information	No time constraints for respondents

Ability to collect more detailed information	Detailed (further) information can not be collected
Possibility to obtain more accurate spatial data	-

In our study we focus on using selected participatory GIS methods, from the above mentioned approaches, to collect information with the aim to aid design and development of campus web map and to study perception of campus area. The target group is the academic community of the Wroclaw University of Science and Technology (Poland). The tasks have been realised within student bachelor diploma theses and activities of student GIS science club. The scope of work consisted of: internet based survey of the usefulness of present-day campus map and the desired thematic content and functionality of campus web map, mapping campus features indicated in the survey with mobile GIS, as well as analysis and identification of most and least likeable places on the campus with spatial statistic methods based on results of second internet-based geo-survey.

The methodology, data processing and analysis and discussion of results have been described in the following sections of the paper.

2 Wroclaw University of Science and Technology Campus and campus maps

Wroclaw University of Science and Technology (WUST) is one of biggest and best technical universities in Poland with approx. 34,000 students being educated by 2000 academic teachers in 16 faculties. There are 157 student scientific clubs and 22 student organizations operating within the University. WUST has been dynamically developing in recent years with new investments on its campus, e.g.: Bibliotech Library building, which is the seat of the Centre of Knowledge and Science and Technology Information, Technopolis, Student Cultural Zone, as well as cable car named "Polinka" across the Odra River connecting main campus of the University with the new Geocentrum building - an educational and research complex where a new Geocentrum III extension is under development [18].

Most of the WUST facilities in Wroclaw are located on the campus located between Wybrzeze Wyspianskiego and Cuire-Sklodowska Streets and Plac Grunwaldzki (Fig. 1). There are several other smaller campuses hosting Faculty of Architecture, sport facilities, laboratory and administrative buildings.

The present interactive map of the University available on its web site has been prepared with Google Maps API and offers functionality and content limited to locations of buildings marked with point symbols. Map user can access information, in a pop-up window, limited to building name and geographical location by clicking symbol representing a given building.

The map is available at the following url address [19] and its visualisation has been shown in Fig. 1.

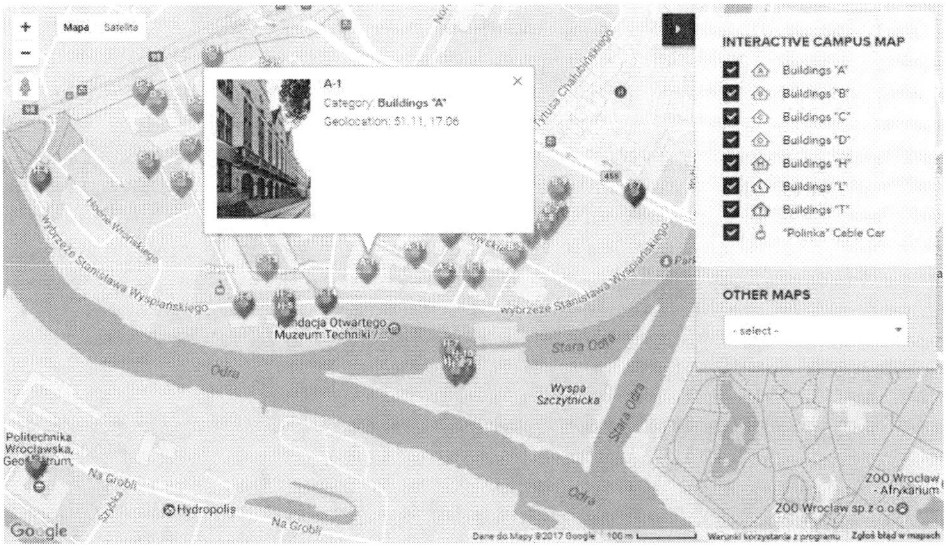

Fig. 1. Present web map of the Wroclaw University of Science and Technology Campus [19].

This is the first public interactive map of the campus. In addition there are numerous paper versions of campus plans included in promotional materials such as factfinders or academic calendars. These plans show location of buildings and selected descriptive information (i.e. symbol of building). One of such plans is shown in Fig. 2.

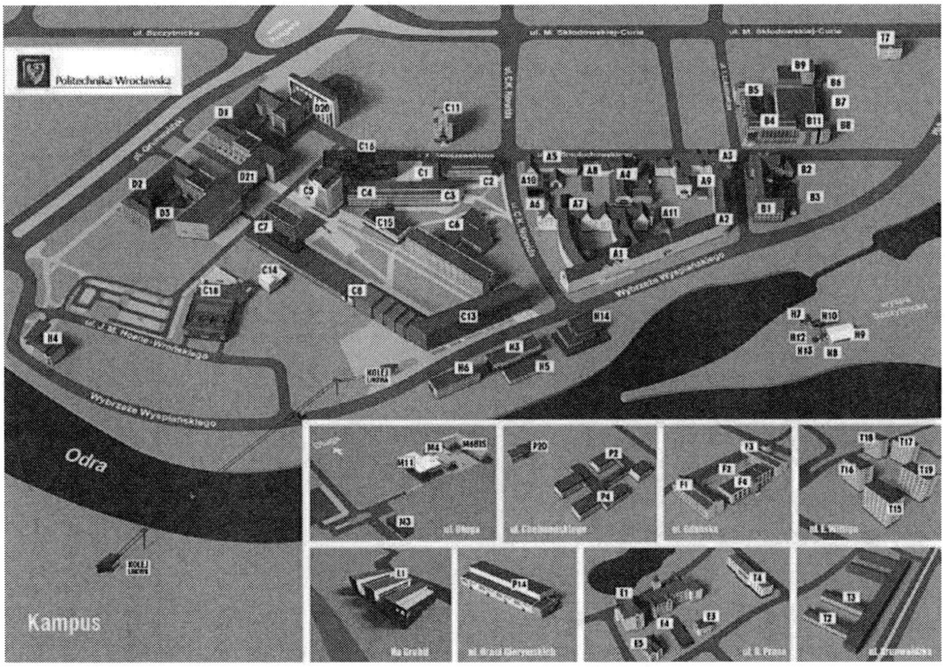

Fig. 2. Campus plan published in promotional materials of 2015/2016 [20].

Project of the first GIS-based campus web map has been developed in the course of three diploma theses realised in succession by students of geodesy and cartography bachelor programme at the Wroclaw University of Science and Technology [21-23]. The web map has been developed with ESRI Web App Builder. In the process of web map design authors' experiences from [24] have been used.

The first edition included 15 thematic layers representing information (point or polygon geometry depending on scale) such as: campus buildings, faculties, dean offices, libraries, pharmacies, photocopy services, food points, as well as cable car "Polinka". The next two editions added information on organisational structure of the University such as thematic layers representing: departments, institutes, as well as student associations and organisations.

In terms of functionality the web map enables: zooming and panning, printing and various queries including, building, department, student association searches. Additional functionalities are time slider and swipe functions. At the moment the map hosts floor plans of the main campus building A-1 and Geocentrum building L-1. Example view of map's graphical interface is shown in Fig. 3.

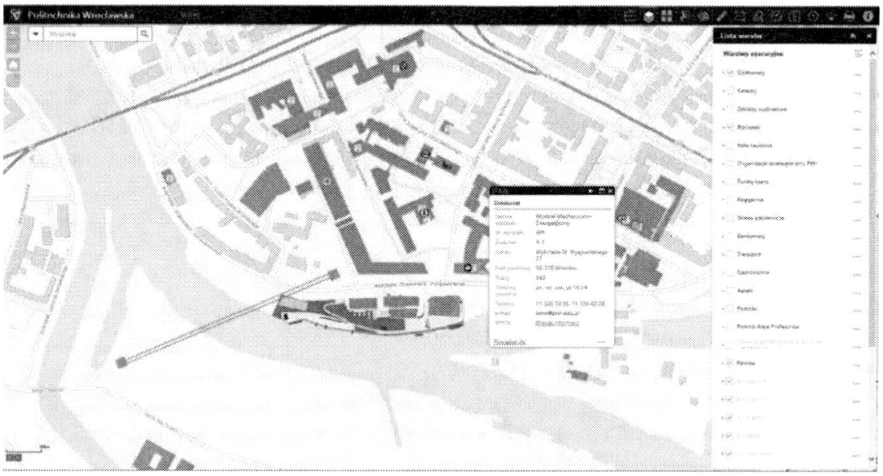

Fig. 3. Project of campus web map layers visible above: buildings including cable car, dean offices, recreational areas [23].

With the aim to enhance applicability and appearance of this web map, surveys of expected usefulness and functionality, as well as perception of campus area have been conducted among academic community.

3 Materials and methods

The methodology of the study consisted of the following main steps: (1) research of the functionality and thematic content of WUST campus web map expected by its potential users, (2) mapping of elements in the campus area with mobile GIS technology, (3) study of visual and functional attractiveness of campus area with web based geoform, (4) analysis and interpretation of results of the study. The procedure is illustrated in Fig. 4.

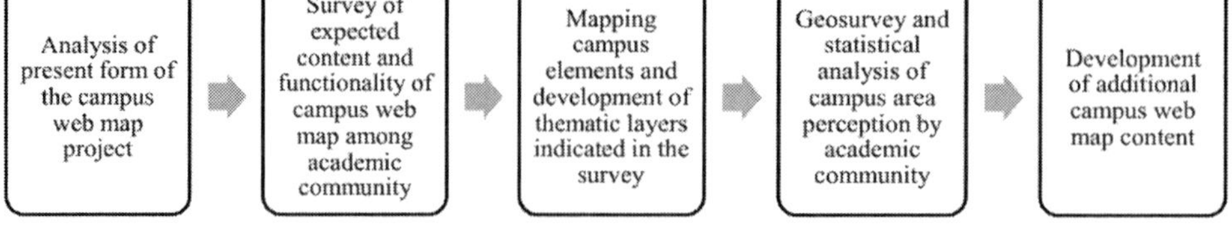

Fig. 4. Methodology applied in the WUST campus web map study.

Step 1 has been accomplished by web survey and analysis of its results. The survey has been done with the use of Google Forms. It included 10 questions, 8 closed and 2 open and optional. The first part of the form consisted of 5 questions to test knowledge of web mapping and familiarity of respondents with WUST maps available on the Internet, e.g.: *Do you think web based map of the campus is necessary (answers YES/NO)? Have you used maps of campus available on the web (answers YES/NO)? How do you rate these maps (grading scale 0 – useless to 5 – excellent)?* The second part of the study focused on gathering data on what kind of information potential web map users seek. The responders could have indicated importance of 26 thematic layers on a scale of 0 – useless to 5 – very useful and propose new thematic layers in open question. Analogously responders could have chosen between 14 functionalities (multiple choice question) and propose additional tools to be added to the web map application. The last question concerned home faculty of respondent.

Step 2 consisted of a practical part where members of the student GIS club created mobile application to collect geographical and attribute information (location and description) on features of the campus area indicated in Step 1. For this purpose Collector for ArcGIS application has been used. In the first step a geodatabase feature class with appropriate attribute field definitions is created in ArcGIS Desktop and shared as a service to ArcGIS Online platform. There, it is used to create a map and interactive form that can used by authorised users who connect to ArcGIS Online with smartphones. The Collector application allows to use selected forms to input data to the database while in the field. The complete procedure is described in [25].

Step 3 utilised web based geographic form application created with template available within the WUST ArcGIS Online SITE licence subscription. The form utilised two editable layers used to collect geographically referenced data of parts of the campus that are attractive and those that are disliked by the survey respondents. The graphical interface of the form has been presented in Fig. 5. The questions in the first part of the survey included queries related to: gender, status (student, staff, other), level of studies (in case of students) and home faculty. The second part included map window where respondents indicated places favoured and disfavoured by them and a field where additional and optional description detailing why a given place is liked or disliked could have been added. The geoform has been shared on popular social media platforms. However, most of the data have been collected by members of the GIS student club with tablets during WUST Student Activity Days that took place on the 9th of March 2017. Results of this survey have been used to prepare heat and hot spot maps of campus area and were described and analysed in part 4.

Fig. 5. Graphical interface of web survey utilising geoform application (*in Polish*).

4 Results and discussion

With the aim to enhance project of campus web map developed with GIS tools [26] prior to its publication two studies have been performed. Survey of expected functionality and content of campus web map and analysis of campus area reception in terms of visual and functional quality. The results have been presented and discussed below and used to map new elements of campus and expand current version of web map.

4.1 Study of the expected functionality of campus web map

The web based survey has been filled by 508 respondents, both students and academic staff, representing 14 faculties at the University [23]. The results have been presented and commented below. In the assessment of the functionality of present interactive map (question 6) 34,3% participants said that it is useful or very useful and 35,2% that it is helpful. Whereas, 20,5% decided that this map is useless or unhelpful (Fig. 6).

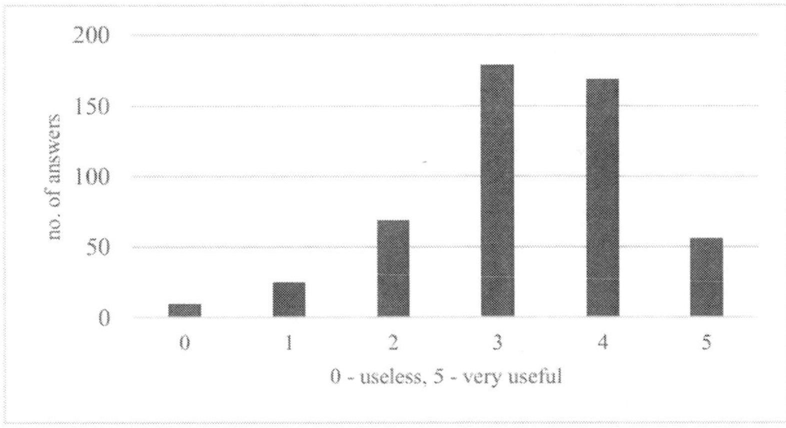

Fig. 6. Assessment of present interactive web map of WUST campus.

In the survey of the proposed 26 thematic layers representing map content (question 7) the following have been the most sought after, i.e. declared as helpful or useful (%):

- buildings (94%),
- libraries (88%),
- photocopy services (88%),
- food points (86%),
- dean offices (83%),
- faculties (78%),
- parking (78%),
- floor plans of buildings (76%),
- stationery shops (74%),
- atm's (73%),
- bicycle stands (68%).

Student associations (55%) and organisations (57%), other organisational units of the WUST (49-56%), pharmacies (56%), shops (56%) and monuments (18%) have been less recognised in the survey.

In the following question (no 8) respondents could provide their own ideas and suggestions for additional map layers. The following have been mentioned most often:

- building entrances,
- wi-fi zones,
- disabled access friendly facilities,
- bookshops,
- places of relaxation (benches, study areas, etc.).

In the examination of expected functionality and content of WUST campus web map (question 9) the following, among 14 proposed, have been found to be the most attractive (in brackets percentage of answers finding a given functionality helpful or useful):

- building search (93%),
- room search (81,5%),
- route planning (73%),
- map zooming and panning (62%),
- map printing (56%),
- other language version (56%),
- distance measurement (55%).

Other functions of web map such as: change of coordinate system, transparency of map layers, time slider (showing progress of campus development), base map gallery, photo gallery, blog/forum, and 3D visualisation have not been considered as important by the survey respondents.

The last, open type, question (10) concerned what else is expected from campus web map. The most frequent answers suggested: simplicity, clarity, intuitiveness, as well as mobile version of the application.

4.2 Field mapping of campus area

Based on the results of survey described in part 4.1. GIS database structure of the following campus area features has been prepared and collected in the field using procedure given in section 3. Location of the following features on the campus have been collected:
• location of building entrances (98 features),
• wheelchair accessible facilities (18 features),
• location of bike racks (288 features),
• benches (261 features),
• trash bins (238 features),
• street lamps (266 features).
• monuments (8 features),
• hydrants (8 features).
Table 2 contains information on additional attributes recorded during mapping.

Table 2. Attributes of collected features.

Feature class	Attributes
Location of building entrances	Emergency exit: yes/no
Wheelchair accessible facilities	Type: entrance, lift, ramp
Location of bike racks	Type: rack, city bike station, storage
Benches	Material: concrete, metal, wood, Back of bench: yes/no
Trash bins	Type: bin, container Trash type: mixed, glass, paper, plastic
Street lamp	Type: lamp, other

The collected data have been used to create new thematic layers of campus web map application.

4.3 Study of the campus area quality

In the second geo-survey of academic community 335 answers have been collected, 203 concerning most attractive and likeable parts of the campus and 122 concerning locations considered to be unattractive. 51% of respondents were female and 49% were male. Based on these results two heat maps have been created showing most (Fig. 7) and least (Fig. 8) attractive areas of the campus. To create these maps Kernel Density function of the ArcGIS software licensed to Wroclaw University of Science and Technology has been used. Statistical significance of identified sites has been analysed with Hot Spot function using the Getis-Ord Gi* statistic [27].

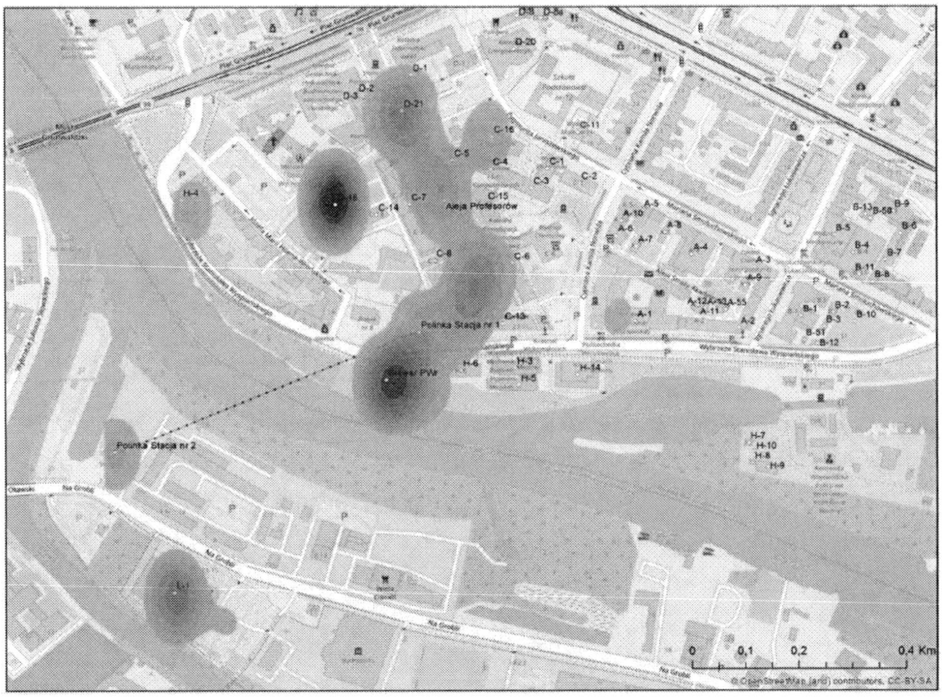

Fig. 7. Heat map showing most attractive areas of WUST campus (background OSM).

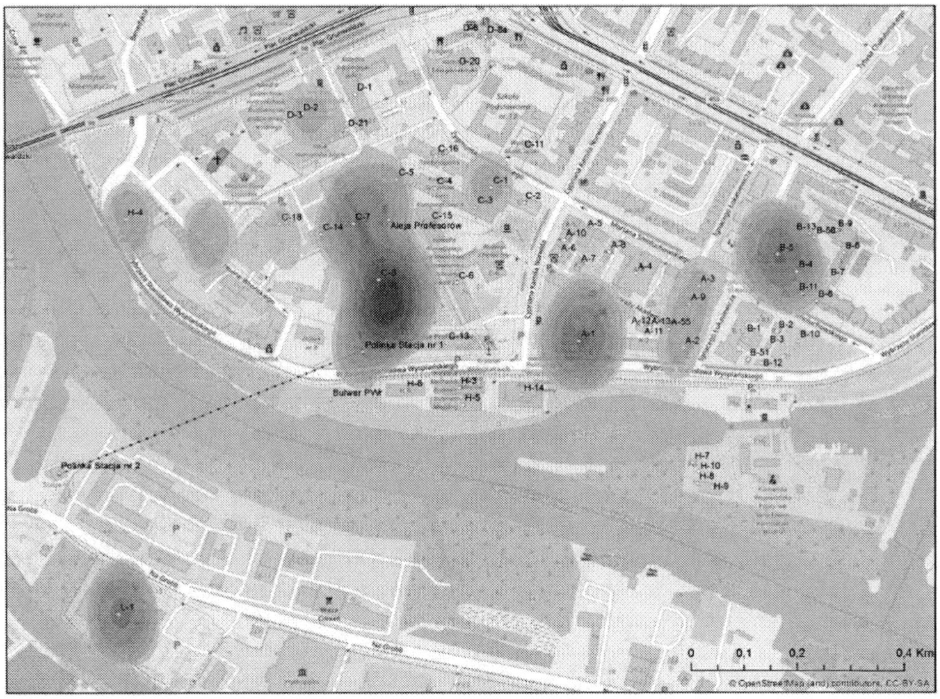

Fig. 8. Heat map showing least attractive areas of WUST campus (background OSM).

The most liked areas included: building C-18 – seat of the Zone of Student Culture, boulevard on the Odra River and cable car "Polinka", while as the most disliked ones the following have been pointed out: C-13 building - seat of the University's Integrated Student Center, L-1 building - Geocentrum, as well as the main building of the campus A-1, building

B-4 – seat of the Mechanical Faculty and D-2 buildings. In addition to location of popular and unpopular sites of the campus information on the causes of these feelings have been collected. In the case of attractive areas, 63% of respondents selected these sites because of their functional qualities, and 37% have chosen the sites based on visual appearance (Fig. 9). Unlikeable areas have been indicated due to their functional properties in 76% of votes and visual appearance in 24% of votes. There has been general agreement between answers provided by male and female respondents. The only significant difference has been observed in answer concerning visual aspect causing features to be regarded as unlikeable. The most common causes have been uncomfortable temperatures (too cold or too hot) pointing to defective air-conditioning systems, dirty or out of use toilets and difficulties with obtaining something to eat due to lack of nearby food points and shops.

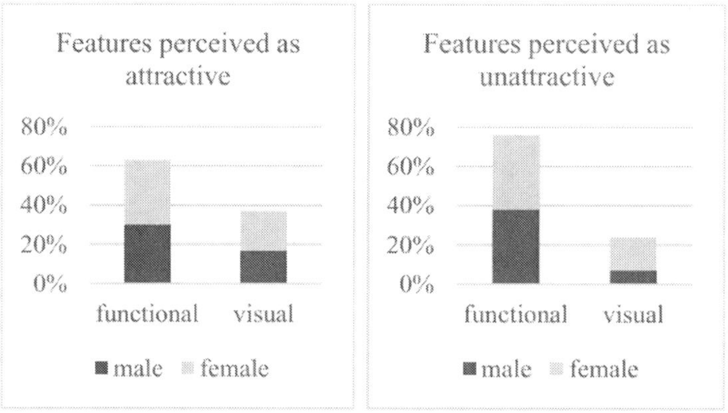

Fig. 9. Comparison of number of answers indicating functional and visual aspect of features, total and in male and female respondent groups.

These maps representing results of functional and visual perception of campus area have been added as thematic layers of the proposed campus web map.

5 Conclusions

In the presented study methodology for community participation in design and development of campus GIS web map has been proposed. In the result of web-based survey of academic community the following new feature classes have been prepared: location of building entrances, wheelchair accessible facilities, location of bike racks, benches, trash bins, street lamps, monuments, hydrants. In an accompanying study web-based geoform has been used to collect information on perception of campus area. In the result of GIS processing heat and hot spot maps have been prepared and most and least attractive areas of WUST campus have been located and the functional and visual reasons for such perception was determined. The most attractive features are: boulevard on the Odra River bank, building of the student culture zone and cable car "Polinka". According to the respondents the least attractive sites of the campus are buildings with poor air-conditioning systems, unclean facilities and lacking food services.

The focus group has been represented by over 500 people (students and academic staff) in the case of first survey and over 300 in the second one. We regard that the sizes of these groups are satisfactory to provide information representative for the academic community of WUST. One of the noticed drawbacks, and in accordance with [7] is the limited control on the course of mapping (e.g. quality of data introduced into forms).

The study has demonstrated usefulness of mobile and web based GIS for participatory mapping and inclusion of local community opinion on the example of campus web map project and assessment of campus area quality and atmosphere. The proposed methodology and tools will be used in further studies concerning assessment of condition of WUST facilities.

This work was financed by the Grant no. 0401/0128/17.

References

1. R. Sieber, Annals of the Association of American Geographers, **96(3)**, 491–507 (2006)
2. S. Carver, Int. J. Geogr. Inf. Sci., **5(3)**, 321-339 (1991)
3. J. Abbot, R. Chambers, Ch. Dunn, T. Harris, E. de Merode, G. Porter, J. Townsend, D. Weiner, IIED, Sustainable Agriculture & Rural Livelihoods, **33**, 27-34 (1998)
4. J. Forrester, S. Cinderby, *A Guide to using Community Mapping and Participatory GIS*, Retrieved May 16, 2017
5. M. Schlossberg, E. Shuford, URISA Journal **16** (2), 15-16 (2005)
6. I. Wolf, T. Wohlfart, G. Brown, A.B. Lasa, Tourism Management **51**, 112-130 (2015)
7. G. Brown, D. Weber, K. Bie, Applied Geography **53**, 77-89 (2014)
8. G. Brown, D. Weber, K. Bie, Land Use Policy 43, 228-238 (2015)
9. M. Goodchild, Geojournal, **69**, 211-221, (2007)
10. M. Haklay, *Knowledge: Volunteered Geographic Information (VGI) in Theory and Practice*, Springer, 105-122 (2013)
11. L. Palen, R. Soden, T.J. Anderson, M. Barrenechea, M. Proceedings of the 33rd annual ACM conference on human factors in computing systems, 4113-4122 (2015)
12. L. Palen, R. Soden, In COOP 2014-Proceedings of the 11th International Conference on the Design of Cooperative Systems, 311-326 (2014)
13. P. Fu, J. Sun, *Web GIS: Principles and Applications* (2010)
14. A.J. Flanagin, M.J. Metzger, GeoJournal, **72** (3-4), 137-148 (2008)
15. A. May, Ch. Parker, N. Taylor, T. Ross, Transportation Research Part C: Emerging Technologies, **49**, 103-113 (2014)
16. Z. Peng, Environment and Planning B: Planning and Design, **28**, 889-905 (2001)
17. http://linkage.iop.krakow.pl/en, Retrieved May 18, 2017
18. http://pwr.edu.pl/en/university/about-us/general-information, Retrieved May 18, 2017
19. http://pwr.edu.pl/en/university/campus-map, Retrieved May 18, 2017
20. http://pwr.edu.pl, Retrieved May 17, 2017
21. J. Spych, Project of web map of the Wroclaw University of Science and Technology campus, Bachelor Thesis (2015)
22. K. Stec, Update of the project of web map of the Wroclaw University of Science and Technology campus, Bachelor Thesis (2016)
23. P. Zagrodnik, Development of interactive map of the Wroclaw University of Science and Technology campus, Bachelor Thesis (2017)
24. J. Blachowski, A. Nowacka, Mining Science **133**, 37-47 (2011)
25. http://doc.arcgis.com/en/collector/, Retrieved May 15, 2017
26. J. Blachowski, Arcana GIS, **March** (2017)
27. ArcGIS 10.4 Desktop Help, ESRI (2016)

Using raster and vector data to identify objects for classify in flood risk. A case study: Raciborz

Mariusz Porczek[1,*], *Dorota* Rucińska[1], and *Stanisław* Lewiński[2]

[1]University of Warsaw, PolandFaculty of Geography and Regional Studies, 26/28 Krakowskie Przedmieście St., 00-927 Warsaw, Poland
[2]Space Research Centre of Polish Academy of Sciences, 18A Bartycka St., 00-716 Warsaw, Poland

Abstract. The severe flood of 1997, which seriously affected Polish, Czech and German territories, gave impetus to research into the management of flood-prone areas. The material losses caused by the "Flood of the Millennium" totalled billions of Polish zloty. The extent of the disaster and of infrastructure repair costs changed the attitude of many branches of the economy, and of science. This is the direct result of consideration of the introduction of changes into spatial management and crisis management. At the same time, it focused the interest of many who were trained in analysing the vulnerability of land-use features to natural disasters such as floods. Research into the spatial distribution of geographic environmental features susceptible to flood in the Odra valley was conducted at the Faculty of Geography and Regional Studies of the University of Warsaw using Geographic Information Systems (GIS).
This study seeks to examine the possibility of adapting vector and raster data and using them for land-use classification in the context of risk of flood and inundation damage. The analysed area of the city and surrounding area of Raciborz, on the upper Odra River, is a case study for identifying objects and lands susceptible to natural hazards based on publicly available satellite databases of the highest resolution, which is a very important factor in the quality of further risk analyses for applied use.
The objective of the research was to create a 10×10-m-pixel raster network using raster data made available by ESA (Copernicus Land Monitoring Service) and vector data from Open Street Map.

1 Introduction

Changes in population size and increased population density in at-risk areas are directly responsible for the increasing losses resulting from hazardous natural events, including heavy rainfall and flooding. The potential for towns and neighbourhoods to expand into safe areas is becoming increasingly limited. Developers and local governments are attempting to

* Corresponding author: mt.porczek@uw.edu.pl

remedy the housing crisis by developing residential areas in locations naturally unsuited for habitation [1]. Society is also adopting such land for agricultural or industrial requirements, due to the ever-growing needs of various sectors of the economy. There is increasing use of areas such as floodplains, artificial and natural polders, and escarpments, as well as of places either susceptible to landslide or surrounded by environmental features, which would hamper evacuation.

Such areas are particularly at risk. Natural hazards in areas of human activity lead to major losses, both societal and material. They can also cause irreversible damage to features of the natural environment and cultural heritage. Such events can slow local development as a result of falling investment and the frequent inability to rebuild damaged infrastructure and economic resources related to agriculture, industry or services. Therefore, spatial planning, which takes into account natural disasters and which adapts to places exposed to extreme natural events is also a very important aspect of risk management.

Studies of the spatial distribution of susceptible geographic environmental features on the floodplains of the Odra allowed for the creation of, inter alia, flood risk maps showing the extent of threat of material losses in the vicinity of Wrocław and Raciborz. The analysis used Geographic Information Systems and spatial data from scanned raster maps, which limited analysis to a maximum accuracy of a basic field of 250×250 m [2].

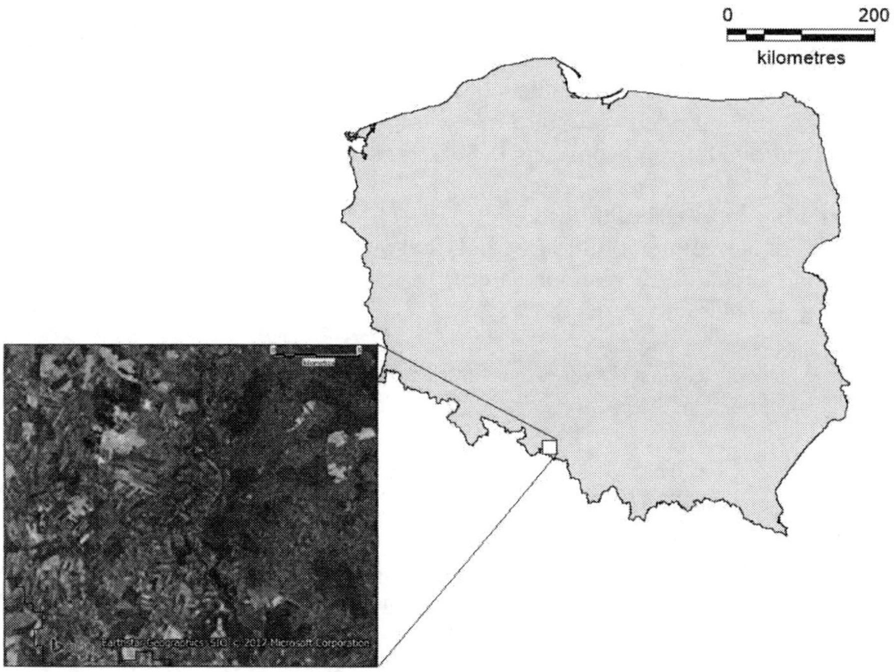

Fig. 1. Research area against the background of Poland.

According to land-use categories adopted in earlier studies [2] in this study has verified the possibility of an analogous analysis based on newer and more accurate spatial data from satellites dating to 2012 at the latest. Source data made available by the Copernicus Land Monitoring Service, Corine Land Cover (CLC) and High Resolution Layers (HR) and Open Street Map (OSM) were processed using GIS software in order to correlate them and obtain an accurate 10×10-m-pixel resolution map of land use determined. This allowed us to check the potential for adapting these publicly available data and using them in further disaster risk reduction studies.

2 Method

Work on the use of GIS and remote sensing techniques to assess a variety of phenomena, including the geographical [3-7] are subject to continuous development from the 90s. These techniques are used in the analysis of various types of hazards [8-12]. Characteristics and discussion on the use of models, vector and raster is still valid [4, 5, 13-19].

The works on indicators of flood losses in Poland was conducting in the 1990s by Chojnacki [20- 23]. He carried out the regionalization of indicators of topographic objects in terms of the estimated value of damages, dividing them into 11 types of investments: arable land, grassland, embankments, regulated banks, buildings, national roads, voivodship and municipal roads, railway tracks, bridges on national roads, on voivodship and municipal roads and on railway tracks. Probably because of the level of contemporary technology, objects were researched for the number of occurrences, not the total area. More recent studies similar to those described above are based on studies of determinants and development strategies of territorial units [24] or databases of vector objects [2, 25] which elaborated on the vulnerability scale of the development objects used in the study below.

At the time, within the framework of the EU Directive 2007/6/EC [26], the Institute of Meteorology and Water Management of the National Research Institute started implementing the ISOK program, under which created flood risk maps [27, 28]. The research on this project uses data derived from satellite imagery. At the same time, such research methods have attracted interest in the scientific community [29]. A new view of the problem is the use of both types of data that are made available to the public.

The analysis used a graded scale of 5 classes, from 0 to 4, where 4 indicates the most negative impact of flooding on people and their lives, and 0 indicates no negative impact [2]. There are:

4 - residential or services buildings
3 - industrial or agricultural buildings, roads and railways
2 - arable land, orchards, recreational land
1 - forests, grasslands and wetlands
0 - standing or flowing water

This scale was used to classify land-use features presented on raster layers from satellite images further enhanced with spatial vector objects (also identified from satellite sources).

3 Study area

The Raciborz region was selected for its location in the Odra valley. The town is located on the upper course of the river among numerous tributaries. The town lands and surroundings have in many places been transformed to meet human needs. This is also true of its watercourses (e.g. the Ulga canal, which drains the area of Raciborz). To the south of the town, a canal connects the Odra basin with the Vistula basin.

The town was greatly affected by the flood of the millennium and is permanently at risk of flood and inundation. Each such event causes serious material losses for local inhabitants, companies and institutions. Raciborz is a medium-sized (pop. 56,000) historic town, which is densely built up and has a well developed urban infrastructure. Meanwhile, it is surrounded by the strictly agricultural lands which are typical of the Polish countryside. There are also forests and wetlands around the town. Together, these represent a typical land-use profile for Poland, making it an optimal area for research into flood risk reduction in the country.

4 Source material

Three public-access online sources were selected. The base layer of the map was the Corine land cover map made available by the European Space Agency at the website Copernicus.eu [30]. The Corine Land Cover 2012 database (CLC 2012) is highly structured; it is divided into 44 categories of land-use features for Europe, of which 41 occur in the study area, including 13, which are significant to the present work. These categories were reclassified to the previously selected 5-point scale of losses (Table 1).

Table 1. Corine Land Cover 2012 classes, reclassified for the study (classes 0–4).

Corine LC		Study classification	
LV 3	Description	Class	Feature
112	Discontinuous urban fabric	4	Residential buildings
121	Industrial or commercial units	3	Industrial buildings
131	Mineral extraction sites	3	Industrial buildings
142	Sport and leisure facilities	2	Recreational land
211	Non-irrigated arable land	2	Arable land
231	Pastures	1	Grasslands
242	Complex cultivation patterns	2	Arable land
243	Land occupied by agriculture with natural vegetation	2	Arable land
311	Broad-leaved forest	1	Forest
312	Coniferous forest	1	Forest
313	Mixed forest	1	Forest
511	Water bodies	0	Standing water
512	Water courses	0	Flowing water

The database has a resolution of 100×100 m, which is insufficient for the current study, particularly in urban areas and areas with highly developed infrastructure. For this reason, data was selected from high-resolution thematic layers published – similarly as with the CLC – by the Copernicus Land Monitoring Service [31]. Raster layers available at 20×20-m resolution indicated Imperviousness, Forest Type, Grassland, Wetlands and Permanent

Water Bodies. Their recency is also set at 2012, and they cover Europe (excluding Russia), Ukraine, Belarus and Moldova (similarly as with the CLC). All listed land-use features occur to at least a minimal extent in the research area (Fig. 2).

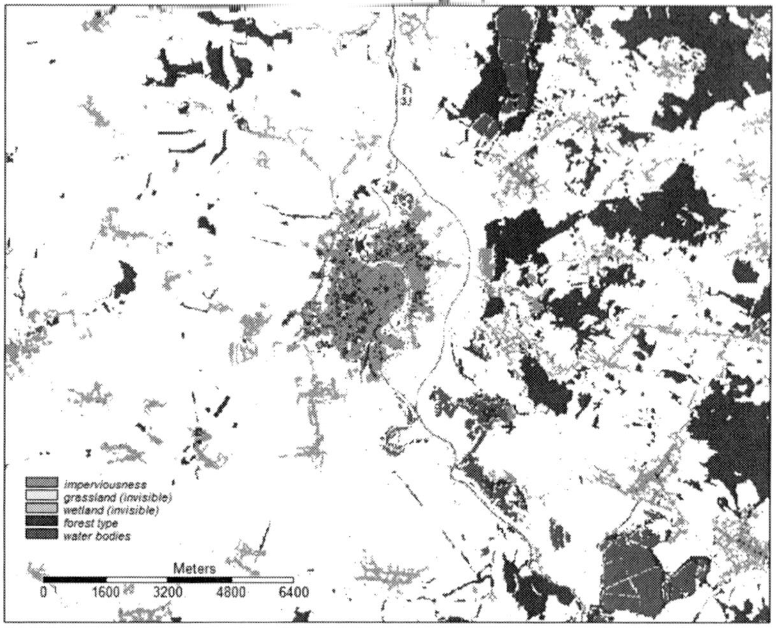

Fig. 2. High Resolution Layers (Copernicus Land Monitoring System).

The High Resolution Layers do not fully correspond to the classification adopted for the study, and hence their correlation to the generally adopted scale was difficult and incomplete (Table 2).

Table 2. High Resolution Layers reclassified for the study (classes 0–4).

HR Layer	Class	Feature
Imperviousness	4, 3	Residential buildings Industrial buildings
Forest type	1	Forest
Grassland	1	Grasslands
Wetlands	1	Wetlands
Water Bodies	0	Water

In order to complete the gaps and reclassification of these selected raster layers for the land use classification study, the key features from the Open Street Map database [32] were made available in vector format. Layers from the OSM project present features identified from satellite imagery from, among others, Yahoo and Bing Maps. Their accuracy depends on the accuracy of the photo, although their resolution is decidedly greater than the resolution of the earlier selected raster layers.

Features that were difficult to identify on the raster images from other sources were selected from the Open Street Map database. Attention was given primarily to the linear feature layers (main roads, railways, and watercourses) not found in raster databases due to pixel resolution, but it was also decided to select residential areas due to the of dispersal of objects of this layer, which is not visible on lower-resolution raster layers (Fig. 3). These features were reclassified according to the aforementioned scale (Table 3).

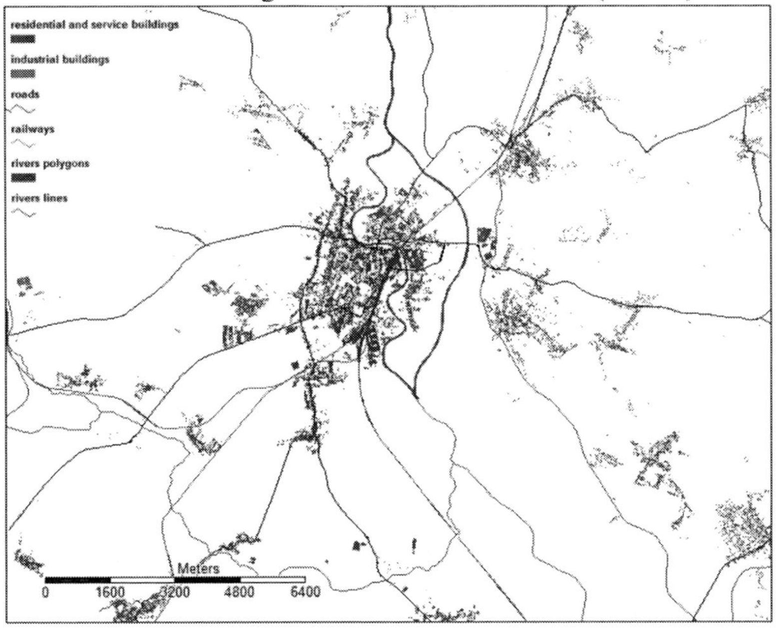

Fig. 3. Open Street Map layers selected for the study.

Table 3. Open Street Map objects reclassified for the study (classes 0–4).

Layer type *OSM attribute*	OSM class	Class	Feature
Polygon *Building*	'apartments' 'chapel' 'church' 'civic' 'college' 'commercial' 'detached' 'dormitory' 'farm' 'farm_auxiliary' 'house' 'office' 'residential' 'retail' 'school' 'service' 'shop' 'train_station' 'yes'	4	Residential buildings
Polygon *Building*	'cowshed' 'garage' 'garages' 'greenhouse'	3	Industrial buildings

 https://doi.org/10.1051/e3sconf/20182900026

	'hut' 'industrial''roof' 'shed' 'stable' 'transformer_tow er''transportation' 'warehouse'		
Line *Z_order*	5 3	3	Railway line
Line *Highway*	'primary' 'secondary' 'secondary_link'	3	Main roads
Line *Waterway*	river 'riverbank'	0	Flowing water
Polygon *Waterway*	'riverbank'	0	Standing and flowing water

5 Data Processing

Due to their varying resolutions and referencing systems (Table 4) all studied sources needed to be converted. The ETRS89 system (EPSG:3035), which corresponds to the Copernicus layer arrangement and a pixel resolution of 10×10 m, was selected. Conversion was performed using the SAGA GIS software according to the scheme of Figure 4.

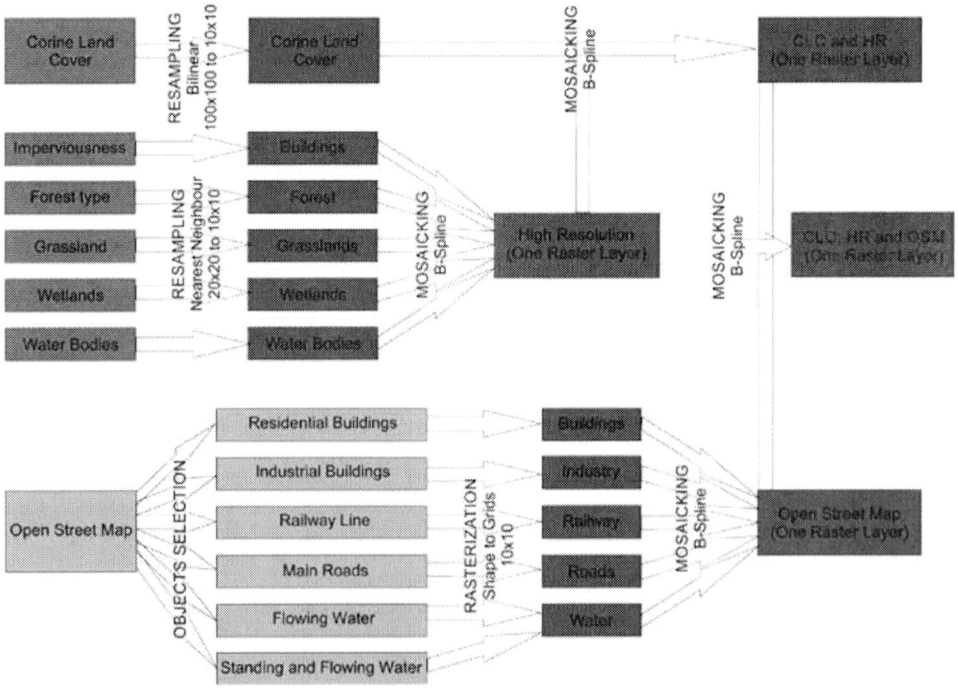

Fig. 4. Scheme of data processing adopted in SAGA GIS[†].

[†]In resampling of HR Layers used a simply interpolation method (Nearest Neighbour), beacause this grid have only one atribute (value exist or not). In other more complicated layers adopted a B-Spline method, which is optimal in situation, when more value are adjacent.

In the case of raster layers, a simple resampling was performed, which consisted of changing pixel size. The resultant maps were unchanged. Meanwhile, the Open Street Map layers, which remained in vector form were rasterised, maintaining the same parameters (Fig. 5).

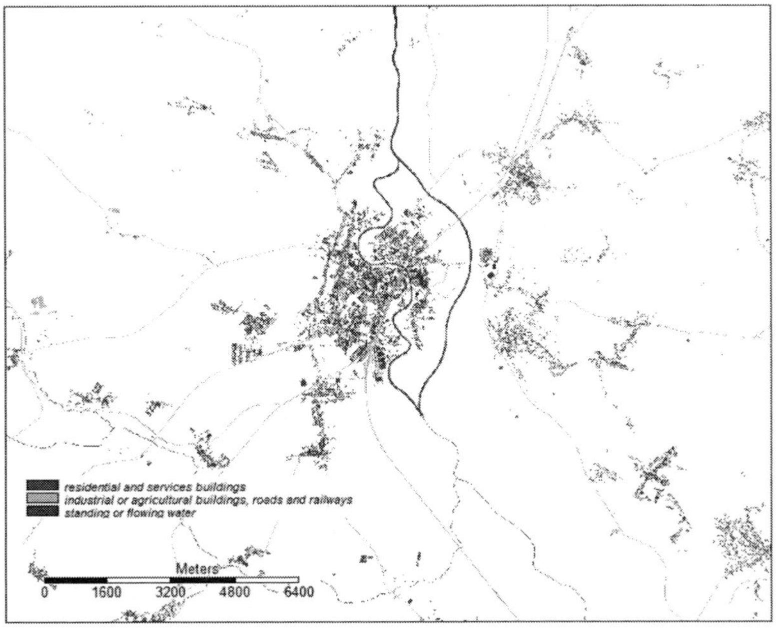

Fig. 5. Open Street Map layers reclassified to the study scale.

The next stage involved combining the layers while retaining the key features for study For reclassification for the study, the High Resolution layers were mosaicked, to combine five layers into one grid. The result contained features in the 0–4 scale of land use classes adopted in the study. A similar process was used for the Open Street Map layers to combine six layers. The reclassified Corine Land Cover data were left unchanged as a single ready layer.

In this way three land-use maps were obtained at the same resolution, of which only the Corine map is complete. Individual layers do not agree with each other in terms of land-use features. This is due to the resolution and methods of identifying features in the source layers.

In order to correctly identify features in the basic unit of study, which is the 10×10-metre pixel, a layer of hot spots, i.e. units with conflicting results, needed to be created. For error correction, combined grids were produced for CLC and HR layers. The resultant combined layer was a grid, which showed only the part common to the two layers, and which was reclassified so that the result values corresponded to the assumptions of the correlation of two overlapping classes (Table 4).

Table 4. List of CLC and HR layer classes.

HR class (reclassified)	CLC class	New study class
4	112	4
4	121	3

https://doi.org/10.1051/e3sconf/20182900026

4	142	2
1	243	2
1	311	1
1	312	1
1	313	1
1	231	1
0	511	0
0	512	0

The resultant reclassified grid of common parts was combined with the High Resolution Layers, again using Mosaicking, assuming the layer with the common parts as the main layer, and the HR layer as only supplementary (Fig. 6).

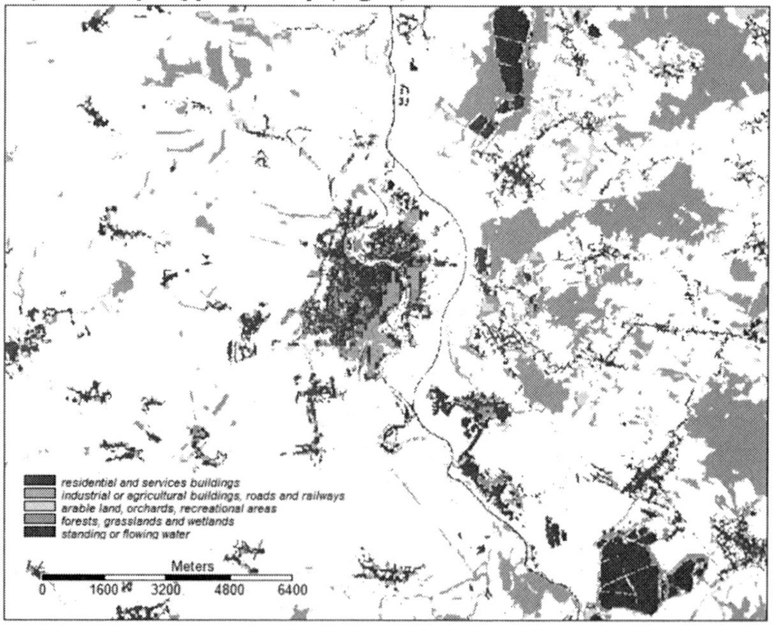

Fig. 6. Land use based on High Resolution Layers (Copernicus Land Monitoring System).

The method was repeated, adding the CLC grid to the resultant layer, this time assuming that this layer was supplementary. In this way, a full picture of land use was created, which did not identify linear features or individual buildings (Fig. 7).

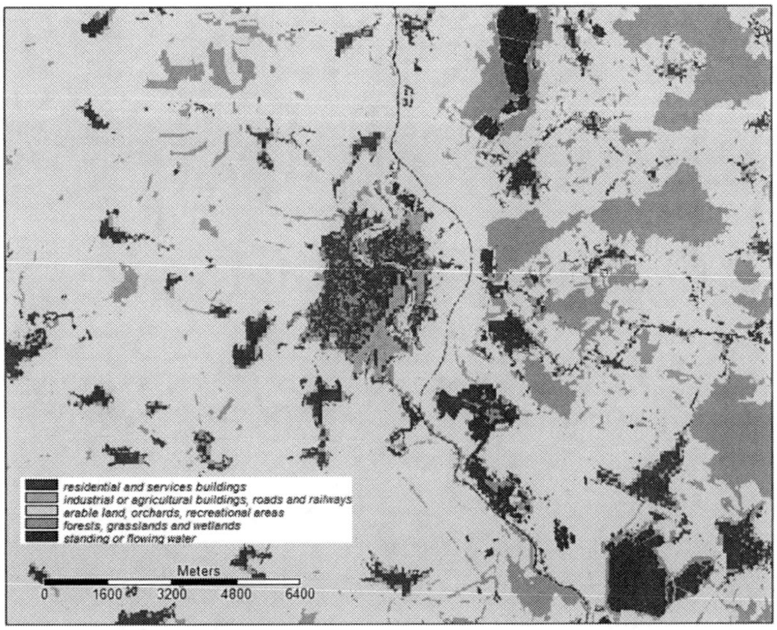

Fig. 7. Land use based on High Resolution Layers and Corine Land Cover (Copernicus Land Monitoring System).

The final research stage involved adding in Open Street Map objects to this newly created layer. Because of the insufficient building descriptions in the attribute table of the OSM layer it was assumed that buildings in an area designated as an industrial built-up area by previous analysis are industrial buildings and should therefore be identified as class 3. To this end, combine grids was repeated to combine the previously indicated common part of the High Resolution and Corine layer with the OSM layer, according to the assumptions mentioned above (Table 5).

Table 5. List of CLC/HR and OSM layer classes.

HR/CLC Class	OSM class	New study class
3	4	3

The layer thus obtained was combined with the Open Street Map using Mosaicking (Fig. 7), and then with the previously obtained layer created from the Corine Land Cover and High Resolution Layers.

6 Results

The end product is a grid layer of land-use according to the classification adopted in the study.

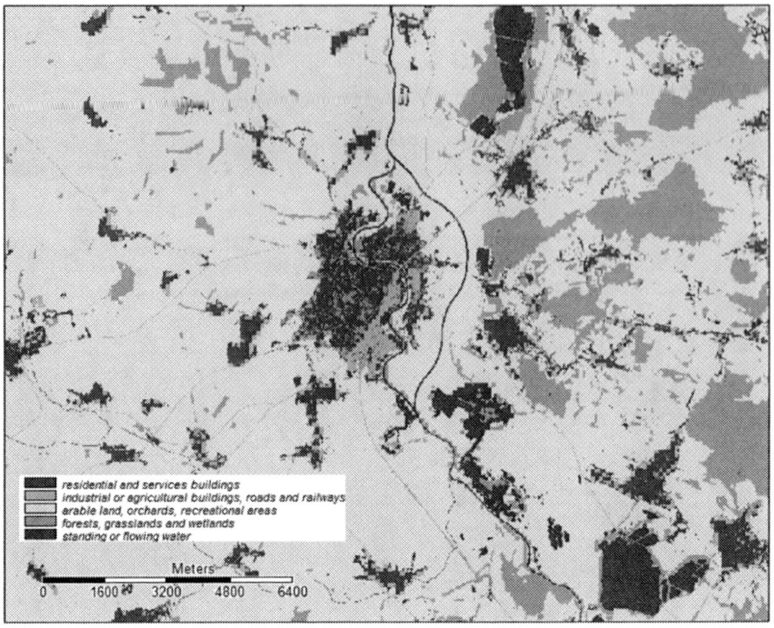

Fig. 8. Land use based on High Resolution Layers, Corine Land Cover (Copernicus Land Monitoring System) and Open Street Map.

All mosaicking steps were performed by B-Spline interpolation method to reduce errors when linking grids. Thus, despite the use of open source software and public data, the most accurate results could be preserved with possible methods [33].

7 CONCLUSIONS

During the processing of individual sources, it was noted that none of the tested databases individually met the criteria to allow their use in further studies. This is because each grid (CLC, HR and OSM) was incomplete, or the classification of features did not align with that used in flood-related research.

The Corine Land Cover 2012 layer is built on a scale, which deviates from the research standard. Although this database has almost all the desired features (except for line objects – infrastructure and watercourses), its resolution does not allow for the exact identification of feature location.

High Resolution Layers, meanwhile, are a collection of grids that represent only some of the sought features, i.e. forests, grasslands, wetlands and water. The Imperviousness layer represents several classes of features, meaning it cannot be used without comparison against other sources.

The Open Street Map's incomplete description of features limits the degree of recognition of individual land-use features. The only linear features, which can be indentified without major obstacle are roads, railways and rivers. Surface features, such as buildings, were classified using numerous key words, and their attributes were frequently insufficient (e.g. question « building », answer « yes »). The remaining land-use features either do not occur or are insufficiently described.

The only solution is to mosaic all the described sources in order to obtain a single grid layer. This study paid close care to ensure that the results were precise and reflected reality. This allowed greater imaging quality to be obtained for land-cover components relevant to flood-risk management.

The results layer (Fig. 8) shows land-use features for Raciborz and surrounding area as surface features from one of five possible classes to be used in further analyses. The 10×10-m resolution proposed in this study allows features to be identified, which are not accounted for in other studies due to their size but which are important to risk assessment.

The produced grid map is therefore the most accurate depiction of land use produced on the basis of data from widely available public sources. It is also a case study similar to studies on other regions of Poland and Europe.

References

1. Z.W. Kundzewicz, P. Kowalczak, *The Climate Change and the Effects* (Poznan: Kurpisz S.A., 2008)
2. D. Rucińska, Risk Analysis **35/2** (2015)
3. D.J. Wright, M.F.Goodchild, J.D. Proctor, Annals of the Association of American Geographers **87**, 2 (1997)
4. J. Solan, *Ocena różnorodności krajobrazu na podstawie analizy struktury przestrzennej roślinności* (Prace Geograficzne, IGiPZ PAN, Warszawa 2002)
5. B. Hejmanowska, Archiwum Fotogrametrii, Kartografii i Teledetekcji **17b** (2007)
6. M. Goodchild, Journal of Spatial Information Science **1** (2010)
7. C. Wang, J. Johnston, D. Vail, J. Dickinson, D. Putnam, Land **4** (2015)
8. A. Magnuszewski, *Woda w badaniach geograficznych*, T. Ciupa, R. Suligowski (ed.) (Instytut Geografii Uniwersytet Jana Kochanowskiego, Kielce, 2010)
9. M. Ogryzek, Acta Sci. Pol., Administratio Locorum **13**, 4 (2014)
10. P. Pokojska, Prace i Studia Geograficzne **57**, Warszawa (2015)
11. E. Woźniak, Teledetekcja Środowiska **51** (2014)
12. P. Werner, A. Magnuszewski, D. Rucińska, M. Porczek, Prace i Studia Geograficzne **57** (2015)
13. H. Goyal, C. Sharma, N. Joshi, International Journal of Computer Applications **169**, 11 (2017)
14. W. Widacki, *Wprowadzenie do Systemów Informacji Geograficzne* (Wyd. Test, Kraków 2017)
15. M. Kistowski M. Iwańska, *Systemy Informacji Geograficznej. Podstawy techniczne i metodyczne. Przegląd pakietów oprogramowania i zastosowań w badaniach środowiska przyrodniczego* (Bogucki Wydawnictwo Naukowe, Poznań 1997)
16. P. Longley, M.F. Goodchild, D. Maguire, D. Rhind, *Geographic Information Systems and science* (Jon Wiley & Sons, Chichester, 2005)
17. R. Kot, *Georóżnorodność – problémy jej oceny i zastosowania w ochronie i kształtowaniu środowiska na przykłądzie Fordońskiego odcinka Doliny Dolnej Wisły i jej otoczenia* (Towarzystwo Naukow Toruń, UMK w Toruniu 2006)
18. B. Kuraś, Archiwum Fotogrametrii, Kartografii i Teledetekcji **17b** (2007)
19. H. Gong, M. Simwanda, Y. Murayama, ISPRS Int. J. Geo-Inf. **6**, 257 (2017)
20. J. Chojnacki, J. Rataj, Gospodarka Wodna **9/1975**, Warsaw, SIGMA NOT (1975)
21. J. Chojnacki, Gospodarka Wodna **1/1993**, Warsaw, SIGMA NOT (1993)
22. J. Chojnacki, Gospodarka Wodna **10/1994,** Warsaw, SIGMA NOT (1994)
23. J. Chojnacki, H. Zawada, Gospodarka Wodna **3/1998,** Warsaw, SIGMA NOT (1998)

24. M. Borowska-Stefańska, Biuletyn Uniejowski **4/2015**, Uniejów (2015)
25. M. Borowska-Stefańska, *Zagospodarowanie terenów zagrożonych powodziami w województwie łódzkim* (Łódź, Uniwersytet Łódzki 2015)
26. Directive 2007/60/EC of the European Parliament and of the Council of 23 October 2007 on the assessment and management of flood risks (2007)
27. *Raport z wykonania Wstępnej Oceny Ryzyka Powodziowego* (*Report of the Implrmentation of hazard risk maps and risk map of flood),* Project ISOK (*Informatyczny System Osłony Kraju przed Nadzwyczajnymi Zagrożeniami*), title of the task 1.3.1: *Wstępna Ocena Ryzyka Powodziowego* (WORP), IMGW i KZGW (2011)
28. Z., Kurczyński, Archiwum Fotogrametrii, Kartografii i Teledetekcji **23**, Wrocław (2012)
29. E. Drożdżal, M. Grabowski, K. Kondziołka, J. Olbracht, M. Piórecki, R. Radoń, A. Ryłko, Gospodarka Wodna **1/2009**, Warsaw, SIGMA NOT (2009)
30. *Corine Land Cover 2012*, Copernicus Land Monitoring Service, copernicus.eu
31. *High Resolution Layers*, Copernicus Land Monitoring Service, copernicus.eu
32. *Open Street Map Project*, openstreetmap.org
33. S. Lee, G. Wolberg, S.Y. Shin, IEEE Transactions On Visualisation And Computer Graphics **3/3** (1997)

Analysis of post-mining excavations as places for municipal waste

Justyna Górniak-Zimroz[1,*]

[1]Wroclaw University of Science and Technology, Faculty of Geoengineering, Mining and Geology, 27 Wyb. Wyspianskiego St., 50-370 Wroclaw, Poland

Abstract. Waste management planning is an interdisciplinary task covering a wide range of issues including costs, legal requirements, spatial planning, environmental protection, geography, demographics, and techniques used in collecting, transporting, processing and disposing of waste. Designing and analyzing this issue is difficult and requires the use of advanced analysis methods and tools available in GIS geographic information systems containing readily available graphical and descriptive databases, data analysis tools providing expert decision support while selecting the best-designed alternative, and simulation models that allow the user to simulate many variants of waste management together with graphical visualization of the results of performed analyzes. As part of the research study, there have been works undertaken concerning the use of multi-criteria data analysis in waste management in areas located in south-western Poland. These works have proposed the inclusion in waste management of post-mining excavations as places for the final or temporary collection of waste assessed in terms of their suitability with the tools available in GIS systems.

1 Introduction

Despite the desire to reduce the amount of waste produced and to intensify the disposal processes in waste management, landfill sites are necessary for placing waste that cannot undergo waste treatment in the present state of the art. The need to build new landfills results from the fact that existing ones are being closed due to overfilling or failure to meet stringent legal, sanitary or environmental requirements [1, 2]. Therefore, research was carried out to identify new sites for the location of waste management facilities. Designing these sites is a complex issue in which many factors have to be taken into account in determining the best location for a landfill. Therefore, multi-criteria data analysis can be used to determine this location. This analysis aims at selecting the optimal solution according to different criteria that are difficult to compare with each other and affect the implementation and operation of the solution. It is used as a decision support tool when multiple criteria are available. The selected examples described in [3-8] constitute the applications of multi-criteria data analysis in the studies concerning the support in the

* Corresponding author: justyna.gorniak-zimroz@pwr.edu.pl

processes of making planning decisions. It was proposed to calculate the suitability of land in waste management by means of multi-criteria data analysis based on the equation:

$$S = \sum_{i=1}^{n} w_i \cdot p_i \quad (1)$$

where:
S - land suitability
w_i - the weight of the criterion
p_i - value of the parameter
i - criterion
n - number of criteria

$$\sum_{i=1}^{n} w_i = 1 \quad (2)$$

$$i \in <1, n> i \quad (3)$$

The research offered the inclusion of post-mining excavations in waste management, which can be used to prolong mining activities and at the same time contribute to the development of mining services in the waste market. However, the location of the waste management facility in each excavation is not possible due to numerous spatial, social and environmental constraints [9]. Therefore, it is necessary to assess the suitability of post-mining excavations in multi-criteria waste management using data analysis methods.

2 Materials and methods

The analyses were carried out for municipal waste management as well as hazardous and radioactive waste, for which an appropriate area of investigations with precise limits was chosen. These boundaries include spatially linked settlements that are the source of waste generation and mining areas, together with the common solutions for waste management and waste management, and the rational development of mining areas in this territory. The decision concerning the location of the facility related to the management of hazardous and radioactive waste is made by the Minister competent for the economy, the Minister responsible for the environment and the Department of Waste Disposal. Such an object should be a regional object. Therefore, the region of Lower Silesia has an area of 19 936 km^2 with localized deposits of hard coal, copper ore, barite, white-burned clay, rock salt and gypsum and anhydrite (Figure 1a). Despite the fact that municipal waste management is a responsibility of the municipality itself, according to a rational waste management, newly created municipal waste management facilities should perform over-the-counter functions. Therefore, the pilot areas are limited to the administrative boundaries of the Milicz, Wrocław and Swidnickie districts (Figure 1b). The limitations of the study area do not affect the analysis method itself and do not limit the applicability of the developed algorithms and analysis tools. Extending the research area requires the introduction of new empirical data into the system.

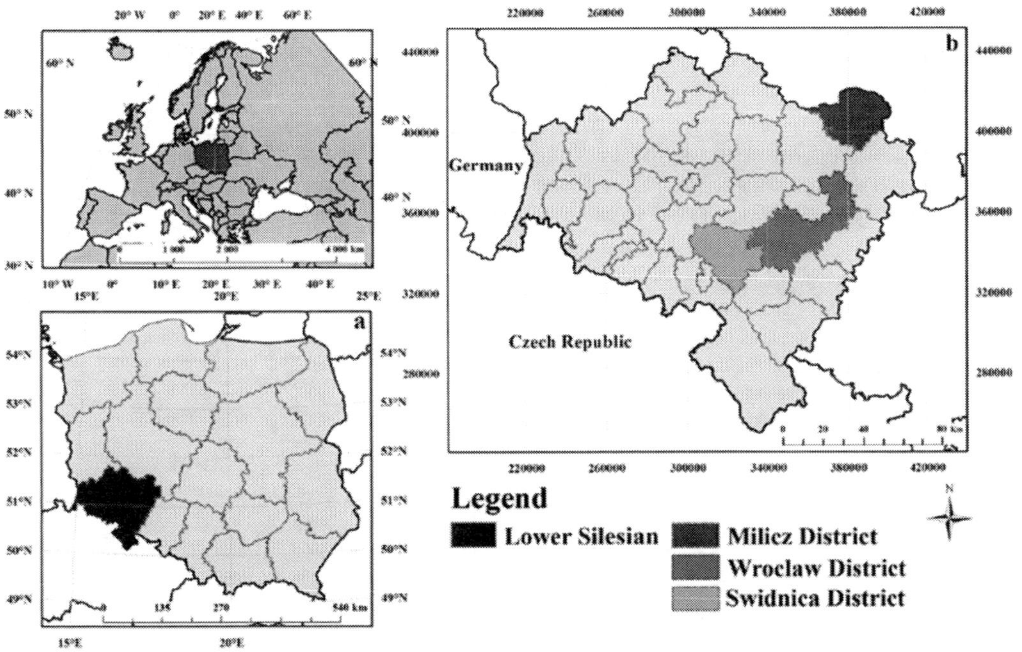

Fig. 1. Localization of selected areas for analyses against the background of Poland (a) and of Lower Silesia (b).

The suitability of the post-mining excavation is described in [10-12] as a function of many variables of unknown form divided into four main groups of factors, i.e. spatial factors (v), environmental factors (p), technical factors (t) and social factors (s):

$$jw = f(x_1, x_2, \ldots, x_n) \tag{4}$$

$$jw = f(v, p, t, s) \tag{5}$$

where:
x – the factor describing the quality of the excavation
v - spatial factors
p - environmental factors
t - technical factors
s - social factors

The function describing the state of the excavation, the so-called "quality of the excavation" depends on different variables in different ways, leading to the following dependence:

$$jw = \sum_{i}^{n} w_i \cdot x_i \tag{6}$$

where:
w_i - weight determining the value of the i-th factor describing the quality of the excavation
x_i - the i-th factor describing the quality of the excavation

Designing weights is a complex issue, but it is possible through empirical research. Determination of the value of the "excavation quality" function is necessary to verify in which of the ranges of variability of this function, where ranges correspond with the suitability classes of the excavation, the evaluated excavation site is found. In the presented research, the factors and their weights were selected on the basis of the literature on the subject, legal references and surveys of experts dealing with spatial planning, waste management, environmental protection and mining. The number of factors describing the suitability of the post-mining excavation in the conducted research was selected for the analyzed case. Data and knowledge of the content needed for analyzes were collected in databases developed for mining objects and for selected research areas.

Research in the described cases was made by means of selected spatial analyzes available in GIS systems. Their final result was to obtain a map of the assessment of the suitability of excavations for locating a waste management facility. Figure 2 shows schematically how the research was conducted.

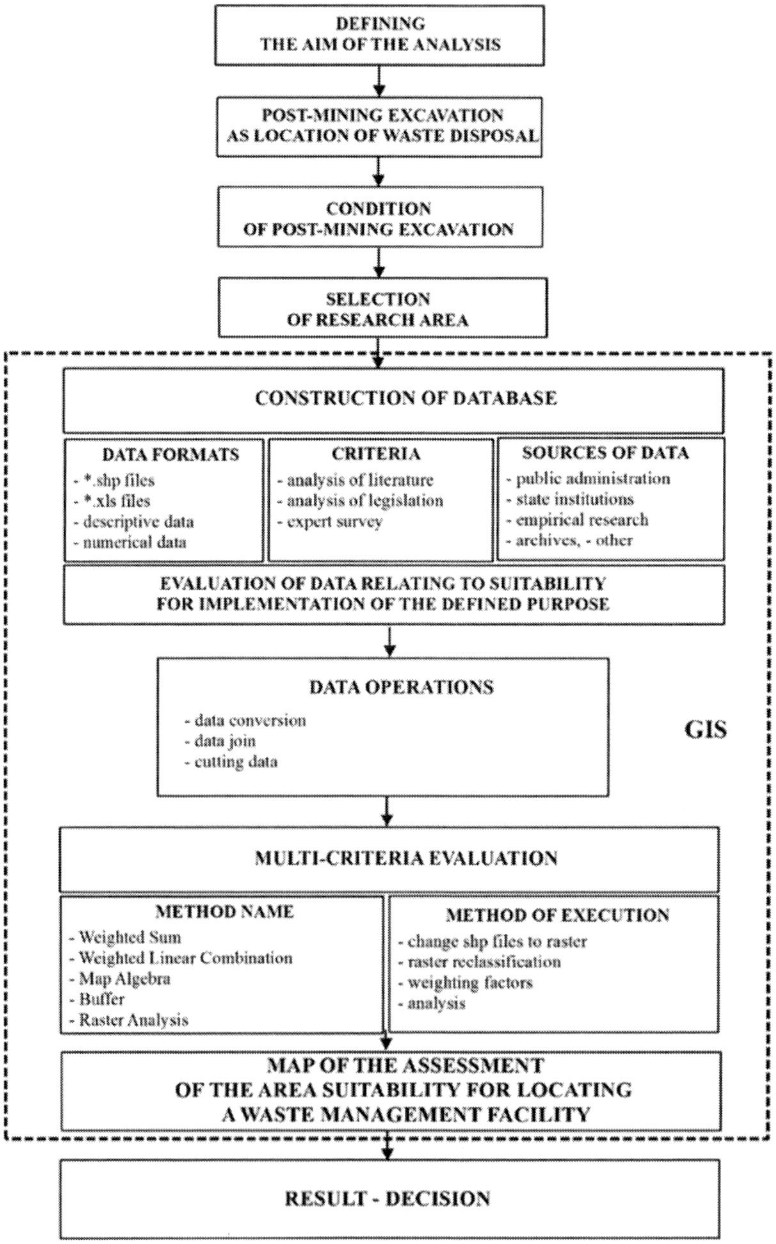

Fig. 2. Methodology for the evaluation of the suitability of post-mining areas in waste management.

3 Results

3.1 Milicz District

A database of 98 factors was developed for the Milicz District based on the analysis of the literature on determining the factors defining the direction of reclamation, the factors determining the use of post-mining areas for the disposal of municipal waste, the factors taken into consideration during the selection of the site for the location of the municipal landfill and the analysis of the Polish and EU law on environmental protection and the management of municipal waste. These factors are described by quantitative and qualitative variables that were converted into numerical variables. The numerical value for the i-th variable describing the quality of the post-mining excavation was determined according to the following principle: the advantageous factor for the location of the site was 1, neutral or no factor was 0, and the disadvantageous factor for the location of the site was -1 [10-13].

Figure 3 shows the result of the analysis of the suitability of the site for the location of the municipal waste disposal landfill made for 16 rock raw material deposits from the Milicz District. This analysis takes into account the influence of other structures located within the analysed land on the location of the landfill, i.e.: the proximity of post-mining areas in relation to waste generating sources and roads, and the remoteness of groundwater intakes, rivers, reservoirs and nature reserves. The suitability of the area is depicted in colour, the brighter the colour, the better place for the location of the landfill. The analysis was done in ArcGIS using multi-criteria data analysis by the Weighted Sum Method. The obtained result was compared with the deposits located in the Milicz District and it was found which is located in the highest rated area with regard to its waste management suitability. On the obtained map of results, this area was marked with the location of the clay raw material deposits of Zduny-Cieszków building ceramics [10-13].

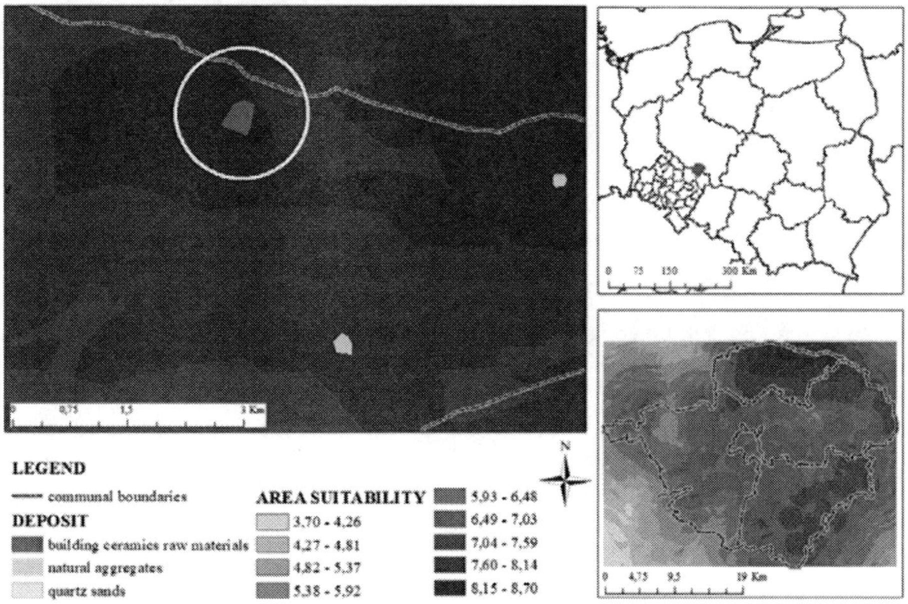

Fig. 3. Analysis of the suitability of the site for the location of a municipal landfill in the Milicz District.

3.2 Wroclaw District

For the Wroclaw District, 59 rock raw material deposits were analyzed in terms of their suitability in the management of municipal waste. A buffer construction tool was constructed for five criteria based on the literature for the following parameters: 500 m for municipal waste generation sources, 300 m for the communication network, 90 m for the hydrographic network, 500 m for inland surface water bodies, and 300 m for landscape parks with buffer zones. On the obtained restrictions there was a layer applied with the location of deposits and analysis of their suitability [14].

Figure 4 shows the results. In the Wroclaw District only 4 deposits are located outside of the restrictions in question. These are the deposits of natural aggregates Jezierzyce Wielkie and Jezierzyce Wielkie I as well as deposits of road and building stones Jordan and Jordan 1 marked in the picture.

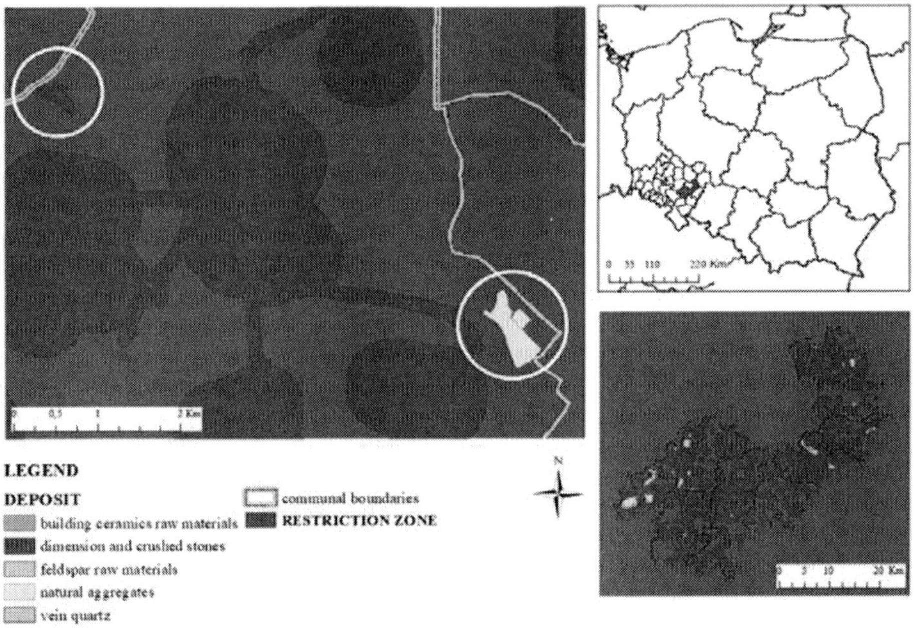

Fig. 4. Analysis of the suitability of the site for the location of a municipal landfill in the Wroclaw District.

3.3 Świdnica District

In the Świdnica District, 94 rock raw material deposits were analyzed. The factors defining the suitability of the post-mining area in waste management were determined on the basis of the analysis of literature and legal references. 48 factors were divided into six groups: development, communication network, water and wetlands, forests, protected areas, soils, grasslands and landslides. Weights for factors were identified during the survey of mining, waste management and environmental experts [15]. After preparation of the database, multi-criteria data analysis was performed by means of buffering and map algebra consisting in performing algebraic operations on rasters [16]. A layer of deposits was applied on the resulting raster to analyze which of them could qualify as a location for a municipal waste disposal site. In the Świdnica District, 40 rock raw material deposits were identified. The resultant map is shown in Figure 5. The dark-green colour means the areas with the best suitability for the location on the map. The map shows the deposits located in these areas. These are the Gniewków I granite deposit, the Grodziszcze gneiss deposit, the Julia kaolin deposit and the Wiry-Gogołów magnesite deposit [15].

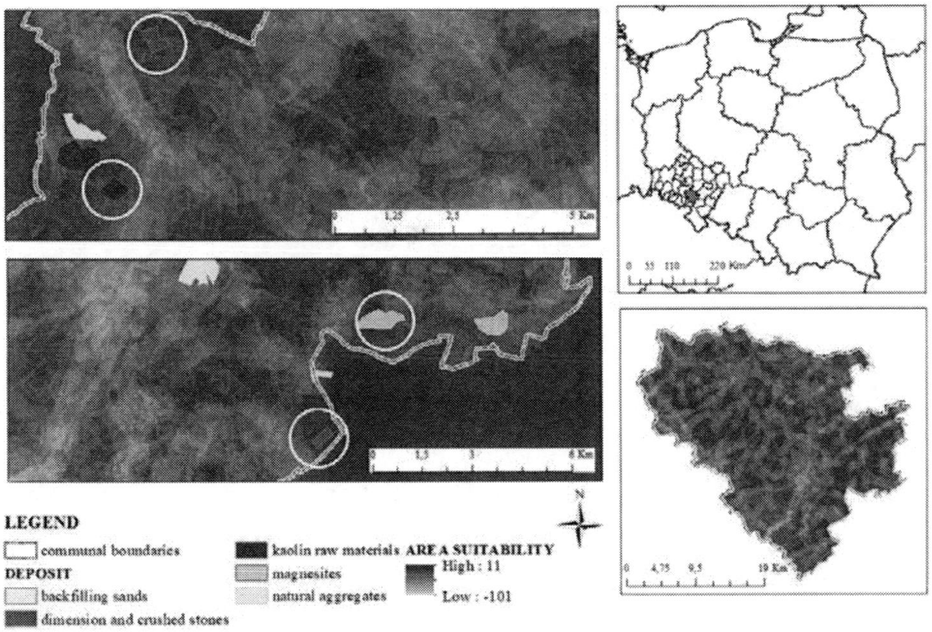

Fig. 5. Analysis of the suitability of the site for the location of a municipal landfill in the Świdnica District.

3.4 Lower Silesia

In the Lower Silesia area, underground excavations of hard coal deposits, copper ores, gypsum and anhydrite, barite, white-burned clays and rock salt deposits were selected. These deposits were analyzed for the location of hazardous and radioactive waste disposal sites. The selection of factors was made on the basis of literature and legislation regulating issues related to the management of hazardous and radioactive waste. The geobase with factors was divided into five groups: factors that define areas of nature protection, factors describing soils, factors concerning hydrographic structures, factors related to afforestation, as well as building and infrastructure factors. As a method of multi-criteria spatial analysis, a method of Weighted Linear Combination was chosen. WLC consists in a simple combination of layers based initially on multiplication of parameters and estimated weight of a given criterion and then on the addition of multiplication results of all criteria [18]. Weights for factors were based on the analysis of literature and legislation related to the studied subject matter. For weighting, the scale from 0 to 10 was assumed, where 10 is the best location for the factor and 0 is the exclusion of the location [17].

Figure 6 shows the result of the assessment of the suitability of the site for the location of the hazardous and radioactive waste landfill. The barriers strictly limiting the suitability of the site and the deposit were applied on the resultant map and it was analyzed which of these deposits could be used as a location for the site. As a result of the analyzes, it was obtained that there are Nowy Ląd deposits of gypsum and andyrethrite, Jeżów Sudecki barite deposit and deposits of Sieroszowice and Bądzów salts, which are the accompanying minerals for copper ores.

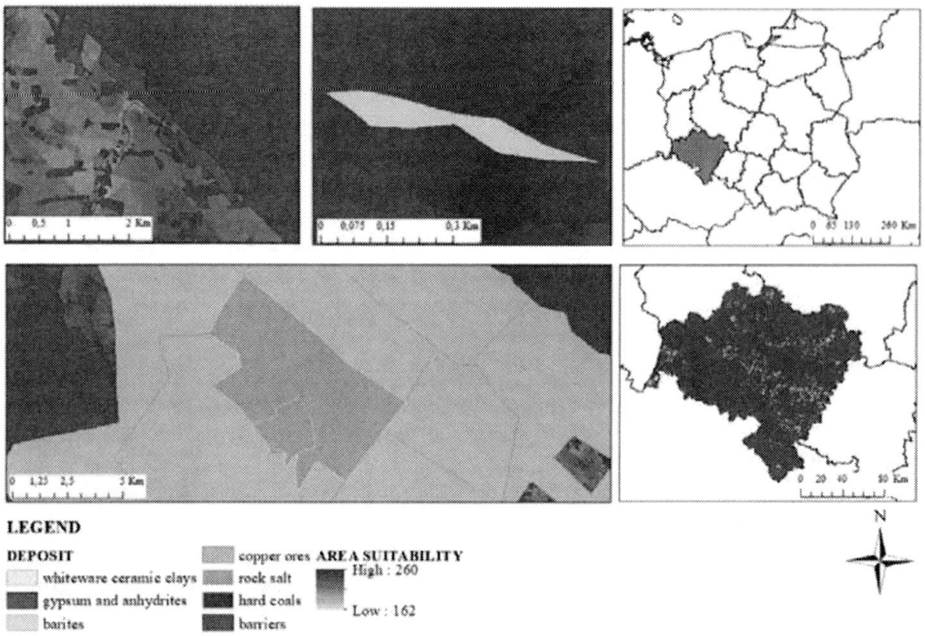

Fig. 6. Analysis of the suitability of the site for the location of the hazardous and radioactive waste disposal landfill in Lower Silesia.

4 Conclusion

The research showed that multi-criteria data analysis used to assess the suitability of a site for the location of a waste management facility can be applied as a useful tool in spatial management to support decision-making processes in waste management. Choosing a location involves analyzing many factors that influence one another. The suitability maps generated by multi-criteria spatial analysis show the information on the site and the requirements that should be met by the location of the municipal waste disposal facility and the hazardous and radioactive waste site. It is important that the use of GIS tools in the decision-making process also makes it possible to incorporate in the analysis the legal aspect understood as the content of legislation, regulations and directives translated into spatial data used in the analysis. However, as with most automated processes, it is necessary to verify the results with a reality that do not explicitly exclude areas covered by factors that exclude the location of the research objective, namely, determining the location of the site, but only reduce their availability. Therefore, multi-criteria data analysis in GIS environment can be considered as a decision support tool. Further studies can use the results obtained during the analysis and provide a detailed analysis of the geological and hydrogeological structure of the areas designated as the most favourable for the location of the landfill.

This article was financed with the research project no. 0401/0174/16 *Data mining in modelling environmental changes* and Grant no. 0401/0128/17.

References

1. *Basic statistics of the European Community,* EU, Eurostat, Brussels, http://ec.europa.eu/eurostat/en, March (2017)

2. American Environmental Protection Agency, https://www.epa.gov/landfills, March (2017)
3. J. Blachowski, Environ Earth Sci., Springer **71** (2014)
4. J. Chen, Environmental Systems Research **3/13** (2014)
5. J.R. Eastman, M.E. Van Fossen, L.A. Solorzano, *Transition potential modelling for Land-Cover change*, in *GIS, Spatial Analysis and Modelling*, editors: Maguire D.J., Batty M. and Goodchild F., ESRI Press, 380 New York Street, Redlands California (2005)
6. J. Górniak-Zimroz, Applied Mechanics and Materials **683** (2014)
7. G.A. Kiker, T.S. Bridges, A. Varghese, T.P. Seager, I. Linkovjj, Integrated Environmental Assessment and Management **1 (2)** (2005)
8. J. Malczewski, International Journal of Geographical Information Science **20 (7)**, August (2006)
9. B. Radwanek-Bąk, R. Patorski, *Possibilities of using post-mining excavations on landfills*, work edited M. Kistowskiego *Landscape-ecological studies for sustainable development programming review of polish experiences on the eve of European Union accession*, The Problems of Landscape Ecology **XIII**, Gdańsk (2004)
10. J. Górniak-Zimroz, *Integrated municipal waste management and post-mining excavation,* doctoral thesis, unpublished work in Polish (2004)
11. J. Górniak-Zimroz, J. Malewski, *Application of the Kohonen neural network for classification of mining voids*, AI-METH 2004, Symposium on Methods of Artificial Intelligence, Gliwice (2004)
12. J. Górniak-Zimroz, J. Malewski, *Decision support system for management of municipal waste and mining voids*, MPES 2004, the Thirteenth International Symposium on Mine Planning and Equipment Selection, Wrocław (2004)
13. J. Górniak-Zimroz, J. Malewski, J. Woźniak, *The methodology of development of decision support system for management of municipal waste and mining voids*, Sustainable post-industrial land management, SPILM 2004, The international conference, Kraków (2004)
14. J. Górniak-Zimroz, J. Malewski, *Logistics aspects of municipal waste management*, Management of municipal waste in the Wroclaw District in the light of solutions implemented in the Upper Rhine Department (France) and the Borken District (Germany), International Conference, Wroclaw (2005)
15. S. Sewina, *Analysis of post-mining excavations as potential landfills of municipal waste - analysis for the selected district,* master thesis carried out under the supervision J. Górniak-Zimroz at the Faculty of Geoingeneering, Mining and Geology Wroclaw University of Science and Technology, unpublished work in Polish, (2016)
16. C.D. Tomlin, Elsevier, Landscape and Urban Planning **30** (1994)
17. A. Dymacz, *Determination of potential locations of underground hazardous and radioactive waste dumps in Lower Silesian Voivodeship*, engineering work carried out under the supervision J. Górniak-Zmroz at the Faculty of Geoingeneering, Mining and Geology Wroclaw University of Science and Technology, unpublished work in Polish (2017)
18. J. Eastman, *Guide to GIS and Image Processing*, **2**, IDRISI 32 (2001)

E3S Web of Conferences **29**, 00028 (2018) https://doi.org/10.1051/e3sconf/20182900028
XVIIth Conference of PhD Students and Young Scientists

Analysis of environmental-social changes in the surrounding area of KWB Turow in the historical context

Tobiasz Cieślik[1], and *Justyna* Górniak-Zimroz[1,*]

[1]Wroclaw University of Science and Technology, Faculty of Geoengineering, Mining and Geology, 27 Wyb. Wyspianskiego St., 50-370 Wroclaw, Poland

Abstract. Opencast mining of large-area lignite deposits impacts the environment, and the health and life of people living in the vicinity of the conducted mining activity. Therefore, the attempt was made to develop a methodology for identification of environmental and social changes in the Bogatynia municipality (south-western Poland), resulting from functioning of Turow lignite mine within its area. During the study of changes occurring over the years, the development of mining pit was noticed, as well as the transformations of this area and impact of the mining plant on the selected elements of environment and surrounding areas. Analogue and digital data were used for the preparation of cartographic compilations, the usefulness of which was analyzed in accordance with the guidelines contained in the standard [1]. The conducted cartographic studies allowed to learn the history of the mine together with identification of changes taking place in the municipality Bogatynia. The obtained results show the form and condition of the objects in the analyzed year, allowing for the interpretation of changes that occurred in the surrounding areas of the Turow mine. Due to the conducted activity of the mine and Turow power plant, both negative and positive aspects were noted in connection with the carrying out of mining activity in the Bogatynia municipality.

1 Introduction

The operation of large-area mining companies that exploit the deposits using opencast method has a significant impact on the environment and society. The consequences associated with the operation of such companies are visible in the landscape, affect the elements of environment located in or near the opencast area and impact the lives of residents in villages and cities located in the surrounding area. Large-area opencast mines in Poland include the following lignite mines: Adamow, Belchatow, Konin, Turow and Sieniawa. Within this study, the works have been undertaken to identify and analyze environmental and social changes in the Bogatynia municipality, which is located in south-western Poland (Figure 1), during the operation of lignite mining from KWB Turow within

* Corresponding author: justyna.gorniak-zimroz@pwr.edu.pl

its area. The conducted study covered the years between 1890 to 2016, divided into several periods due to the availability of source materials.

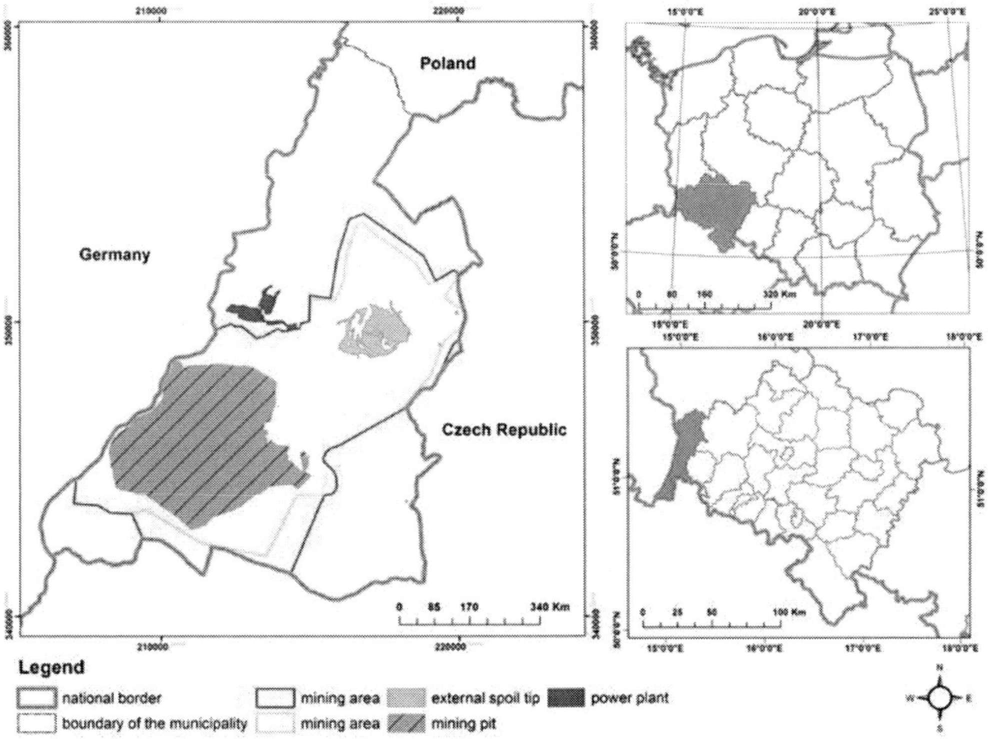

Fig. 1. Location of Turow Lignite Mine (own elaboration).

2 Materials and methods

In analysing environmental-social changes against historical background, it is important to define the time range examined. It is important to preserve temporal continuity, in addition to the data from the beginning and the end of the time frame, it is necessary to present transitional states in order to better analyse the changes taking place in the area analysed. Time frames were determined after analysing the data obtained in analogue and digital form. Accepted time forms should be referenced to one reference level, allowing identification of transformations that occurred in the time interval. When presenting historical data, the uniform boundaries of the surveyed area should be determined for each development period. This is complicated because the boundaries of the administrative units have changed, so determining the area to be investigated should determine the extent of impact of the investigated phenomenon. The solution to the problem of non-uniform borders may be the adoption of a radius of influence and, on that basis, the designation of the field under investigation or the acceptance as an area under investigation in each period of current administrative boundaries. The administrative boundaries of the municipality of Bogatynia were accepted. The impact of large-scale mining activity on particular elements of the environment and on society was characterized in the next step. Based on this characteristic, the structure of the geodatabase with objects changing under the influence of mining activity was determined. In order to fill the geodatabase, the derived source data was processed (scanning, calibration, vectorization of the analysed objects, calculating their surface area, calculating surface changes) and cartographic studies were performed for the analysed time frame. The developed geodatabase and maps allowed to identify environmental and social changes caused by the exploitation of brown coal from the large-scale open-pit mine. Figure 2 shows the proposed methodology.

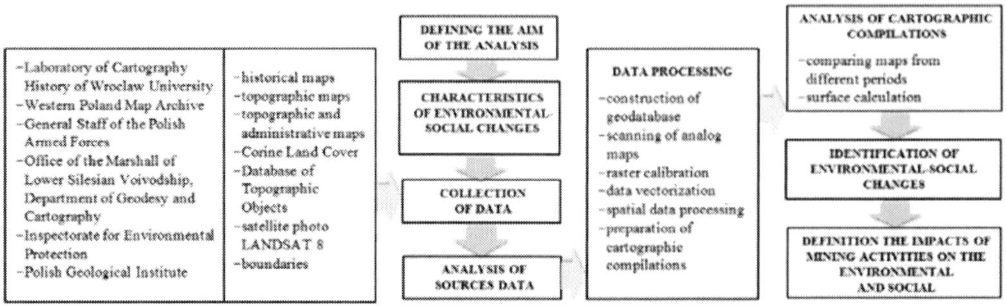

Fig. 2. Scheme of conduct in the methodology of analysis of environmental-social changes in the surrounding areas of KWB Turow in the historical context [2].

2.1 Identification of environmental-social changes

Each stage of deposit management interferes with the environment and affects the health and life of people. The character and size of these changes vary and depend on the type of exploited raw material, size of its extraction, applied mining technologies, transport and processing of minerals. The identification of environmental and social changes resulting from the large-area opencast mining of lignite was carried out on the basis of subject literature [3-8] and it's described in detail in Table 1.

Table 1. Environmental-social changes resulting from lignite mining using the opencast method.

Stage of mining activity	Environmental changes	Social changes
Searching and identification of the deposit	-change of ground surface -emission of communication pollution -emission of dust pollution -noise and vibrations	-emission of communication pollution -emission of dust pollution -noise and vibrations
Preparation works, supplementary works and mining	-transformations of terrain -change of geological structure -transformations of soils and lands -destruction and pollution of plant and animal systems -hydrological and hydrogeological transformations -emission of communication and dust pollution -noise and vibrations -exclusion of areas from the previous use	-transformations of terrain -change of land development -liquidation of crops and breeding -liquidation of residential infrastructure, transport and industrial infrastructure -change of real estate values -change of place of residence and lifestyle -emission of communication and dust pollution -noise and vibrations
Transport	-emission of communication pollution -emission of dust pollution -noise and vibrations	-emission of communication pollution -emission of dust pollution -noise and vibrations
Liquidation, reclamation and management	-transformations of terrain -transformations of soils and lands -transformations of flora and fauna -hydrological and hydrogeological	-transformations of terrain -emission of communication and dust pollution -noise and vibrations

	transformations -emission of communication and dust pollution -noise and vibrations	-change of land development

2.2 Characteristics of the study subject

The history of first attempts to extract lignite from the Turow deposit dates back to 1740-1790. In XIX century, as a result of the flourishing of industry in these areas, the first mineshafts were created allowing for extraction of coal, which was supposed to become the new material used for energy production. Unfortunately, the water in the ground caused that the windshields were flooded in a short period of time and most of them stopped coal mining. In 1879, many changes were introduced to the mining system, because a lignite-fired power plant started its operation in Hirschfelde. Systematic extraction using the opencast mining method in the industrial scope started in 1904, at the time of "Hercules" company. Production of new mining machines was started in order to increase the efficiency of mining of the deposit, while the briquetting plant and railway line allowing for transport of coal to the power plant were created in 1911. In 1947, after the II World War, the mining plant was taken over by the Polish State Treasury. Further development of the mine took place in 1958, when the expansion of existing open pit was started, in order to create a modernized mining plant "Turow". At the beginning of 1958, the decisions were made to expand the plant and the district of Bogatynia became famous on a national scale. The next stage in development of the mine was the commencement of construction of the Turow Power Plant. At the beginning of 1990s, the tactics of modernization of the mining plant was introduced. Innovative machines and devices were introduced in the mine, which significantly contributed to the efficient extraction of lignite. The mining activity associated with environmental protection programs has been successively expanding. In 2001, the "Turow" Mining Plant was removed from the list of 80 companies that are the most harmful to the environment. Due to the concern for the environment, the mine implemented a nature protection program in 2004-2008, which resulted in the afforestation of the external spoil tip. In the subsequent years, as a result of policy associated with power supply sector and merging of companies engaged in energy production, KWB Turow became a part of the PGE Mining and Power company. Currently, the mining plant in Turow is one of the largest opencast mining companies in Poland. Coal extraction is estimated at 12 million tons/year with an overburden of 32 million m^3 [9-13].

2.3 Construction of geodatabase

Usefulness of the collected source materials in analogue and digital form was analyzed according to guidelines in the standard [1], which defines the basics of geographic data description, defines the description components of their quality and structure of the register contents containing data quality measures. This standard determines the rules for presenting data quality in further applications and describes the general procedures for their assessment. During the analysis of source data, it was assumed that the compilations, which lack parts of the municipality area, would not affect the outcome of the work. Areas that were not shown on analogue maps are located at the outskirts of Bogatynia municipality and the greatest impact of the mining plant is visible in its immediate vicinity. KWB Turow is located in the central part of the municipality, which was included in the entire collected source materials. It was also found that the information contained in the Corine Land Cover from 2006 is not very accurate, because they show real changes above 5 hectares, which can significantly affect the results of conducted analysis. Therefore, data from this database were used only for the preparation of cartographic compilation from 2006 and were omitted in the analysis of changes. After analyzing the source data, the geodatabase [2] was prepared for individual years covered by the analysis, the structure of which is shown in Table 2.

Table 2. Geodatabase prepared within the framework of research [2].

Object	**Year**									
	1890	**1919**	**1936**	**1963**	**1981-1988**	**1999**	**2006**	**2010**	**2013**	**2016**
road infrastructure	V	V	V	V	V	V	V	V	V	V
railway infrastructure	V	V	V	V	V	V	V	V	V	V
hydrographic network	V	V	V	V	V	V	V	V	V	V
surface waters	X	V	V	V	V	V	V	V	V	V
built-up areas	V	V	V	V	V	V	V	V	V	V
arable lands	V	V	V	V	V	V	V	V	V	V
forests	V	V	V	V	V	V	V	V	V	V
mining pit	V	V	V	V	V	V	V	V	V	V
external spoil tip	X	X	X	V	V	V	V	V	V	V
power plant	X	X	X	X	V	V	V	V	V	V

Where: V - means the occurrence of an object in a geodatabase, X - means no object in geodatabase for a given year.

2.4 Preparation of cartographic compilations

During preparation of cartographic compilations, the definition of presentation principles of individual objects presented in the compilations is of great importance. It's essential to use appropriate symbols, fonts and colours in order to present the studied phenomenon in the best possible way. The study [2] determines the style of presentation of objects for cartographic compilations prepared in order to analyze the environmental-social changes caused by the activity of Turow mine, depicting the state of development of the Bogatynia municipality in the years: 1890, 1919, 1936, 1963, 1981-1988, 1999, 2006, 2010, 2013 and 2016. Examples of cartographic preparations for 1890 and 2016 are shown in Figure 3.

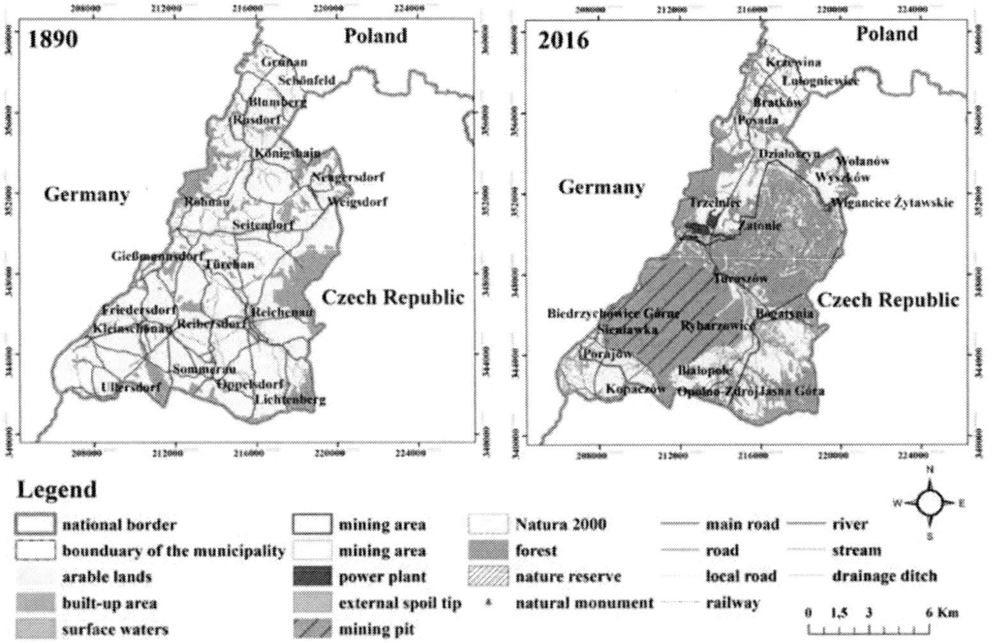

Fig. 3. State of development of the Bogatynia municipality in 1890 and 2016. Based on [2].

2.5 Analysis of the impact of mining activity

The area of analysis adopted for each examined year consists of current administrative borders of the Bogatynia municipality (2016). The state of development for individual years was compared with the state of municipality development from 2016. The following forms of land development were selected for the analysis: mining pit, external spoil tip, power plants, standing surface waters, forests, arable lands and built-up areas. Using the tools available in GIS systems, the areas of land development forms were calculated for analyzed years, which are presented in figure 4.

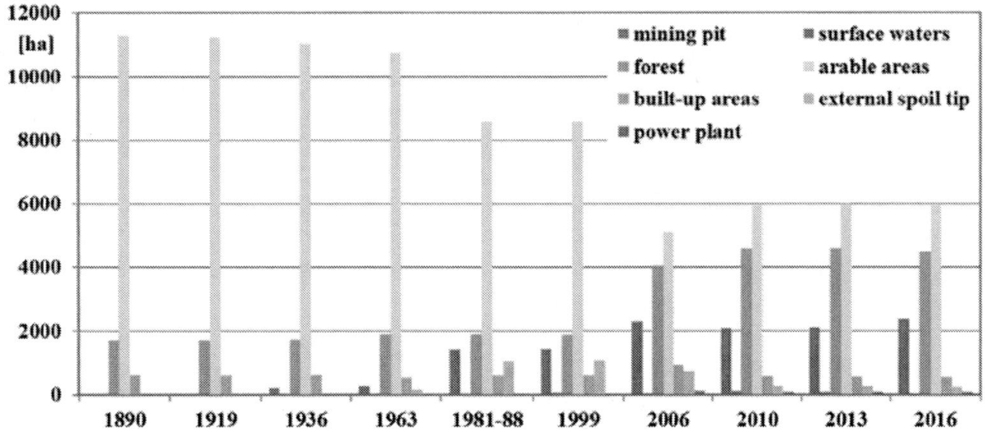

Fig. 4. Changes of land development in the Bogatynia municipality in analysed years.

By analyzing the data obtained from calculations, it can be noticed that over the years, the mining pit has been continuously expanded. The greatest dynamics of changes can be observed between 1963 and 1980s. This may result from the fact that during this time interval, the lignite-fired "Turow" power plant was created. The first stage of its construction was completed in 1965 and in the following years, it was expanded with subsequent blocks in order to produce greater amount of energy. In 1971, the second stage of power plant expansion was completed. The subsequent years are associated with modernization of the facility, ensuring more favourable and more efficient generation of electricity. The third phase of development was implemented in 2000. The high demand for lignite by the Turow power plant resulted in increased amounts of extracted raw material and, consequently, expansion of the mining pit. Such method of information analysis shows the difference between the state of mining pit in 1890 and the area occupied by it in 2016. The activity of large-area opencast mine caused great changes in the landscape of Bogatynia municipality. The expanding opencast mine took up further parts of arable lands, reducing their number. The greatest dynamics of changes in arable lands can be observed between 1963 and 1980s, as well as 1999 and 2010. Also, the greatest differences in the surface of mining pit occurred within those years. On the basis of these changes, it can be stated that the expansion of mining pit was a direct factor impacting the reduction of arable lands. However, the conducted mining activity did not significantly contribute to the reduction in forest areas. Changes between the individual periods are small and there can be observed a large dynamic of changes only between 1999 and the state of forest in XXI century. Increase in the number of forest areas results from the works associated with forest reclamation carried out by the mine. Young forest was created on the area of external spoil tip within few dozen years, which covers the eastern part of the Bogatynia municipality. The works associated with the restoration of these lands to their original state have been ongoing since the 1960s. Since 2010, the number of forest areas has been decreasing. This phenomenon is related to the expansion of mining pit in the direction of the village of Białopole, where the mining activities cause degradation of the forest occurring in these areas. By analyzing the values of surface occupied by the built-up areas, there can be seen a downward trend. Due to the fact that the expanding mining pit is occupying further parts of the land, many households were degraded. A slight increase in the number of these areas can be seen only in 1936. This may be associated with immigration of people, who have been employed in the mine, which was developing during that period. In 1963, there can be seen a rapid reduction in built-up area, caused by the interpretation of built-up areas by authors of the source data. Based on comparison of maps from different periods, it was found that other elements were recognized during these years as built-up area. As a result of expansion of the mine, many local residents were displaced and such towns like Rybarzowice, Biedrzychowice-Gorne ceased to exist. Many residential areas were degraded and the best example is the current state of Zatonie village and Wigancice Zytawskie village, which reduced their area. In the early years of operation, the overburden was tipped by the mining company. Since 1963, there has been created an external spoil tip in the north-west direction from the mining pit. The spoil tip occupied the largest area in 1980s and 1999. In the later periods, this area has been already reclaimed, therefore in Figure 4, it is visible that there was a decrease in the area occupied by the overburden. The area, on which the spoil tip was located in the previous years, still occupies a similar surface (2016), but instead of the tipped material, there's a young forest now.

The occurrence of large-area mining activity within the municipality affects the society living in the municipality, in which this activity is located. Each mine is obliged to pay fees related to the performance of exploitation works. The budget of municipality receives 60% of the value of paid exploitation fee. The study [2] includes calculation of the amount of exploitation fees received by the budget of the Bogatynia municipality. The greatest amount of the proceeds from exploitation fees was paid to the municipality budget in 1977 and it amounted to approx. 24,92 million. This amount significantly contributed to the improvement of economic conditions of the municipality, which affected the living comfort of population residing in this area. These funds were spent on own needs of the municipality. The mine is also one of the largest employers of local residents.

In connection with activity of the mine, both negative and positive aspects associated with the mining activity in the Bogatynia municipality were noted. By comparing the number of these aspects, there can be seen a balance between these activities. Nevertheless, not all factors affect the life quality of society and the state of natural environment in the same way. In order to carry out a more detailed analysis, it was proposed to assign weights to the effects of mining activity of Turow mine, according to the key proposed in the study [2]. The results of this analysis are presented in Table 3.

Table 3. Analysis of impact of the KWB Turow mining activity [2].

Positive aspects	Weight	Negative aspects	Weight
reclamation of post-mining areas	37	emission of pollutions	42
provision of jobs	25	emission of noise	40
possibility to acquire energy	23	degradation of forest and agricultural areas	39
proceeds to the municipality budget	18	change of water ratios	35
lower energy costs for municipality residents	19	landscape changes	31
opportunity for development of smaller companies	17	degradation of built-up areas	29
subsidizing public investments	15	displacement, emigration of population	20
Sum of weights:	**154**		**236**

3 Summary

The presence of large-area opencast mining activity in the Bogatynia municipality in the form of Turów lignite mine caused many changes in the existing landscape, contributed to changes in land development and affected the everyday life of residents of this area. By studying the phenomenon of changes that took place over the years, the development of mining pit, transformations of land and impact of the mining plant on the nearest surrounding area can be observed. The source data used for the analysis allowed to acquire knowledge about history of the mine and identification of changes occurring in the Bogatynia municipality. The maps prepared in the framework of research present the form and state of identified areas in the analysed years, allowing for the interpretation of changes that occurred in the vicinity of conducted mining activity.

On the basis of conducted analysis, the predominance of negative effect of the mining activity was found. Such outcome results from the fact that most of the positive activities of the mine are economic factors and in this comparison, the economic-market conditions were assigned the weight of the lowest value range. The highest value of weights was assigned to the factors related to expansion of the mine by occupying forest areas, agricultural areas and build-up areas, along with the emission of pollution and noise. This contributed to the predominance of negative effects of large-area opencast mining in the Bogatynia municipality.

The cartographic compilations done in the framework of the work allows to know the history of the Bogatynia municipality and to identify changes that have taken place in the forms of land development over the years. Geodatabases from the analysed years with vector data created on the basis of archival analogue maps allow for dissemination of the cartographic records describing the history of analysed area and enable the analysis of environmental-social changes.

This article was financed with the research project no. 0401/0174/16 *Data mining in modelling environmental changes* and Grant no. 0401/0128/17.

References

1. PN-EN ISO 19157:2014-04, *Geographic information – Data quality* (2014)
2. T. Cieślik, *Identification of environmental and social changes of the Turów Quarry on a historical background*, engineering work carried out under the supervision J. Górniak-Zmroz at the Faculty of Geoingeneering, Mining and Geology Wroclaw University of Science and Technology, unpublished work in Polish (2017)
3. J. Badera, Mineral Resources Management **21 (1)** (2010)
4. J. Badera, P. Kocoń, Energy Policy **66** (2014)
5. H. Izydorczyk, *Turów - economically and socially responsible*, Lignite **2** (2012)
6. Z. Kasztelewicz, M. Zajaczkowski, Ecological Policy **13/2** (2010)
7. A. Maciejewska, *Reclamation and protection of the environment in opencast mining*, Publishing House of Warsaw University of Technology, Warsaw (2000)
8. E. Pietrzyk-Sokulska, R. Ubermann, J. Kulczycka, Mineral Resouces Management **31** (2015)
9. Z. Dobrzyński, *The lignite is flowing*, Turów Lignite Mine, Bogatynia-Zgorzelec (2002)
10. H. Izydorczyk, Lignite **2 (91)** (2015)
11. Z. Kasztelewicz, *Polish brown coal mining*, The Association Agreement of Brown Coal Manufacturers, Opencast Mining, Bełchatów-Wrocław (2004)
12. Z. Kasztelewicz, *Reclamation of post-mining areas in Polish open-pit mines*, Foundation for Mining Science and Traditions, Kraków (2010)
13. M. Łukjanowicz, Lignite **2 (59)** (2007)

Map of the terminal moraine of the Werenskioldbreen glacier (South-West Spitsbergen) from 2015

Laura Downar-Zapolska[1], and *Tadeusz* Głowacki[1,*]

[1]Wroclaw University of Science and Technology, Faculty of Geoengineering, Mining and Geology, 27 Wyb. Wyspianskiego St., 50-370 Wroclaw, Poland

Abstract. The first direct field measurement of the entire terminal moraine of the Werenskioldbreen glacier on Spitsbergen took place at the end of July and the beginning of August 2015. The results of the measurements were the basis for the development of the large-scale terminal moraine map. The article presents the stages of map creation and editing based on measurement points established by the GNSS method. The map editing required knowledge of the morphologically complex terrain. Key documentation was field notes and documentary photographs, enabling the cartographic interpretation of the varied topographic features of the terrain's surface. Based on the documentation, the water bodies on the moraine were located, the outline of the structure and the boundary points were executed and the latter were excluded from the triangulation process. The glacial river was also excluded from triangulation, which allowed a DTM to be developed. An important step in point cloud mapping was to generate a topologically correct digital elevation model of satisfactory accuracy. On the basis of the DTM, contour lines were generated showing the topographic features of the terrain's surface. The printable resultant map's scale is 1: 5000; it is in the UTM coordinate system, in the 33X zone. Complementing the content of the map, a grid of geographic coordinates, a kilometer grid, and map key descriptions were added.

1 Introduction

Global climate changes are the most visible in polar regions. Surface changes are noticeable with the naked eye; maps as cartographic studies help one to document the changes that have occurred on the surface. Most studies in the polar regions focus on assessing climate changes and the impact of the changes on the environment. The interest in the polar regions, as well as greater accessibility to them, make them be a growing topic of debate about Earth and the environment. Polish scholars engaged in polar studies primarily focus on Spitsbergen [1], where international projects are implemented, including workshops and conferences. Research within the scope of various scientific disciplines is conducted in this

* Corresponding author: tadeusz.glowacki@pwr.edu.pl

area. There are 6 Polish polar bases in the Svalbard archipelago, which are supervised by scientific centers such as Maria Curie-Skłodowska University in Lublin, Institute of Geophysics of the Polish Academy of Sciences, Nicolaus Copernicus University in Toruń, Adam Mickiewicz University in Poznań, Jagiellonian University and Wroclaw University [2]. With the growing interest in polar subjects, and the possibilities of using geodetic methods in the polar regions, cartographic studies began to appear. Analogue topographic maps were once the only source of information on topographic features of the terrain. By analyzing the collected material, researchers have been able to see changes in both relief, location of terminus of the selected glacier, or height changes in the glacier area [3, 4]. New technologies and methods of data collection enable the development of polar research and the resulting possibilities. Currently, the choice of technology itself depends on economic and logistic aspects. Today, satellite technology is the leading one. The GNSS measurement method is unrivalled in small areas where measurement has to be done with great accuracy, while in case of large areas, they are mapped by means of photogrammetric and remote sensing methods. Software and hardware have contributed to the development and increase in a number of cartographic documents concerning polar areas. The development of measurement techniques and the possibility of obtaining accurate data have led to an increase in the number of polar regions that are considered both interesting and dangerous. In addition, geodetic measurements are part of the non-invasive polar research without disturbing or interfering with the environment.

2 Scope of the study

The entire area of Svalbard is protected, so scientists must comply with the rules imposed by the Governor of the archipelago. Measurements were made on Spitsbergen - the largest island of the Svalbard archipelago between 74° and 81°N and 10° and 35°E. In global terms it is approx. 800 km north of Norway and approx. 1100 km south of the North Pole [2]. The measurement facility is located in the southern part of Spitsbergen, in Wedel Jarlsberg Land, in the Hornsund Bay. The measurement area is shown in Fig.1a. On fig.1b is shown the Werenskioldbreen glacier [5, 6] on TopoSvalbard web page.

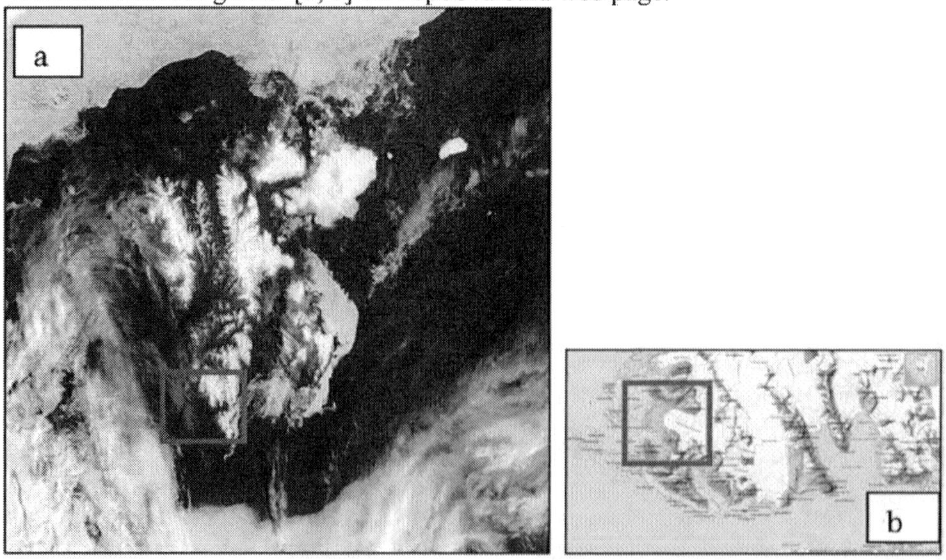

Fig. 1. a) Svalbard - NASA photo from August 8th, 2015 with measurement area marked [7]; b) Wedel Jarlsberg Land with the area of research [8].

 https://doi.org/10.1051/e3sconf/20182900029

3 Svalbard maps

3.1 Historical background

The first work on Svalbard's mapping began more than 100 years ago. Then participants in the Swedish-Russian expedition of 1899-1902 conducted measurements to determine the shape and size of Earth. They measured the length of the meridian arc on the east coast of Spitsbergen. During this period, the first cartographic documents were drawn that met the criteria of carto metrics [9]. Historical maps created at the turn of the 17th century depict the development of cartography of the northern regions. The first document dates back to 1596. Willem Barents, a Dutch cartographer, during his third expedition, developed a map showing the first outline of Spitsbergen coast. Historical studies are characterized by the fact that they were not based on any measurements but on the observations of people engaged in the expansion of the polar regions. Graphic forms such as animals or ships were designed to fill in empty spaces that were not discovered.

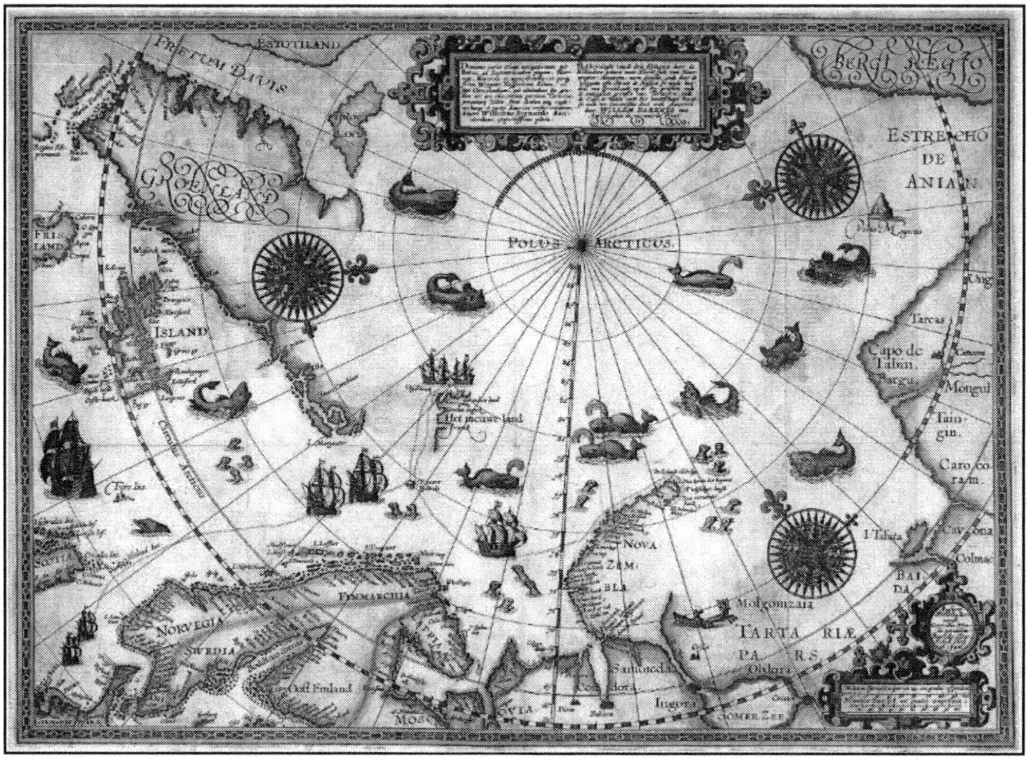

Fig. 2. First cartographic study with the Spitsbergen shores - Willem Barents, 1599 [10].

Fishing, whaling and hunting contributed to the development of cartography of the Spitsbergen area. However, until the development of Svalbard's first cartometric map, other maps were just the results of subjective observations, so many discrepancies in the Spitsbergen coastline between the studies can be observed. The 20th century was a time of scientific research and numerous expeditions. Thanks to technology and measurement possibilities, the first world's most famous scientists arrived in the Arctic. The main subjects were geology and glaciology. Numerous thematic maps - geological and geological - of Spitsbergen were first developed. The researchers from England, Sweden, Norway, Germany and Austria led the way in these expeditions [10].

After the First World War, the development of measuring and cartographic technologies took place. The demand in the extraction industries for the cartographic studies of the selected area increased. In 1936 Svalbard mapping started using photogrammetric methods. The area of measurement covered 40,000 km, 2. 3800 photos were taken then. Photos were taken on a scale from 1:30 000 to 1:50 000. After the Second World War, they were developed with a working scale of 1:50 000, and published on a scale of 1: 100,000 in the Gauss-Krüger transverse cylindrical projection. After that mapping, the maps were used as base maps for a long time [9].

3.2 Contemporary cartographic works

Modern technology, which develops at a rapid pace, enables both precise measurements and quick results development and visualization. Thanks to topographic maps and aerial photographs, numerous scientific research and studies on glaciers of various types have been made, and the glaciers cover almost 60% of Svalbard. With the GNSS geodetic surveying technology, the speed of glacier movement is currently being monitored, and then the position of their snouts is mapped onto the basic map.

There are maps of each region of Svalbard available. Topographic maps were scaled as follows: 1: 500,000, 1: 250,000, 1: 100,000. 1: 100,000 scale maps are a series of 62 sheets representing the entire archipelago divided into parts. Most of the sheets were made in the WGS-84 reference system. The exceptions are G7 and H7 sheets - they were made in ED-50 system (reference: Hayford ellipsoid). The assumed contour interval of these maps is 50m. The sheets were developed on the basis of photogrammetric photographs made in 1990. The exception is sheet 6 because it was also developed on the basis of the flight pass over this area made in 2009. The advantage of photogrammetric methods in Svalbard is undoubtedly the fact that a large area can be measured in a short space of time. In addition, the measurement is also conducted in hard-to-reach areas. In their works, Małgorzata Blaszczyk and Leszek Kolondra [4, 11] claim that photogrammetric methods are among the best methods of obtaining data in polar areas. An orthophotomap is often used in glaciated areas. Its greatest advantage is that the data for its development can be obtained quickly and the measurement covers a large area of land, even hard-to-reach one. The number of measurement points is very high, and thus the relief is well represented on the map. What is more, the method enables further work with data based on the generated digital terrain model. A DTM is a model that is a set of information about the topographic features of the land in the form of points with specific XYZ coordinates. In order to get an accurate orthophotomap it is important to make sure that a DTM is of the highest quality. Therefore, manual point height correction should be performed if necessary. The DTM on this orthophotomap was presented with a grid of squares of 20 m side. In order to maintain a clear representation of the field situation, additional points and break lines were added. Currently, software and hardware make it possible for cartographers to produce accurate maps, and most editorial operations allow for automatic generation of structures that fall into the map content. An additional advantage of an orthophotomap is that it is not interpreted by the editor [6].

The main administrator of the Svalbard maps and their publisher is the Norwegian Polar Institute (Norsk Polarinstitutt) in Tromso (Norway). They provide both printed and digital maps.

Norsk Polarinstitutt provides digital maps via the following websites:

a) TopoSvalbard - a topographic map portal developed by the Norwegian Polar Institute, where one can see the details of the selected area and view aerial photographs. The user has the possibility of reading coordinates (however, their accuracy, which is represented on the map, is not useful for detailed geodetic surveys), search for a town name, measure distance and area. This portal also provides a 3D view.

b) Svalbardkartet - Svalbard interactive map collection. It is possible to include thematic layers on the environment, glaciers and hunting grounds [12].

3.3 Polish cartographic documents concerning Spitsbergen

Polish scientists have contributed to the development of scientific research on Spitsbergen, including the development of cartographic studies. The substantive contribution to Polish and world's polar research is appreciated worldwide. The first Polish scientific expedition was organized in 1934 by Stanisław Siedlecki, geologist. The study area was the south-western part of Torell Land. The main premise of this expedition was to map the previously unmeasured area using the triangulation method and photogrammetric record. The result of the study was a 1:50 000 map, issued by the Polish Military Geographical Institute in 1936. The map was bicolour, and the contour interval was 50m. Its authors were Antoni Zawadzki and Sylweriusz Zagrajski.

Cartographic documents are produced for research purposes. The increase in the number of maps covering the Hornsund area was accompanied by the development of research facilities, i.e. the construction of polar stations. The main station in Wedel Jarlsberg Land is Stanisław Siedlecki Polish Polar Station managed by the Institute of Geophysics of the Polish Academy of Science in Warsaw. Interest in Werenskiold glacier and research conducted in its vicinity contributed to the establishment of Stanisław Baranowski Polar Station of Wroclaw University in 1971. The possibility of using the station by Polish scientists made the majority of Hornsund area be mapped by Poles. Polish research teams have published a number of cartographic documents covering the areas of Svalbard. Typical map scales range from 1: 2000 to 1: 100000.

The following maps are produced in the polar areas [13]:

a) topographic,
b) geological,
c) glaciological,
d) geomorphological,
e) orthophotomaps,
f) Other (bathymetric, hipsometric and DTM).

In the group of topographic maps, the documents in scales 1:20 000 and 1:25 000 are prevailing. Large-scale maps were formerly developed using analogue measurement methods. They covered the relief of the glacier area (e.g. Werenskiold Glacier) and the marginal zones. Topographic middle-scale maps are basic maps for other documents and are used for comparisons of changes in relief and glacial surface height in time.

The interest in Werenskiold glacier came in 1957 when the 3rd International Geophysical Year started. The expeditions were attended by specialists from the Topographic Military Service of Polish Army. One of the participants of these expeditions was Lieutenant Colonel Cezary Lipert, who carried out measurements to produce a Werenskiold glacier map [14]. The measurement method was terrestrial "terrophoto" photograps that were made by the Zeiss-Jena theodolite. C. Liper and M Bałdyga were responsible for autogrammetric processing. The result of the study was a map developed by T. Naumienko in two sheets: Werenskioldbreen Glacier - terminus zone, Werenskioldbreen Glacier - central zone. The map was produced on a scale of 1: 5,000. The applied contour interval was 1.25m. The Gauss-Krüger mapping for $\lambda = 15°$ was adopted. Due to the method of measurement of terrestrial photographs, the map was made in local system, with the starting point of $x = 8500$ and $y = 5400$. In 1960 Jan Szupryczyński used a topographic map developed by T. Naumienko as a basemap for geomorphological mapping of marginal zone forms while conducting geomorphological surveys in the forefield of the Werenskiold Glacier [15, 16]. The scientist evaluated the map as the most detailed representation of the Hornsund area. Therefore, he decided to conduct research in this area due to the fact that he had the accurate cartographic material.

The next development was an orthophotomap. As a result of cooperation between Leszek Kolondra, and scientists from Norsk Polarinstitutt in Tromso, 17 images on a scale of 1:50 000 were used to produce a 1:25 000 orthophotomap (Fig. 3) [17]. It was developed in a standard coordinate system used in Svalbard - UTM - and the adopted reference ellipsoid was WGS-84. The development methods of the orthophotomap of Werenskiold glacier and its surroundings had their limitations. It was necessary to have accurate coordinates of the photo control points in order to correctly carry out this development. These points had to be high-precision, so they had to be measured by direct methods. GNSS receivers are currently used for this purpose. The identification of photo control points is difficult [11]. The glacier surface often does not have any characteristic points. In addition, varied relief, characterized by considerable height differences of the land, creates shadows that are difficult to correct.

Fig. 3. Orthophotomap "Werenskioldbreen's neighborhood" [17].

4 Method of data collection for -producing a map of the terminal moraine of the Werenskioldbreen glacier

Many of the published maps are based on photogrammetric methods, based on terrestrial imagery processing or photogrammetric record. On the other hand, scientists from Wroclaw University of Technology within the framework of the III Polar Expedition in 2015 have measured the entire terminal moraine of the Werenskioldbreen glacier in south-west Spitsbergen. The measurement was conducted from 5 July to 16 August 2015. Measurements were made within the scope of the project "Geodetic measurements of the Werenskioldbreen glacier (South Spitsbergen)", which was registered in the database of The Research in Svalbard (RIS) with number 10257 [18]. Field work enabled GNSS receivers to obtain data using the FastStatic method. Three GNSS receivers were used to measure points. The aim of the study is to present the terminal moraine of the Werenskiold Glacier in the form of a topographic map of 1: 5000 scale. The map was made on the basis of the coordinates of postprocessing measurement points in reference to the nearest EUREF reference stations: Ny-Alesund (NYAL) and Hornsund (ASTR). A total of 20 721 station poles were measured, which were characteristic features of the very varied relief. Measurement sessions for 1 point lasted 31". The coordinates of the points were calculated in the UTM system (33X zone) and in the WGS-84 mapping. The average accuracy of the location of the points in the horizontal plane falls within 5 and 25 cm and in the vertical plane it is within 40 and 60 cm. Because of the time limitation, no field sketches were made, only field notes were used to identify point groups and point numbering sequences.

5 Results

5.1 Map editing

Editing the content of the map started with water bodies. The lakes were easy to identify because before the measurement it was assumed that station poles at the edges of the water body were measured one by one, at the water surface. So, the presence of the lake was indicated by consecutive point numbers and an empty spot in the cloud of points. In the case of ambiguity in the interpretation of selected measurement areas, satellite imagery available on the Norsk Polarinstitutt website [19] was used (Fig.4). The pictures clearly show large lakes on the moraine, and the comparison of the shape and location with respect to other structures allowed for unambiguous identification of lakes as enclosed areas. (comparison of Fig. 4a and Fig. 4b). Figure 5 shows a fragment of a vector map with the lake inserted into a point cloud.

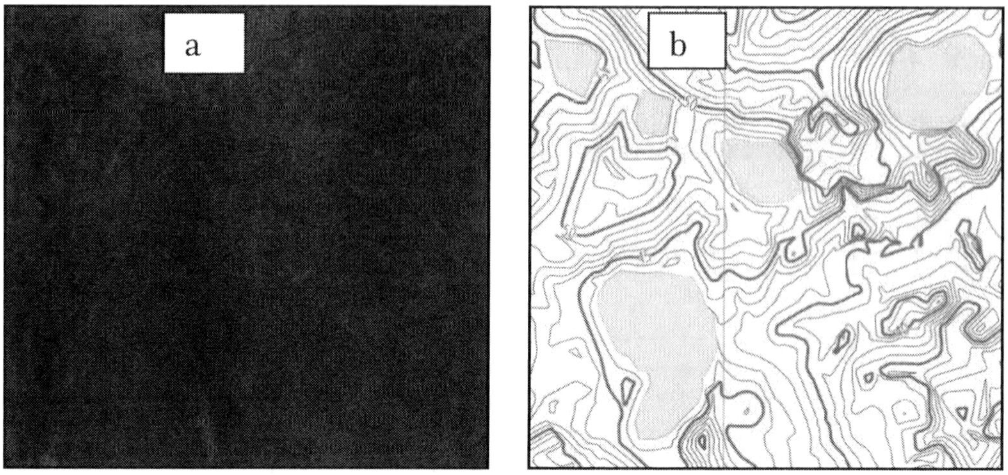

Fig. 4. a) Fragment of the satellite image - lakes on the moraine [19]; b) Fragment of the developed map - lakes located on the moraine.

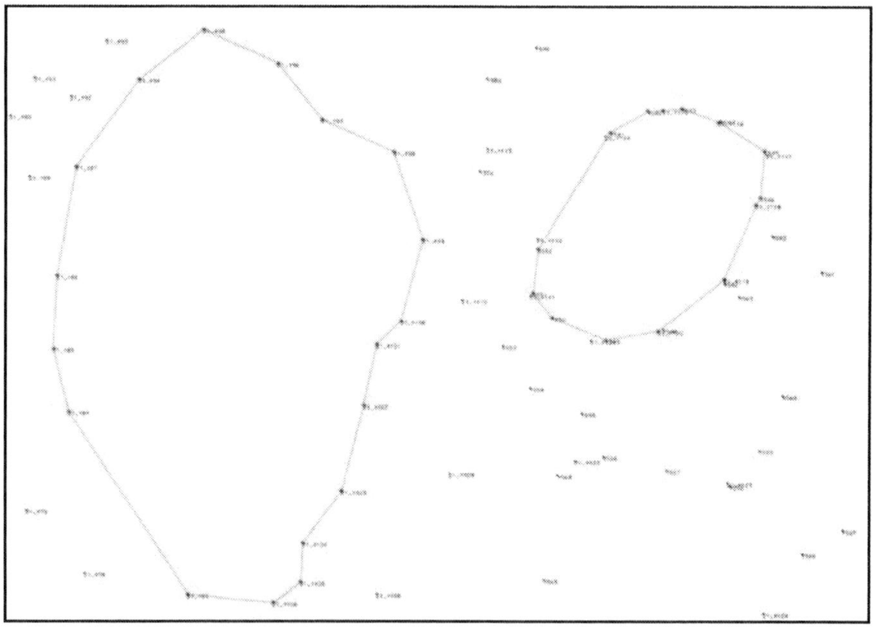

Fig. 5. Identification of water bodies.

The next stage was the development of the glacial river bank line. The basis for identification was, like with the water bodies, the numbering of measurement points, and the information recorded in the field notebook. Points of the riverbank were joined by a line that formed the outline. In addition, the points of the riverbank line limited the measurement area from the east. The limitation from the western, northern and southern sides was the foot of the glacier's terminal moraine escarpment. The measurement ended at the foot of the moraine escarpment. In order to indicate the flattening, apart from the escarpment, there were still 1-2 additional points located along the entire south, west and north sides.

The photographic documentation was a valuable source of information when identifying water bodies. Approximately 500 images of terminal moraine along with the surrounding area were taken. Thanks to them, it was possible to determine the shape of the lakes, the bankline of the glacial river, the location of the hills, peaks, saddles, valleys, ridges, skeletal lines, break lines like cliffs and slopes, characteristic points and other terrain forms, as well as their position relative to each other. An example of a photo is shown in Figure 6.

Fig. 6. Photo of the terminal moraine's fragment with the lakes and glacier river in the central part of the photo (photo. T.Głowacki, 8/8/2015).

5.2 Digital Terrain Model

Digital Terrain Model is a set of data in a form of points of XYZ coordinates, which, based on interpolation algorithms, recreates the shape of the terrain surface. The boundaries of DTM were determined by the measurement area described in the preceding paragraph.

The next stage of the development was the triangulation (applying a grid of triangles) between all measured points for the whole area. The lakes and the glacial river were excluded from the calculations, and the defined characteristic points were included in the structure. As a result of the automatic operation of connecting the nearest neighbours, a digital terrain model in a form of a triangulated irregular network (TIN) was created according to the rule of nearest-neighbour connection (Fig. 7). The model was saved in a universal .dtm format. The vertices of the generated triangles were measurement points directly measured. This model was the working version, because each model should be subjected to control. Initial verification of the resulting model allowed for the elimination of gross errors and errors resulting from measurement imperfections. The points that stood out from the model were deleted. Another surface triangulation process no longer contained points that were significantly different from the surface model. In total, about 110 points were removed which definitely did not match the generated model.

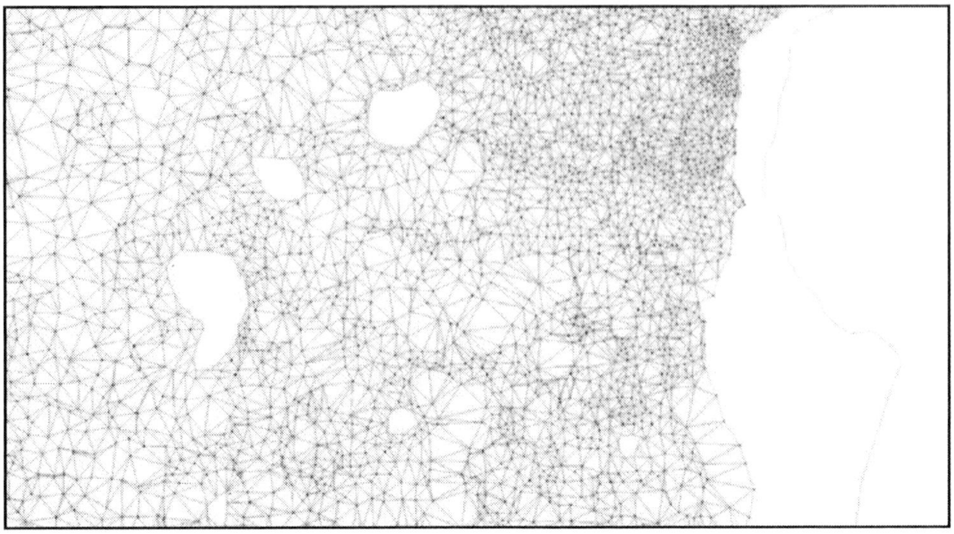

Fig. 7. Fragment of the generated Digital Terrain Model in TIN form - top view.

5.3 Contour lines

Contour lines are lines joining points of the same height on the map at a reference level. They depict the topographic surface and enable one to control the previously developed DTM.

In this study, contour lines of 1 m basic interval were generated in order to eliminate the wrong heights that were applied to the measuring points model. Contour lines were automatically generated using the linear method embedded into the functionality of the application used. The main contour lines were every 5 m, and the auxiliary contour lines were every 1 m. Along with the interpolation of the contour lines, the descriptions constituting the elevation features of the contour lines and the labels in the characteristic points were generated. The result of automatic generation is shown in Fig. 8.

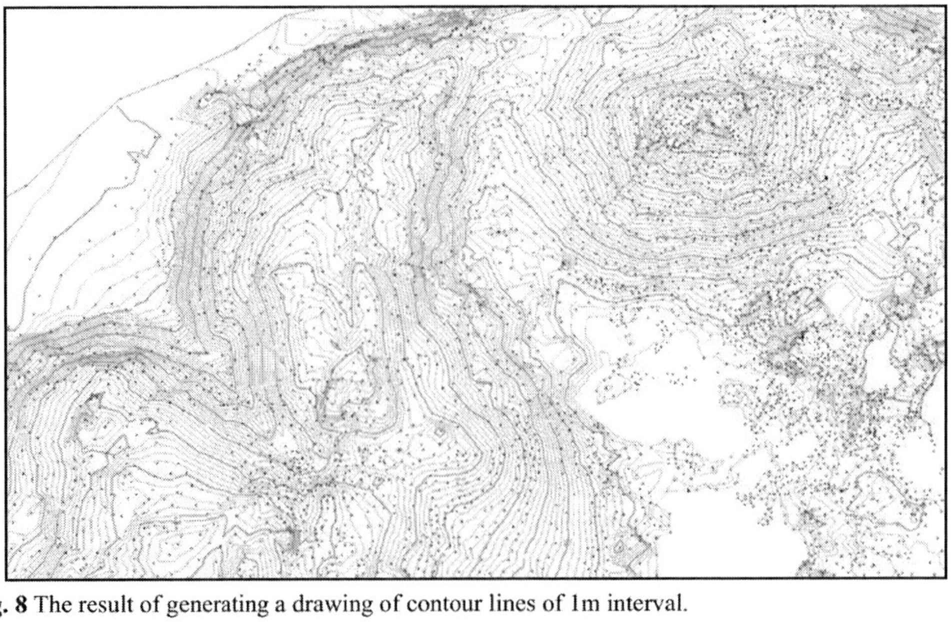

Fig. 8 The result of generating a drawing of contour lines of 1m interval.

After drawing out the contour lines, it turned out that among the set of points there were those that deviated from the average height values occurring in a given measuring area. An example of the wrong contour lines is shown in Figures 9a and 9b. The error could be caused by measurement inaccuracies, such as a temporary loss of GPS signal in an area with a covered horizon. Comparison of moraine form drawn out on the basis of contour lines with satellite imagery and documentation photographs of the moraine made it possible to definitely exclude wrong measurement points. Analysis of the accuracy of measurement of vectors to the reference points confirmed the validity of the decisions taken. Throughout the project, the analysis of contour lines and satellite imagery allowed for the location of about 260 outlying points that were removed from the DTM.

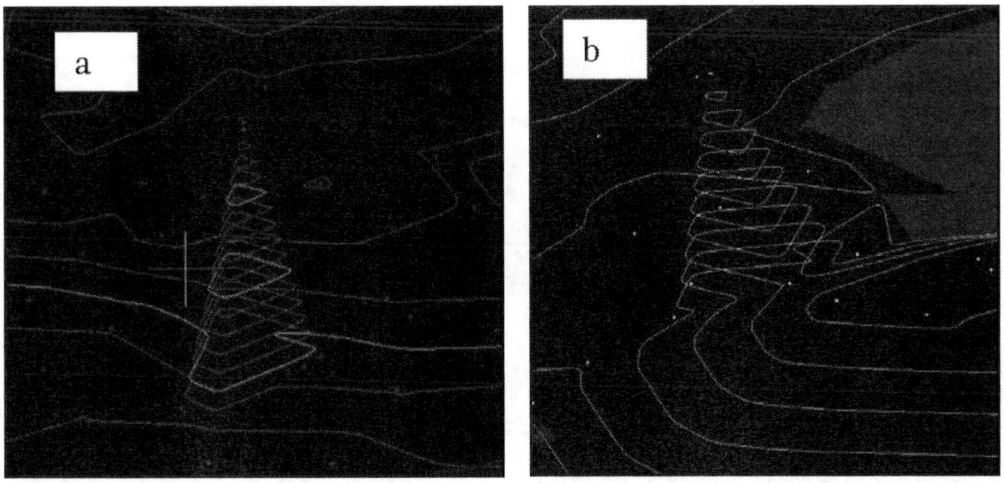

Fig. 9. a) An example of wrongly generated contour lines; b) An example of erroneously generated contour lines.

The mutual arrangement of the contour lines allowed one to re-check the layout of the land. New, smaller lakes were observed in land depressions. The boundary points had another numbering on the map. Another comparison with satellite imagery and field photographs confirmed the correctness of the procedure. Unfortunately, some of the water bodies identified on the moraine could not be found in the photographs, which indicates the surface changes in the layout of the glacier's terminal moraine. After total correction, approximately 355 points (1.7%) were removed from the initial set of 21,000 points. A new digital terrain model was then generated from the remaining set of points. On the basis of the moraine model, the contour lines of 1 m basic interval and thickened contour lines every 5 m were automatically drawn out.

5.4 Editing map content

The process of map development and editing requires the editor to be familiar with the specifics of the area as well as the individual approach to the topic of the development. Although there is currently access to technology supporting cartographic developments and most operations can be done automatically, the limitations of the automation implemented in the programs must be taken into account, and thus the need to manually correct the resulting map.

After developing the digital terrain model, a map was to be made more readable. The first stage of editing the content of the map was to analyze the course of the contour lines. As a result of interpolation, contour lines were generated with "sharp" contours, contour lines in contact or looping. Such a situation may be a consequence of automatic interpolation and smoothing algorithms, which do not work in a complex and varied terrain, generating a lot of errors. The course of the contour line on the map should be smooth. In the first close-up, not all contour shapes were smoothed out. It was necessary to carry out a manual correction of this type of contour lines.

The elevation values of the basic contour lines on the map were added automatically. Only the font, size, and repetition distance of the contour feature in the same line were specified. In areas with a large fall of the land, not only the value contour but also the nearest auxiliary contours are interrupted (Fig.10). Therefore, it was necessary to close the auxiliary contour lines, and in case of difficulties related to the size of the font, it was also necessary to move the inscription to another place on the map.

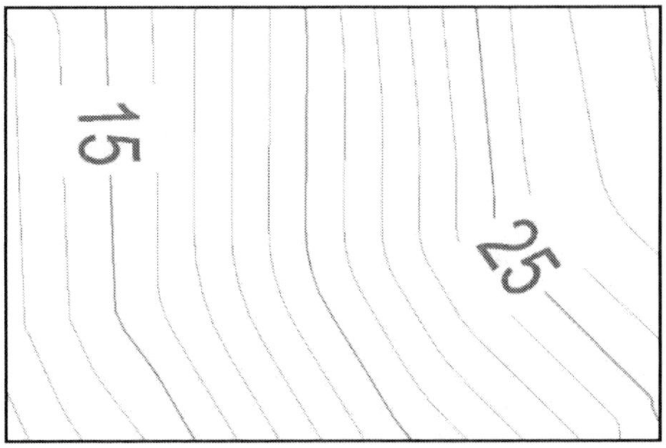

Fig. 10. An example of interrupting neighbouring contour lines in the automatic addition of a height feature.

In addition, defining the distance after which the contours are broken again causes the inscriptions to be inserted into the wrong places, limiting the readability of the map significantly. In such situations, during the editing and making the map more readable the height features were shifted to the most appropriate locations, simultaneously describing the map content and character of the terrain.

After editing structures on the map, two grids were manually drawn in the form of a box: the first kilometer one referring to the UTM coordinate system in zone 33X and the geographic one with latitude and longitude. Lines separating minutes were added to the frame. Longitudes and latitudes on the box were converted on the basis of the corner points of the frame.

5.5 Content outside the frame and map printout

Content outside the frame is an integral part of every map. It contains important information such as title, scale, development information (measurement date, measurement method), coordinate system used and mapping, legend and additional information.

Cartographic development was entitled "Terminal Moraine - South Spitsbergen". 1: 5,000 scale map was developed for printout. Information was included below the content of the map along with a linear scale. The project adopted the UTM coordinate system, zone 33X, binding for the whole Svalbard. The content outside the frame contains important information about the method of measurement and the date of fieldwork. In addition, the names of the people performing the measurement and the map editor are listed. Due to the layout of the prepared map, it was possible to place the legend in a vacant part of the map within the frame.

Because of the large area and elongated shape of the moraine, the map was printed on an A1 format sheet. The vector map was converted into a PDF format using the application, rasterizing the map to a pixel format. The resulting drawing guarantees that each printout is without loss in the drawing part and content of the map. Unfortunately, the raster is not scalable, which means that the printout is only possible in one, previously assumed, scale. The map, reduced to A4 format, was added to the article on the final page.

6 Summary and conclusions

The article presents the stages of map creation and editing based on approximately 21 thousand measurement points measured by the GNSS method. The objective assumed at the beginning of the work was achieved.

The map editing required a special approach to the development and good knowledge of morphologically complex terrain. The key documentation were field notes and documentary photographs taken by the expedition's participants. The notes, together with terrestrial and satellite imagery available on the Norwegian website [19] provided a faultless multiple interpretation of the location of the structures on the very complex surface of the moraine.

An important step in the editing of the map was the development of the digital terrain model. The DTM was developed in two stages to eliminate outlying points. The first of the cloud points allowed to eliminate a part of points. The control of the course of contour lines, allowed the elimination of consecutive points with the wrong height coordinate. The rest of the points was the basis for the development of the final DTM. The contour lines were developed from the created model, whose course required manual correction.

For enhanced map readability, the automatically generated height features of the contour lines were enlarged and moved to the correct location. During map editing, attention should be paid to the generated objects as they may adversely affect map readability.

After the correct presentation of the moraine on the plane, the drawing was given the form of a map. Grids of geographic and kilometer coordinates were added. The last stage of the study was an outside the frame descriptions constituting the source of information about the map.

The map developed is a source of knowledge about the layout of the terminal moraine. Comparing the map developed in 2015 with previous cartographic studies, one can see how many processes and changes took place in the selected area. Map "Terminal moraine - Werenskiold glacier. South-West Spitsbergen" is another Polish cartographic compilation, which fits into the canon of the maps of Hornsund. The map developed is the result of measurement work conducted within the III Polar Expedition of Wroclaw University of Technology in 2015 and financed from a statutory order S50051 at the Faculty of Geoengineering, Mining and Geology, Wroclaw University of Technology.

The study was carried out thanks to statutory order No. S50067 and Grant no. 0401/0128/17.

References

1. Z. Zwoliński, A. Kostrzewski, Polish geomorphological research on Spitsbergen, Z. Zwoliński, A. Kostrzewski, M. Pulina (eds.), *Ancient and modern geoecosystems of Spitsbergen* (Bogucki Wydawnictwo Naukowe, Poznań, 2013)

2. W.T.J. Stankowski, S. Bartoszewski, K. Birkenmajer, J. Borsiak, E. Bukowska-Jania, A. Karczewski, S. Skiba, Z. Zwoliński, Geographical environment of Spitsbergen. Introduction, Z. Zwoliński, A. Kostrzewski, M. Pulina (eds.), *Ancient and modern geoecosystems of Spitsbergen* (Bogucki Wydawnictwo Naukowe, Poznań, 2013)
3. E. Bukowska-Jania, Glaciological Workshops, Spitsbergen 2004, Glaciology, geomorphology and sedimentology of the Spitsbergen polar environment, 100-103 (2004)
4. L. Kolondra, *Werenskioldbreen and surrounding areas, Spitsbergen- orthophotomap 1:25 000.* University of Silesia, Norsk Polarinstitutt-Tromso, Sosnowiec (2002)
5. W. Dallmann, W. (ed.) *Geoscience Atlas of Svalbard,* Tromso: Norsk Polarinstitutt (2014)
6. J. Hagen, O. Liestol, E. Roland, T. Jorgensen, *Glacier Atlas of Svalbard and Jan Mayen*, Oslo: Norsk Polarinstitutt (1993)
7. NASA, National Aeronautics and Space Agency website, http://eoimages.gsfc.nasa.gov/images/imagerecords/86000/86468/widesvalbard_amo_2015213_lrg.jpg, access: 23.12.2015
8. TopoSvalbard, http://toposvalbard.npolar.no/, access: 23.12.2015
9. L. Kolondra, Geoinformatica Polonica, **7**, 45-73 (2005)
10. Canadian Geographics: https://www.canadiangeographic.ca/article/10-fascinating-historic-maps-arctic access: 23.12.2015
11. M. Błaszczyk, Archiwum Fotogrametrii, Kartografii i Teledetekcji, **7**, 53-62 (2007)
12. Norwegian Polar Instytute http://www.npolar.no/en/services/maps/interactive/,access: 21.12.2015
13. L. Kolondra, Mapy Svalbardu [In:] Z. Zwoliński, A. Kostrzewski, M. Pulina (eds.), *Ancient and modern geoecosystems of Spitsbergen*, Bogucki Wydawnictwo Naukowe, Poznań, 427-456 (2013)
14. E. Sobczyński, www.geoforum.pl, access: 23.12.2015 (2014)
15. T. Naumienko, *Werenskiold glacier 1: 5000 - terminal zone.* Military Topographic Service, Warsaw (1961)
16. J. Szupryczyński, Geographical works of the Institute of Geography of the Polish AcaDTMy of Science, No. **39**, 56-92 (1963)
17. L. Kolondra, Polish Polar Studies, 28th International Polar Symposium, Poznań, 173-186 (2002)
18. Research In Svalbard Database, https://www.researchinsvalbard.no/project/7644, access: 22.12.2015
19. The Nothern Lights Route, http://www.ub.uit.no/northernlights/eng/index.htm, access: 23.12.2015

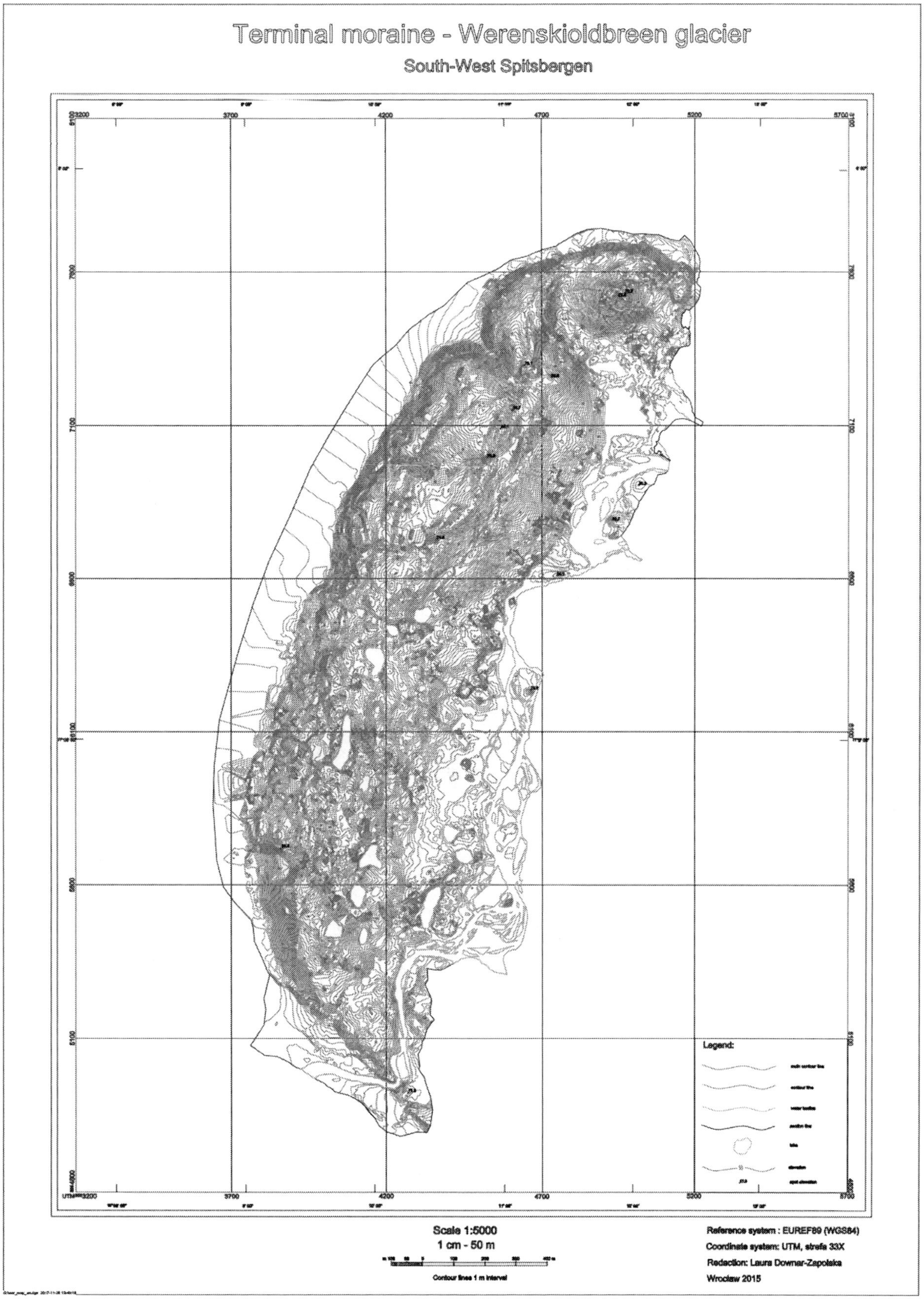
Terminal moraine - Werenskioldbreen glacier
South-West Spitsbergen
3200
3700
4200
4700
5200
5700
Legend:
Scale 1:5000
1 cm - 50 m
Contour lines 1 m interval
Reference system : EUREF89 (WGS84)
Coordinate system: UTM, strefa 33X
Redaction: Laura Downar-Zapolska
Wroclaw 2015

E3S Web of Conferences **29**, 00030 (2018) https://doi.org/10.1051/e3sconf/20182900030
XVIIth Conference of PhD Students and Young Scientists

Front of the Werenskiold Glacier (Svalbard) – changes in years 1957–2013

Wojciech Ciężkowski[1], *Tadeusz* Głowacki[1], *Katarzyna K.* Grudzińska[1], *Damian* Kasza[1,*], and *Paweł P.* Zagożdżon[1]

[1]Wrocław University of Science and Technology, Faculty of Geoengineering, Mining and Geology, 27 Wyb. Wyspiańskiego St., 50-370 Wrocław, Poland

Abstract. Werenskiold Glacier in the Isle of Spitsbergen is one of the polygons on which evaluation of the rate of glacier recession is performed. Location of the glacier front is precisely presented in the specialist literature since the mid. 30. of 20th century. Article presents results of studies about changes in the position of Werenskiold Glacier based on selected archival data and own research conducted in 2012-2013. The initial location of the glacier front was described by the topographical map in the scale 1:5000 elaborated in the year 1961 on the basis of data from the years 1957-1959 by the Polish Army Topographic Service. Moreover, as comparative data on location of glacier front in the later years there were applied results of photogrammetrical images from the year 1973 and a orthophotomap from 1990. These data together with the author's GPS measurements were transformed into the uniform coordinate system. Thus, prepared data made it possible to evaluate the rate of recession of Werenskiold Glacier front in the three epochs from the years 1957-2013. It was found that during the 56 years the glacier front was moved by ca. 1200 m, which gives the mean recession value of 25 m/year as well as the mean yearly loss of the glacier surface of the order of 5 ha.

1 Introduction

Werenskiold Glacier (Werenskioldbreen) is situated in the south-western part of Spitsbergen which is the largest island of the Svalbard Archipelago belonging to Norway. The Isle of Spitsbergen is situated between 80°48' and 76°28' of northern latitudes and between 10°28' and 28°50' of eastern longitudes. The research area is situated in the south-western part of the island in Wedel Jarlsberg Land, not far from the entrance of Hornsund bay (Fig. 1).

The climate conditions in Spitsbergen are extremely variable and the access to the research area is largely limited in winter. Direct measurements of the glacier front are possible during summer months only. The choice of measurement technology depends on area conditions. Discussion of variations of glacier front course in the periods of 1957 to 1973, in 1990 and also in the period of 2012 to 2013 allowed for precise and quantitative presentation of recession phenomenon of Werenskioldbreen in the above-mentioned

* Corresponding author: damian.kasza@pwr.edu.pl

periods. The linear and spatial values of recession were calculated. Comparison of images in photos taken from the end moraine in the years 1973 and 2013 reveals the scale of glacier recession (Fig. 2) [1, 2].

Similar studies were performed earlier but they did not comprise so vast temporal range. It is also important to note that the analyses concerned only the process of glacier front recession [3].

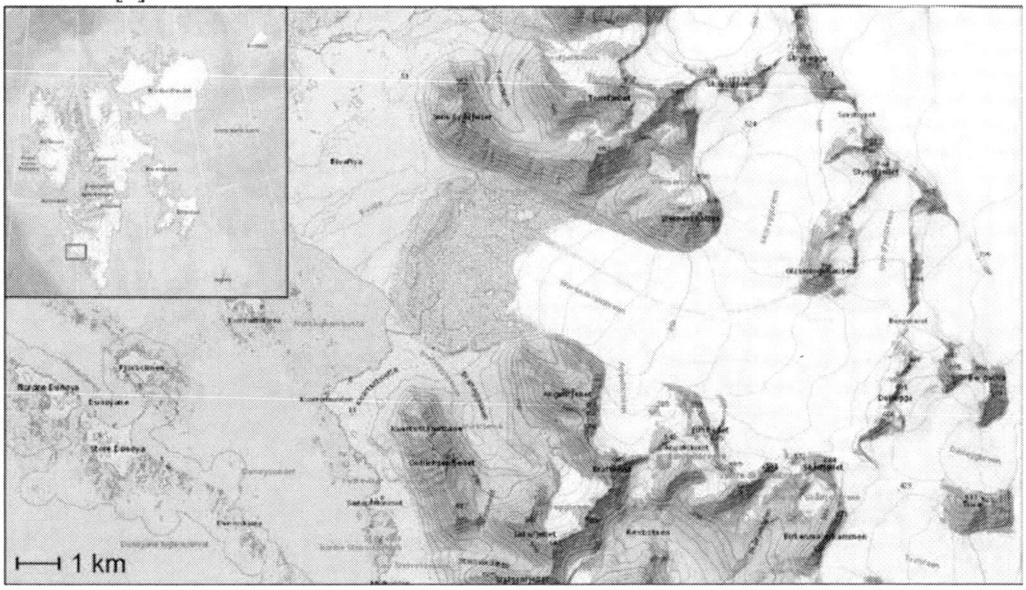

Fig. 1. Location of the research area [4].

Fig. 2. A) The photo from 1973 [1], with a view of Werenskioldbreen front, and B) the photo from 2013 (K. Grudzińska) from the same position.

2 Werenskioldbreen's surroundings and bedrock

Werenskioldbreen is a glacier of valley type with generated structure of connected basins of complex thermics of B type (glacier may have slight cold areas in the zone of dry snow; thickened layer of cold ice down the ablation zone up to thinness freezing front) and with majority of deformations [5]. The total length of glacier is 7 km (state of 1990 - middle of the analyzed area; total length was measured in a straight line to the end of the glacier without side firn fields). Its firn fields are located on the height 350–550 m a.s.l., among the 630–790 m a.s.l. elevations of summits: Wernerknatten, Strypegga, Glasjologerknausen and Deilegga. In its higher parts the glacier is divided into three clearly extending components:

Skilryggbreen (in the north), Slyngfjellbreen (in the east) and Angellisen (in the south). The lower part of glacier extends along the NWW–SEE direction and it is limited by the mountains: Jens-Erikfjellet (576 m a.s.l.) from the north and Angelfjellet (591 m a.s.l.) from the south. In this zone of the glacier area there is a perfectly formed middle moraine separating the ice masses generated in the fields of Skilryggbreen and Slyngfjellbreen. The glacier front is situated at present on the ordinate of ca. 100 m a.s.l. The end moraine is located about 2 km west-wards, its height reaches 90 m a.s.l. Along the slopes of Angelfjellet and Jens-Erikfjellet there are developed lateral moraines.

Until the seventies of 20th century the glacier front was drained in two directions: westwards through the Kvisla river and south-westwards by Werenskioldelva. Later on the first of the two flows broke the barrier of central moraine and therefore presently the total volume of melting water is remitted SW.

3 Source data and methods

The initial material used to determine the change of Werenskioldbreen's scope was a topographical map in the scale 1:5000 as prepared by the Polish Army Topographic Service in 1961 [6] (Fig. 3A). The photos and terrain measurements were performed by C. Lipert with a ground-based photogrametry in the years 1957–1959 with help of Zeiss (Jena) photo-teodolite. The map was elaborated by C. Lipert and M. Bałdyga using analytical plotter Wild-A7 in Gauss-Kruger transformation with central meridian 15°. The print of colour map was executed in the Army Cartographic Plant in 1961. One of the results of polar expedition of Wroclaw University on Spitsbergen in 1973 was elaboration of another map of Werenskioldbreen front (Fig. 3B). The transformation in 1:5000 scale was produced with the use of ground-based photogrammetric measurements also with the use of Zeiss phototheodolite. The author of photographs, analytic elaboration and map editing was J. Żyszkowski. This cartographic elaboration was performed, just as the earlier elaboration, in Gauss-Kruger's transformation with 15° central meridian. The map printout was black and white and formed an Annex to the published article [1]. The next analytical elaboration is a topographic map in 1:25000 scale performed by a staff from the Silesian University in cooperation with the Norwegian Polar Institute in Tromsø, on the basis of aerial photos transformed to the central projection (orthophotomap; Fig. 3C). It was published in 2002 [7] but it was based on photographs from the year 1990. Elaboration of the orthophotomap was based on a series of 14 stereoscopic models calibrated on 7 terrain points determined by GPS technique. The orthophotomap was depicted with Mercators transformation in 33X zone with the 15°central meridian on the ellipsoid WGS 84.

The polar expedition in July 2012 organized by the Faculty of Geoengineering, Mining and Geology of Wrocław University of Science and Technology measured the location of Werenskioldbreen's front with GPS Trimble GeoXT receiver. The measurements were performed in GPS data registration system with latitude and longitude in the global coordinates frame presented on reference ellipsoid WGS 84. The inventory of glacier front comprised 31 measurement points measured within one day (July 18). Supplementation of GPS measurement was a routing map made on the plotting paper in scale 1:5000 and elaborated on the basis of measurements with help of compass and measuring tape. The next expedition organized by the Wrocław University of Science and Technology organized in July 2013 performed the most relevant GPS measurements of glacier front. GNSS Trimble R6 receiver with registered data from satellites GPS and Glonass were used. The measurement was performed similarly as in 2012 in the system of WGS 84 with registration of measurement data with help of Fast Static method. The measurement of the glacier front was performed in July 5 to 7 and comprised 32 measurement points. The data elaboration was made with post-processing method with adjustment of the values of measured coefficients with reference to the GPS station of EUREF network in Ny-Ålesund, located ca. 220 km northwards from the investigated area. This station functions in the continuous mode with stable registration of changes of location in relation to the global reference frame of geocentric coordinates (the data are accessible in the internet page of GPS stations) [8].

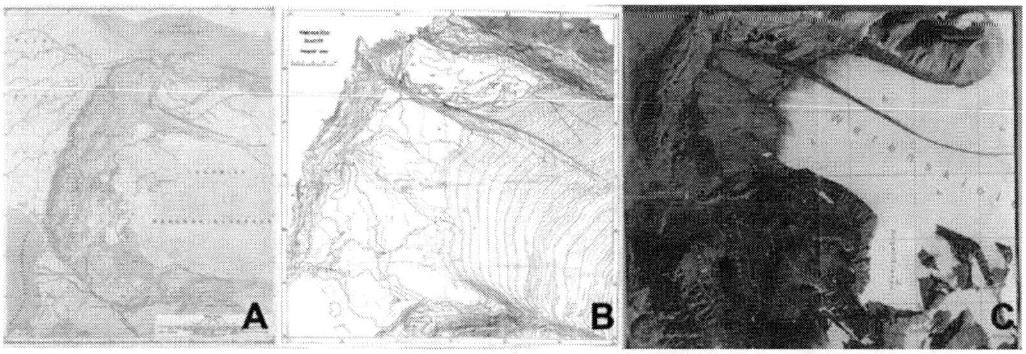

Fig. 3. Source maps for determination of Werenskioldbreen's front: A) Naumienko [6], B) Żyszkowski [1], C) Jania et. al. [7].

4 Elaboration of cartographic and measurement data for analysis of glacier front displacement

Evaluation of velocity of glacier front recession was possible to carry out on the basis of uniform system of coordinates only for all analysed maps and measurement results. Tophographic maps from the years 1957 and 1973 together with the orthophotomap from 1990 were transformed to the single system of coordinates. The system of rectangular coordinates was defined as projection on the surface of reference ellipsoid WGS 84. There was applied crosswise transformation: cylindrical, equiangular projection of Gauss-Kruger's with 15° meridian tangency as applied in Poland projection system "2000" which is used for small scale maps of small areas [9, 10]. Because the investigated surface of glacier covers only few square kilometers, application of the defined projection allows for determination of glacier recession without loss of precision as caused by transformation of coordinate system. The points of accommodation for carrying out transformation were intersection lines of kilometre network and well defined characteristic terrain points such as mountain tops on separate maps. The advantage of the selected projection is its angular fidelity and small deformations of distances on the whole mapping area (7.5 cm in the central part of zone and in the verges of the belt). The geographical longitude and latitude measured with GPS method in 2012 and 2013 was recalculated into the flat system as defined above. The possibility of spatial correlation of all the available archival data and the authors own measurements was obtained as its shown in the resulting map (Fig. 4).

Assessment of the recession of Werenskioldbreen's front was carried out with two methods. The linear values of recession were determined based on analyses of situation on 6 profiles determined generally perpendicularly to the glacier front (Fig. 4). The surfaces by which the glacier was diminished in the analysed time sections also were evaluated.

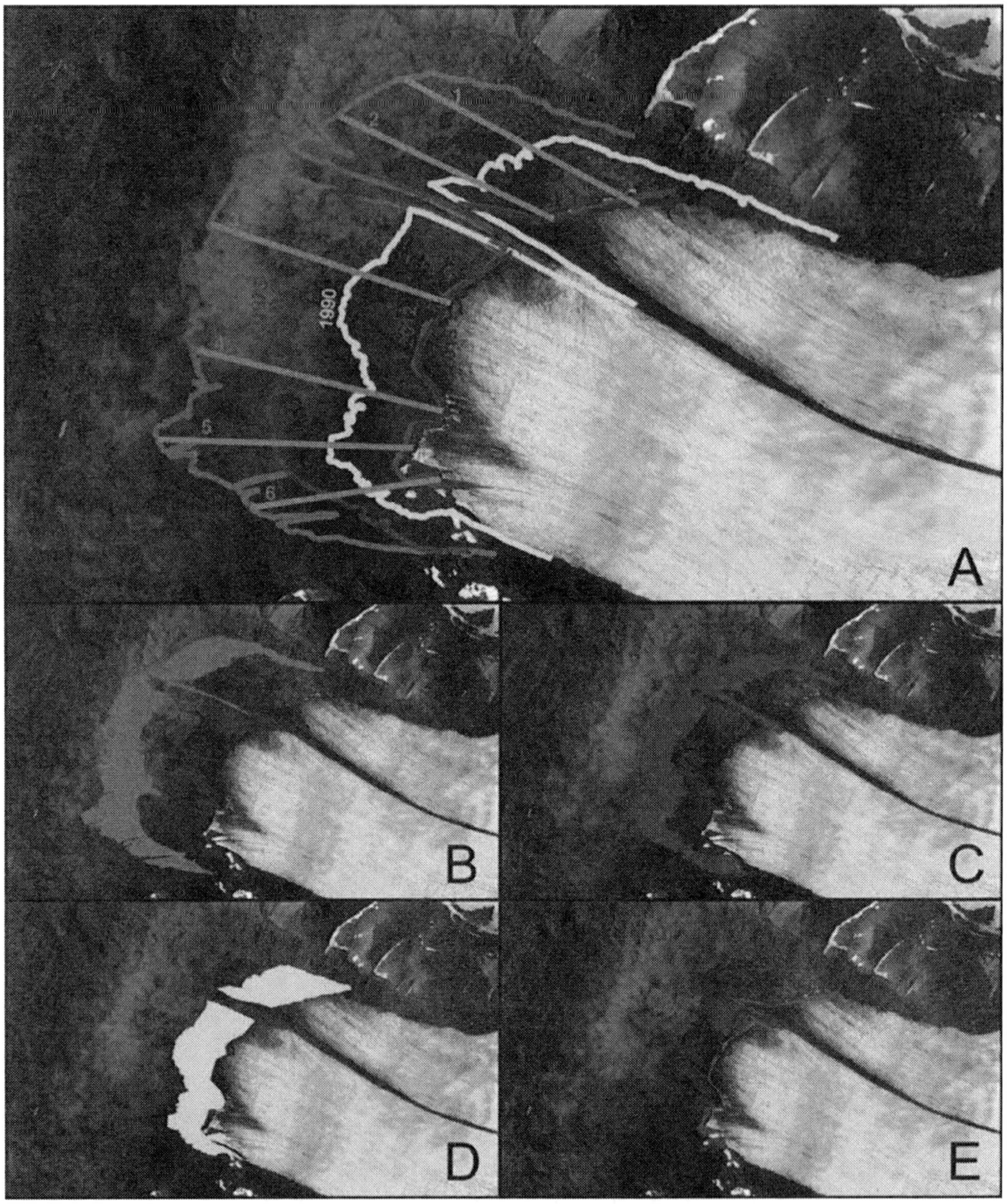

Fig. 4. Location of Werenskioldbreen's front in the years 1957-2013 and longtitudinal profiles where recession was analyzed (A), glacier's surface recession between 1957-1973 (B), glacier's surface recession between 1973-1990 (C), glacier's surface recession between 1990-2012 (D), glacier's surface recession between 2012-2013 (E); background: satellite photos, exemplary, uncalibrated [11].

4.1 The linear method

The linear values of front recession were calculated on the above-mentioned profiles which were determined between the lines of glacier range in the years 1957 and 2013 (Fig. 4). The numeric data concerning the magnitude of the glacier front recession in separate time sections calculated for separate profiles are shown in Tab. 1 and Tab. 2.

Table 1. The values of Werenskioldbreen's front recession in the analysed time sections for separate profiles [m].

Number of profile	The value of the recession in time intervals			Total recession along the profile	The average annual recession along the profile
	1957–1973 (16 years)	1973–1990 (17 years)	1990–2013 (23 years)		
1	375	334	577	1286	23
2	365	300	437	1102	20
3	385	375	443	1203	21
4	528	345	455	1328	24
5	560	270	378	1208	22
6	385	190	360	935	17
Average recession for period	433	302	442	1177	21
Average annual recession for period	27	18	19	21	

Table 2. Mean yearly recession for separates profiles in the analysed time sections [m].

Number of profile	Average annual recession for the profile in a given period		
	1957–1973	1973–1990	1990–2013
1	23	20	25
2	22	18	19
3	24	22	19
4	33	20	20
5	35	16	16
6	24	11	16

Elaboration of the rate of glacier recession in detail is possible based on analysis of vertical sections of glacier as shown in Fig. 5. They show approximate form of the glacier front surface in separate years with marked predominance of sections (the ratio of vertical and horizontal scales is 1:2).

The analysis of presented data enables us to show correctness of the observed process of the glacier recession in the period of the last 56 years. The glacier front retreated most spectacularly – by ca. 1300 m – in its southern part, in the area of profile no. 4 (1328 m) and in the northern part, as it is shown in profile no. 1 (1286 m). For maps of each analyzed period also the length of glacier front was determined as its changes also indicate the rate of recession. Simplified field measurements were performed for the sections of glacier rim in the analysed maps and in the measurements data from the glacier valley. The broken lines marking the border of ice from the north moraine to the central moraine and from the central to the southern moraine were measured.

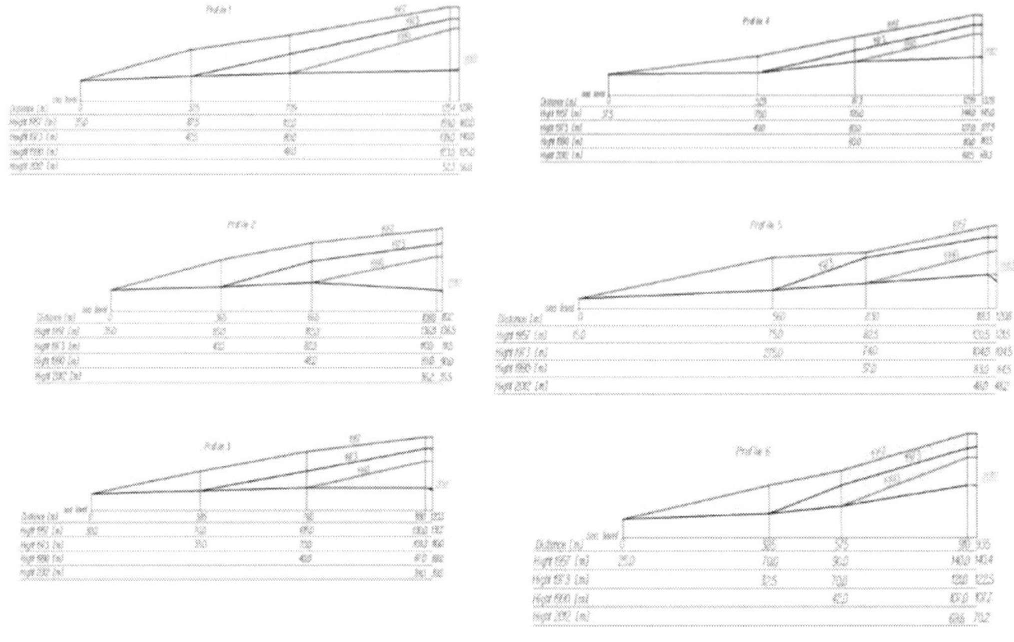

Fig. 5. The cross sections depicting the Werenskioldbreen's recession and front lowering along analyzed profiles (cf. Fig. 4).

The sum of the measured lengths was recognized as the length of glacier front. This value was subject to significant recession – from 6430 m in 1957 to 3480 m in 2013. For the year 1973 this value amounted to 3980 m, whereas for the year 1990 – 3060 m only. This effect is a result of narrowing of the glacier valley to the east and distancing or "escape" of the glacier from the lateral moraines located along the slopes of Angelfjellet and Jens-Erikfjellet.

4.3 The surface method

The second method which was applied, making use of the analysis of changes of the glacier surface, enabled us to determine independently the scale of Werenskioldbreen's recession. In order to obtain this objective with the use of orthophotomap of Werenskioldbreen area the total glacier surface was measured with the Gauss method [7, 12]. The coordinates of the points defining the border of deposition of the ice cover were determined. Based on the determined rectangular coordinates the surfaces of separate discussed areas were calculated. At the first stage the glacier surface in 1990 was calculated because the initial map from 1957 was not the whole glacier which was charted but only its frontal part, which was repeated in 1973. To the surface calculated in 1990 was added the surface between the front lines from the years 1957 and 1990. This way the approximate surface initial for the total glacier in the year 1957 was obtained. The subsequent stages of elaboration consisted in determination of the surfaces of fields between the curves forming the glacier front in the separate measurement periods. The results of these calculations are shown in Tab. 3. The surface is given in hectares and also in percent of the total initial surface of the glacier.

Table 3. The results of calculation of the glacier surface and its recession in different epochs.

Year/period	Surface of the glacier		Decrease of surface		
	ha	%	ha	average ha/year	average % /year
1957	2960	100			
1957–1973 (16 years)			103	6.4	0.22
1973	2857	96.5			
1973–1990 (17 years)			85	5	0.17
1990	2772	93.7			
1990–2013 (23 years)			79	3.7	0.13
2013	2687	90.8			
2012–2013			6	6	0.20

Based on presented data one can formulate several regularities concerning the glacier recession process and scale of baring of sub-glacier surface. The surface of total glacier in the period of latest 56 years of the research decreased by 279 ha, i.e. 9.2% of the surface determined in the year 1957, which gives the mean yearly loss of the glacier surface of the order of 5 ha, i.e. 0.22% of the initial glacier surface. The greatest loss of surface occurred in the initial period of research in the years 1957–1973 and it amounted to 103 ha, i.e. 3.5% of the initial glacier surface. The smallest loss of the surface occurred in the period 1990–2013 and it amounted to 79 ha, i.e. 2.9% of the initial glacier surface which gives the yearly mean loss about 3.7 ha. In the latest measurement period (2012-2013) the loss of surface increased again to the value of 6 ha, i.e. 0.2% of the total glacier surface. The rate of glacier recession determined by the loss of its surface is variable in time. At the beginning it had value of ca. 6.4 ha per year, and finally it decreased to 3.7 ha per year. We obtain the anomalous value of ca. 6 ha also in the case of application of this research method for the latest year (2012–2013) of research.

5 Discussion

In the result of geometrical analysis of the accessible archival cartographic elaborations together with the results of the author's terrain investigations the process of Werenskioldbreen's recession was showed. It was described with the use of results of linear and surface analyses.

The length of the glacier front was subject to reduction by above 2 km i.e. about 30% of its initial length. This is probably a result of morphologic conditions caused by the eastward narrowing of the glacier valley.

The results of linear analyses of Werenskioldbreen's front recession show the mean value ca. 1200 m during the period of 1957–2013, which corresponds to the rate of recession by ca. 21 m per year. However, the rate of the process is variable in time and space. The rate of recession fell from ca. 27 m per year in the epoch 1957–1973 down to 18–19 m per year later on. Initially the glacier retreated most promptly in the south by 528–560 m, i.e. ca. 34 m per year. In the period of 1973–1990 the melting down had more uniform character along the total glacier front with rather indistinctly drawn maximum in its central part, i.e. 375 m and 22 m per year. In the period of 1990–2013 the recession was distinctly fastest in the northern part of the glacier front - 577 m (25 m per year). From comparison of the results of author's elaboration with the analyses presented by Bukowska-Jania and Jania [13], Grabiec et al. [14], Migała et al. [15] and Ignatiuk et al. [16] one can state that the mean values of speed of the glacier recession (as well as calculation of the glacier surface) are in a high accordance with the values cited from the literature, i.e. with 1 to 2 m difference.

The analysis of the changes of the glacier area shows that during the 56 years long research period it decreased by 279 ha, i.e. 9.2% of the initial surface. It corresponds to the mean yearly loss of surface equal to 5 ha, i.e. 0.22% of the initial surface. The most intensive recession of Werenskioldbreen surface occurred in the first analysed period (1957–1973) when it amounted to 103 ha i.e. 3.5 % of the initial glacier surface, that is 6.4 ha per year. In the two subsequent periods i.e. 1973-1990 and 1990-2013 a systematic rate of slowing of this process to the values of 85 ha and 79 ha was observed. Whereas, during the whole research period the slowing of this process reached only 5 ha in period 1973-1990 and average recession 3.7 ha per year in the years 1990–2013. The changes of glacier's areas and their front recession is observed in whole Svalbard, e.g another valley glacier located in southern Spitsbergen – Ariebreen [17] and others [18].

It should be stressed that the values obtained on the basis of author's terrain research made in the last year considered (2012–2013) are anomalous in comparison to the results of analyses from the years 1957, 1973 and 1990. The mean linear recession along the analysed profiles amounts to 26 m, while the surface recession amounts to 6 ha.

It is also important to emphasize that the factor having a significant impact on the rate of retreat of the glacier is rising average value of annual air temperature [19]. This state is confirmed by a number of published research results, especially when it comes to the West Spitsbergen. The increase in average annual temperatures of air results from an increase in average temperatures for each season. Only during the last twenty years the increase in average annual temperatures was set at 1.0 - 1.2°C. This trend is confirmed by direct instrumental observations as well as statistical increase in number of warm days during the year and decrease in number of cold days, especially during winter seasons [20-25].

6 Conclusions

The most essential observation is recognition of the slowing down or stabilization of the process of melting of the glacier in the total analysed period. This trend is visible in the results obtained with both applied methods of the linear and surface analyses. However, for a credible monitoring of the trend of changes it is necessary to implement regular cyclic observations. Accepted in this elaboration irregular measurement periods result from the character of accessible archival data. The periods of 17 and 23 years long do not allow for creation of complete model of glacier recession. It is possible that in these time intervals there occurred a periodic expansion of the glacier which is a natural phenomenon that often takes place in different parts of Spitsbergen.

There was noticed some differentiation of the rate of glacier recession on separate sections of the glacier front caused by the glacier foundation relief and activity of the glacier rivers. The photogrammetric ground and low-ceiling aerial methods as well as the traditional geodetic measurement methods and the GNSS method can be still applied with consideration of often extreme atmospheric, logistic and economic conditions.

This elaboration was financed from research grants of Wrocław University of Science and Technology No. S 200 27/2012, S 200 10/2013, S 50051, 0401/0125/17/Z1, 0402/0137/16 and Grant No.0401/0128/17.

References

1. J. Żyszkowski, Acta Universitatis Wratislaviensis, **525**, 289-298 (1982)
2. W. Ciężkowski, T. Głowacki, K. Grudzińska, D. Kasza, P. Zagożdżon, 35th Polar Symposium, Diversity and state of polar ecosystems (book of abstracts), 4th-7th June (2014)
3. L. Kolondra, 2003. Problemy fotogrametrycznego pozyskiwania danych w badaniach glacjologicznych (studium metodyczne na przykładzie Spitsbergenu), PhD dissertation, University of Silesia in Katowice (2003)
4. Norwegian Polar Institute, http://toposvalbard.npolar.no/ (2014)
5. J. Jania, *Wyprawy Polarne Uniwersytetu Śląskiego 1980–84*, t. 2, Prace Naukowe Uniwersytetu Śląskiego, 12-47 (1988)
6. T. Naumienko (ed.), Mapa topograficzna czoła lodowca Werenskiodbreen w skali 1:5000,Wojskowe Zakłady Kartograficzne (1961)
7. J. Jania, L. Kolondra, H.F. Aas, Orthophotomap 1:25000, Werenskioldbreen and surrounding area, Uniwersytet Śląski and Norsk Polarinstitutt (2002)
8. website of EUPOS Ny-Ålesund, http://www.epncb.oma.be/_networkdata/ siteinfo4onestation.php?station=NYA1_10317M003 (2014)
9. R. Kadaj, *Polskie układy współrzędnych. Formuły transformacyjne, algorytmy i programy*, AlgoRes-Soft (2002)
10. E. Osada, *Osnowy geodezyjne*, UxLan (2013)
11. satellite photo from Quit Buird, https://browse.digitalglobe.com/imagefinder (2014)
12. W. Kosiński, *Geodezja*, Wydawnictwa Naukowe PWN (2012)
13. E. Bukowska-Jania, J. Jania, *Wyprawy Polarne Uniwersytetu Śląskiego 1980–84*, t. 2, Prace Naukowe Uniwersytetu Śląskiego, 65-91 (1988)
14. M. Grabiec, T. Budzik, P. Głowacki, Arctic, Antarctic, and Alpine Research, **44**, No. 2, 164-179 (2012)
15. K. Migała, J. Pereyma, K. Birkenmajer, D. Ignatiuk, C. Kabała, M. Kasprzak, B. Korabiewski, H. Marszałek, J. Matuła, P. Migoń, M. Rysiukiewicz, S. Sikora, S. Staśko, M. Wąsik, A. Witkowski, B.Wojtuń, *Ancient and modern geoecosystems of Spitsbergen*, Z. Zwolinski, A. Kostrzewski, M. Pulina, K. Bogucki (eds.), Wydawnictwo Naukowe, 122-125 (2013)
16. D. Ignatiuk, A. Piechota, M. Ciepły, B. Luks, International Conference of Computational Methods in Sciences and Engineering 2014 (ICCMSE 2014), AIP Conf. Proc. **1618**, 275-280 (2014)
17. J. Lapazaran, M. Petlicki, F. Navarro, F. Machi´o, D.Puczko, P. Glowacki, A. Nawrot, Polar Research **32** (11068), 1-10 (2013)
18. World Glacier Monitoring System, http://wgms.ch/fogbrowser/ (2016)
19. Weather forecast for Longyearbyen (Svalbard), http://www.yr.no/sted/Norge/Svalbard/ Hornsund/ (2016)
20. O. Nordli, Norwegian Meteorological Institute, http://nordicspace.net/2005/01/01/ temperature-variations-at-svalbard-during-the-last-century/ (2005)

21. E. Oseth, *Climate Change in the Norwegian Arctic. Consequences for life in the north.* Norwegian Polar Institute, report series 136 (2011)
22. D. Divine, E. Isaksson, T. Martma, H.A.J. Meijer, J. Moore, V. Pohjola, R.S.W. Van De Wal, F. Godtliebsen, Polar Research **30** (7379), 1-12 (2011)
23. E.J. Forland, R. Benestad, I. Hanssen-Bauer, J.E. Haugen, T.E. Skaugen, Advances in Meteorology, vol. 2011 (2011)
24. O. Nordli, R. Przybylak, A.E.J. Ogilvie, K. Isaksen, Polar Research **33** (21349), 1-23 (2014)
25. H.M. Gjelten, O. Nordli, K. Isaksen, E.J. Forland, P.N. Sviashchennikov, P. Wyszynski, U.V. Prokhorova, R. Przybylak, B.V. Ivanov, A.V. Urazgildeeva, Polar Research **35** (29878), 1-13 (2016)

AUTHOR INDEX

EDP Sciences – Web of Conferences
17, Avenue du Hoggar
Parc d'Activité de Courtabouf
BP 112
F-91944 Les Ulis Cedex A
France

ISBN 978-1-5108-5722-3

9 781510 857223

CURRAN ASSOCIATES INC.
proceedings.com